INNOVATIONS IN GIS

GIS

and Evidence-Based Policy Making

Edited by

Stephen Wise
Max Craglia

CRC Press
Taylor & Francis Group
Boca Raton   London   New York

CRC Press is an imprint of the
Taylor & Francis Group, an **informa** business

CRC Press
Taylor & Francis Group
6000 Broken Sound Parkway NW, Suite 300
Boca Raton, FL 33487-2742

© 2008 by Taylor & Francis Group, LLC
CRC Press is an imprint of Taylor & Francis Group, an Informa business

---

Library of Congress Cataloging-in-Publication Data

GIS and evidence-based policy making / editors Stephen Wise and Max Craglia.
    p. cm. -- (Innovation in GIS)
  Includes bibliographical references and index.
  ISBN 978-0-8493-8583-4 (alk. paper)
    1. Geographic information systems--Government policy. 2. Geospatial data. 3. Information storage and retrieval systems--Geography--Government policy. I. Wise, Stephen. II. Craglia, Massimo. III. Title. IV. Series.

G70.212.W57 2008
910.285--dc22
                                          2007029937

**Visit the Taylor & Francis Web site at
http://www.taylorandfrancis.com**

**and the CRC Press Web site at
http://www.crcpress.com**

GIS
and Evidence-Based Policy Making

INNOVATIONS IN GIS

SERIES EDITORS

Jane Drummond
University of Glasgow, Glasgow, Scotland

Bruce Gittings
University of Edinburgh, Edinburgh, Scotland

Elsa João
University of Strathclyde, Glasgow, Scotland

GIS for Environmental Decision-Making
Edited by Andrew Lovett and Katy Appleton

GIS and Evidence-Based Policy Making
Edited by Stephen Wise and Max Craglia

Dynamic and Mobile GIS: Investigating Changes in Space and Time
Edited by Jane Drummond, Roland Billen, Elsa João, and David Forrest

Contents

Part II Making Policy

Preface

In 1993, the first of the GIS Research U.K. (GISRUK) conferences was held at the University of Keele. Before that, U.K. GIS conferences had been very broad-based, bringing together the entire spectrum of people who created and used spatial data and the software to process the data. While such conferences were very valuable, it was felt that there was also a need for an academic conference where the focus would be purely on research. It was clear from the very first Keele meeting that the GISRUK conference series was fulfilling a useful function in bringing together researchers from all the different disciplines that contribute to geographic information science (GISc) in an informal but stimulating atmosphere. The Sheffield GISRUK conference in 2002 was the 10th anniversary of the conference series, and thus represented a chance to look back at the first 10 years and look forward to the future.

GISRUK can be proud of its achievements in the first 10 years of its existence. Attendance at the conferences has grown and become more international. One of the original aims was to make GISRUK a forum in which young researchers could be encouraged and welcomed, and this has grown to become one of the most distinctive features of the conferences. Every year there is a young researchers' forum, which takes place immediately before the main conference and allows those starting out on research careers in GIS to meet each other and share their experiences and to receive advice and feedback from experienced researchers. There is a special prize for the best paper presented by a young researcher. The success of this approach to inducting newcomers into the GIS research community can be judged by the fact that a previous winner of this prize is now a member of the national steering committee.

One of the interesting things about GISRUK is that there is no formal association behind the conference series. A national steering committee exists, to provide some continuity from year to year, but each conference is effectively autonomous and the local organizing committees have the freedom to run things as they see fit. During the Sheffield conference, one of the invited speakers, Professor Ian Masser, made the suggestion that GISRUK might seek to take on the role of representing the views of the U.K. GIS research community more widely in the way that AGILE does within Europe. This generated a lively discussion both during and after the conference session, which is after all what you want from a keynote talk! It was decided that in order to take on this role, GISRUK would have to constitute itself more formally because presently the steering committee is unelected and thus has no mandate to represent anyone. The consensus was that this was not the way people wished to see GISRUK develop, not least because it was felt that the Association of Geographic Information (AGI) is already constituted in a way that allows it to represent the views of its members, including academics. However, one of the results of the discussions was an agreement to forge closer links between the AGI and GISRUK.

One of the main outputs of the GISRUK conferences has been the annual Innovations in GIS volume. In keeping with the spirit of review, this volume represents something of a change compared with previous volumes. Early volumes in the series presented a selection of the best papers from each conference, covering the full range of GIS. However, more recently the trend has been to produce a volume more focused on one of the main themes of the conference. In this volume we have extended this process. The bulk of the papers collected here are full versions of papers that were presented at Sheffield in 2002. However, in order to produce a book that is a more comprehensive review of the state of the art of research into evidence-based policy making, we have also invited additional contributions from leading researchers in this area.

The papers in the first half of the book are concerned with collection of the evidence that underpins policy making. This has been split into two sections—the collection of data and the methods for analyzing the data to produce new information.

Ian Masser sets the scene with the first paper in this section by stressing the importance of location as a key factor in policy making at the national level. One of the implications of this is that spatial databases need to be compiled and made available at the national level. As Ian demonstrates, the ways in which this issue has been tackled vary widely between countries, with the variations dictated as much by differences in the approach to national government as by differences in the availability of data. One general pattern that does emerge, however, is a distinct difference between the earlier initiatives in this area, which were dominated by the efforts of data producers to complete or extend their data holdings, to current initiatives that are lead by a wider range of stakeholders and are more focused on providing users with access to data, and connecting distributed data repositories.

The increasing availability of data is partly due simply to the passage of time, with various data capture projects increasingly turning a paper archive into a digital one. Moreover, it is also due to technological advances, both in sources of data and in methods for analyzing these sources. Nowhere is this more true than in the area of Earth observation. The resolution of early remote sensing satellites (in the civilian domain at least) meant that while it was possible to distinguish the built from the natural environment, it was not possible to resolve the detailed variation within the built environment. However, recent satellite platforms have changed that and it is now possible to resolve individual buildings and plots of land on satellite imagery. The challenge is to automate the recognition of the individual features and to try and make inferences about land use, as opposed to simply determining land cover. This is the subject of the second chapter by Victor Mesev and Paul McKenzie, in which one possible approach to classifying urban land-use patterns by using relatively simple spatial statistics on the pattern of buildings in an area is reviewed. These statistics are shown to distinguish reasonably well between different patterns of residential developments in a number of cities in the U.K.

Inference is of course an important characteristic of geodemographics, in which neighborhoods are classified by drawing inferences from a series of statistics on the people who live there. The pioneer of this approach, Richard Webber, concludes the first half of Part I with an overview of geodemographics. He points out that this technique was originally developed to help target policy making in the 1970s, and only later "escaped" into the commercial sector, where it is much better known. Originally developed in the U.S. and the U.K., it is now used in 19 other countries around the world. Despite differences in the categories in which they are used, they show some constant factors, such as the emergence of a group in the U.K. who would be called "young upwardly-mobile professionals," or yuppies as they used to be called in the popular press.

The chapters in Section II of Part I are all concerned with developments in the methods that can be used to undertake the analysis of spatial data for the purposes of informing policy. Health and crime are the two areas in which spatial analysis has a long history and it continues to play an important role, and half the papers in this section deal with these two topics. Crime is the subject of the paper by Andrew Newton. It has long been recognized that the occurrence of crime has a strong spatial pattern, governed at the broad scale by the distribution of those who take part in criminal activity and at the local scale by the location of potential targets for criminal activity. Newton's paper extends this focus in two ways. First, new technology such as GPS may allow us to begin analyzing crimes in which the locus of the crime is not fixed but mobile, such as crime on public transport systems, although little work has been done on this to date. Second, Newton argues that certain locations may act as attractors to crime and takes as an example of this, criminal damage to bus shelters. His analysis reveals a tendency for occurrences of damage to be clustered. The clustering shows some connection with parts of the city where crime levels are higher than average, but also shows an additional degree of clustering that seems to suggest that certain locations are the focus of increased criminal activity.

Patterns of crime and poor health are often strongly linked to material deprivation and it could be argued that in a capitalist society, material deprivation is the single most important determinant of social well-being. For reasons of confidentiality, deprivation is normally estimated on an areal basis, which raises all the well-known problems connected with modifiable areas. In Chapter 5, Daniel Exeter, Robin Flowerdew, and Paul Boyle consider the extent to which this approach hides small pockets of deprivation within larger areas. They take Scotland as their example, and by comparing deprivation levels at the scale at which policy decisions are made with the finer scale of the census output areas, they identify a number of pockets of deprivation that are overlooked in policy making. Somewhat surprisingly, these are not in rural areas, where this problem is normally thought to be most acute, but in suburban areas.

Chapter 6 brings us back to crime with a paper by Paul Brindley, Max Craglia, Robert Haining, and Young-Hoon Kim on the development of methods for identifying levels of repeat victimization from crime statistics. There is a technical element to this problem, concerned with how you identify that the same person or property has been the victim of crime, when the reporting of the location of the crime may differ between incidents. What is equally interesting in the context of this volume is that the techniques that were developed to solve this problem had to be appropriate for use by police officers with no expertise in spatial analysis and only minimal training in the new software. This raises an issue that is more thoroughly explored in the second half of the book, which is that if GIS is to make any contribution to evidence-based policy making, such systems must be designed with the potential users in mind.

One of the key elements GIS brings to the analysis of spatial data, of course, is the ability to bring together multiple sets of information, using location as the linking mechanism. There has been a great deal of interest in the U.K. in recent years in the apparently increased incidence of lowland flooding, which many believe to be linked to the first signs of global warming, and a good deal of work on the prediction of which areas are likely to be affected. The focus of much of this work has been on the potential damage to property, the assumption being that as long as this can be avoided or minimized, the inhabitants will be able to take care of themselves. Tom Whittington's paper in Chapter 7 makes the key point that the extent to which this is a reasonable assumption depends on what he terms the social vulnerability of the inhabitants, a term which covers a whole range of factors that determine how vulnerable people are to a crisis, ranging from their mobility to their ability to withstand economic loss. As an example of how these ideas might be carried out in practice, an index of social vulnerability is constructed from readily available data and combined with a simple model of flooding.

Techniques developed in one field of endeavor can often be applied to other fields, as witnessed by the fruitful cross-fertilization between methods for analyzing patterns of crime and health. As GIS users we are all well used to thinking in spatial terms, but we normally interpret the term spatial in the narrow sense of referring to some portion of the Earth's surface. However, analysis that involves spatial relationships exists in other disciplines too, including computational chemistry where the relationships are between the elements of molecules and compounds. Patterns and spatial relationships are important in determining chemical properties, and thus computational chemists have developed powerful tools to describe patterns and match them with one another. Peter Bath, Cheryl Craigs, Ravi Maheswaran, John Raymond, and Peter Willett illustrate how this approach can be applied to geographical space in order to search for patterns. They take the example of searching for patterns in the occurrence of ill health and illustrate how the technique makes it simple to search for patterns that would otherwise be difficult to identify using conventional GIS methods.

The final contribution in this section is the chapter from Peter Bibby which forms a nice junction between the two halves of the book since he describes not only some innovative technical approaches but also considers the broader implications of using GIS to provide evidence for policy making. Peter considers the U.K. government's policy of keeping new housing development away from greenfield sites, by intensifying urban areas, using brownfield sites, and converting existing buildings. The implementation and monitoring of this policy are relatively straightforward using standard GIS techniques—for example, overlaying the location of new building plots onto urban polygons reveals that, in the 1990s, 57% of new buildings fell within urban areas. However, this relies on a simple view of urban areas defined crisply as polygons. By applying natural language-processing techniques to the national address file, Peter was able to create a more subtle representation of the settlement pattern of the U.K., which suggests that even more development is taking place in urban areas than the standard analysis would suggest.

Section I of Part II contains five chapters discussing the relationships between GIS technology and methods and policy making in public administration from different perspectives.

Chapter 10 by Isaac Karikari, John Stillwell, and Steve Carver emphasizes that the implementation of GIS/LIS is not just a technical issue, but primarily involves people, organizations, and adaptation of the technology and working practices to the local context. This applies all the more in developing countries where the projects are often parachuted by donor agencies and then not followed up and maintained. Hence this chapter makes the important point that technology transfer needs to be human-centered and driven by local staff with the necessary technical and organizational skills. Couched within the broad theoretical framework of sociotechnical design, the authors describe the prototype GIS developed for the Accra Land Commission Secretariat as a way to elicit feedback from the staff of that organization and support the process of mutual adaptation between working practices and technology.

The relationship between technology and society is pursued further in Chapter 11 by Sultan Barakat, Adrijana Car, and Peter Halls in the context of postwar reconstruction. The quote by Moore and Davis that they refer to: "Tell me, I forget. Show me, I remember. Involve me, I understand" captures the essence of this chapter focusing on GIS and public participation in the reconstruction effort following natural or man-made disasters, such as war. The authors make the important point that reconstruction is more about people and social relations than aid and material goods. Likewise technology has to fit the social processes and context but can also offer opportunities to foster dialogues and participation, and support the multilayered decision-making process. The authors then outline a methodology that integrates the contribution of public participation GIS, spatial decision support systems, artificial intelligence, and expert systems to develop a flexible and dynamic model of the postwar reconstruction process. This in their view would enable all the stakeholders to explore the potential impacts of alternative scenarios, increase informed dialogue, and support the achievement of consensus.

The organizational theme is pursued in Chapter 12 by Malcolm Borg and Saviour Formosa in describing their major effort in developing the National Protective Inventory of cultural artifacts in the island of Malta. Malcolm and Saviour give an excellent account of this project that is of major significance to improve accessibility and management of the rich historical heritage of Malta, and from which many lessons can be learned for similar projects in Europe and beyond. While the technologies deployed to create a spatial database and make it accessible via Web browser are now mature for this kind of project, and their opportunities are well illustrated in this chapter, what is particularly worth noting is the underlying major effort that has been required to overcome the data hoarding tradition of public administration, and to develop a culture of data sharing. This is a major challenge that applies to public administration throughout the world and certainly not just to Malta.

While the previous three chapters emphasized the organizational and cultural challenges of implementing GIS-based projects in public administration, Andrew Lovett, Julian Parfitt, and Gilla Sünnenberg in Chapter 13 show the opportunities for policy making that arise from the development of long-term partnerships between university-based researchers and local administrations. This chapter presents an interesting application of GIS to assess the likely impacts of different refuse collections and recycling strategies needed to meet government targets. The case-study area is South Norfolk County, which has a well-maintained waste statistics database (what a rarity!) as well as a history of collaboration with the researchers based in Norfolk. From these premises, the methodology described in the chapter is of generic value for other similar types of applications and alternative strategy assessment, and is therefore of significant value.

The final chapter in this section, by Mette Termansen, Colin McClean and Hans Skov-Petersen, looks at how GIS can benefit another widely used analytical approach, discrete choice modelling. This is applied to the modelling of recreational visits to forests in Denmark, using a random utility model (RUM) with numerous parameters calculated using GIS. Interestingly, the researchers found that the effect of distance on people's choice of site is not linear, as is often assumed, but is much stronger when choosing between nearby locations than when trying to decide between locations that are all much further away. A number of characteristics of the area around the forested area, not normally considered in this kind of analysis, were also significant, with people showing a marked preference for forests near the coast and in undulating topography. The researchers also developed a completely GIS-based approach but found that while this was much simpler to implement, it was not able to model the full range of factors as successfully as the RUM approach.

The last section of the book focuses on the opportunities and challenges of involving the public through Internet-based spatial technologies, and other information and communication channels. The development of a shared understanding among multiple stakeholders and the public of local context, problems, and opportunities is the theme of Chapter 15 by Derek Reeve, Erik Thomasson, Steve Scott, and Ludi Simpson. These authors present an excellent example of a Web-based participatory system designed *with* the local community rather than *for* the community in Bradford, England. The key feature of the Maps and Stats system is the ability for users to generate small-area statistics for user-defined areas, which makes it stand out from the traditional approaches based on predefined administrative areas. As persuasively argued by the authors, the increased control by users on what they can get out of this system is a prerequisite to engage the local community and foster active participation. The development of Maps and Stats is another good example of partnership between the local authority, the university, and a not-for-profit community research center, and a model for others to follow.

Chapter 16, by Tan Yigitcanlar, reviews the mix of technologies and methods currently available to support public involvement and participation. The community-based Internet GIS he presents encompasses distance learning modules, analytical modules built around a collaborative GIS approach, and specific modules to support strategic choice. Web-based architectures have now matured sufficiently to make the development of such integrated systems robust enough for wide access and use. The challenge now, as recognized in the discussion of the case study in Tokyo, is to exploit the opportunities opened up by the technology. This is not trivial because it requires overcoming the lack of interest shown by the public, limited vision of public authorities, and skewed availability of the knowledge and skills necessary not only to physically access information and tools, but more importantly to be able to use them to support meaningful participation.

The final chapter of this section by Robin Smith takes a broader view of public participation in the digital age and strives in particular to unpack the often taken-for-granted view of public participation, which as we all know is "good for you." Building on a comprehensive survey of public authority Web sites in 1999, and a series of case studies, Robin makes the point that despite the increased opportunities offered by new information and communication technologies (well illustrated in the preceding chapters), the gap between the rhetoric in support of public participation, stakeholders' expectations, and outcomes is largely the result of our poor conceptualization and shared understanding of public participation itself. To this end, Robin articulates five components of public participation, reflecting differing theoretical constructs (notions), issues, audience, outcomes, and methods. By linking more clearly expected outcomes to the notions of public participation held by the stakeholders, it is then possible to identify the methods most suited to engage the relevant audience on a specific issue. This is an analysis necessary before starting any public participation exercise. Technology comes after theory, not before. An important point always worth keeping in mind.

Stephen Wise
Max Craglia

Editors

Stephen Wise worked in university computer services for 10 years, during which time he chaired the group that negotiated access to ESRI software for the U.K. higher education sector and was part of the ESRC-funded Wales and South-West Regional Research Laboratory. Since 1990 he has lectured in GIS at the University of Sheffield and is the author of a textbook on GIS entitled *GIS Basics*. He is currently a member of the GISRUK national steering committee and an associate editor of the journal *Computers and Geosciences*.

Max Craglia is the research coordinator of the Unit of the Joint Research Centre of the European Commission that has the responsibility for the technical development of the Infrastructure for Spatial Information in Europe (http://inspire.jrc.it). He is also the editor of the *International Journal of Spatial Data Infrastructures Research* (http://ijsdir.jrc.it). Before joining the JRC in 2005, Max was a senior lecturer at the University of Sheffield, teaching GIS for urban planners and researching areas of spatial data infrastructure deployment and use, GIS applications for policy analysis, and data policy.

Contributors

Sultan Z. Barakat Department of Politics, University of York, Heslington, York, United Kingdom

Peter A. Bath Centre for Health Information Management Research, Department of Information Studies, University of Sheffield, Sheffield, United Kingdom

Peter Bibby Department of Town and Regional Planning, University of Sheffield, Sheffield, United Kingdom

Malcolm Borg Ministry for Urban Development and Roads, Valletta, Malta

Paul Boyle School of Geography and Geosciences, University of St. Andrews, St. Andrews, United Kingdom

Paul Brindley Department of Town and Regional Planning, University of Sheffield, Sheffield, United Kingdom

Adrijana Car Centre for GeoInformatics, Salzburg University, Salzburg, Austria

Steve Carver School of Geography, University of Leeds, Leeds, United Kingdom

Max Craglia DG Joint Research Centre of the European Commission, Institute for Environment and Sustainability, Ispra, Italy

Cheryl Craigs Centre for Reviews and Dissemination, University of York, York, United Kingdom

Daniel Exeter Section of Epidemiology and Biostatistics, School of Population Health, University of Auckland, Auckland, New Zealand

Robin Flowerdew School of Geography and Geosciences, University of St. Andrews, St. Andrews, United Kingdom

Saviour Formosa Information Resources, Information and Communications Technology, Malta Environment and Planning Authority, St. Francis Ravelin, Floriana, Malta

Robert P. Haining Department of Geography, University of Cambridge, Cambridge, United Kingdom

Peter J. Halls Computing Service, University of York, Heslington, York, United Kingdom

Isaac Karikari Lands Commission Secretariat, Cantonments, Accra, Ghana

Young-Hoon Kim Department of Geography Education, Korea National University of Education, Cheongwon, South Korea

Andrew Lovett School of Environmental Sciences, University of East Anglia, Norwich, United Kingdom

Ravi Maheswaran Public Health GIS Unit, School of Health and Related Research, University of Sheffield, Sheffield, United Kingdom

Ian Masser Taddington, Buxton, Derbyshire, United Kingdom

Colin J. McClean Environment Department, University of York, Heslington, York, United Kingdom

Paul McKenzie University of Ulster, Coleraine, County Londonderry, United Kingdom

Victor Mesev Department of Geography, Florida State University, Tallahassee, Florida

Andrew Newton The Applied Criminology Centre, Department of Behavioural Sciences, University of Huddersfield, Huddersfield, United Kingdom

Julian Parfitt Waste & Resources Action Programme, Banbury, United Kingdom

John Raymond Department of Information Studies, University of Sheffield, Sheffield, United Kingdom

Derek Reeve School of Computing and Engineering, University of Huddersfield, Huddersfield, United Kingdom

Steve Scott School of Computing and Engineering, University of Huddersfield, Huddersfield, United Kingdom

Ludi Simpson Cathie Marsh Centre for Census and Survey Research, University of Manchester, Manchester, United Kingdom

Hans Skov-Petersen Department of Forest and Landscape, University of Copenhagen, Copenhagen, Denmark

Robin S. Smith Informatics Collaboratory of the Social Sciences, University of Sheffield, Sheffield, United Kingdom

John Stillwell School of Geography, University of Leeds, Leeds, United Kingdom

Gilla Sünnenberg School of Environmental Sciences, University of East Anglia, Norwich, United Kingdom

Mette Termansen Sustainability Research Institute, School of Earth and Environment, University of Leeds, Leeds, United Kingdom

Erik Thomasson Research Unit and Consultation Service, City of Bradford MDC, Bradford, United Kingdom

Richard Webber Centre for Advanced Spatial Analysis, Department of Geography, University College London, London, United Kingdom

Tom Kieron Whittington GIS and Planning Research Department, Savills plc, London, United Kingdom

Peter Willett Department of Information Studies, University of Sheffield, Sheffield, United Kingdom

Tan Yigitcanlar Urban Research Program, Griffith University, Nathan, Queensland, Australia

GISRUK Committees and Sponsors

GISRUK National Steering Committee (as of GISRUK 2002)

Jane Drummond	University of Glasgow
David Fairbairn	University of Newcastle
Bruce Gittings	University of Edinburgh (Chair)
Peter Halls	University of York
Zarine Kemp	University of Kent
David Kidner	University of Glamorgan
Andrew Lovett	University of East Anglia
David Miller	MLURI
Nick Tate	University of Leicester
Stephen Wise	University of Sheffield
Jo Wood	City University

GISRUK 2002 Local Organising Committee

Stephen Wise (Chair)	Chris Clark	Chris Openshaw
Peter Bibby	Max Craglia	Kate Schofield
Paul Brindley	Young Hoon-Kim	Paul White
Rob Bryant	Ravi Maheswaran	

GISRUK 2002 Sponsors

GISRUK is extremely grateful for the generous support of the following sponsors:

Association for Geographic Information	www.agi.org.uk
Blackwell Publishers Ltd	www.blackwellpublishers.com
CadCorp Ltd	www.cadcorp.com
EDINA	edina.ed.ac.uk
Elsevier Science	www.elsevier.nl
ESRI (UK) Ltd	www.esriuk.com
Ordnance Survey	www.ordnancesurvey.co.uk
Oxford University Press	www.oup.co.uk
Taylor & Francis	www.taylorandfrancisgroup.com
Wiley Europe Ltd	eu.wiley.com
RRL.net	
Quantitative Methods Research Group	www.casa.ucl.ac.uk/qmrg

Part I

Collecting Evidence

Section I *Data Issues*

1

National Spatial Data Infrastructure Phenomenon

Ian Masser

CONTENTS

1.1 Introduction

The widespread and rapid diffusion of national spatial data infrastructures (NSDIs) throughout the world during the last 10 years has led to the emergence of the NSDI phenomenon. This chapter deals with the nature of this phenomenon and discusses some of the lessons that can be learnt from this experience. The chapter is divided into four main parts. The first part describes some of the main milestones in the NSDI emergence, whereas the second part examines its nature. The third part evaluates the experience of the first generation NSDIs and discusses the changes that are taking place in the emerging second generation. It also explores some of the changes that are involved in the transition of NSDIs from formulation to implementation and considers some of the main features of the emerging spatial data infrastructure (SDI) hierarchy. The fourth and final section deals with some of the broader strategic issues that are associated with the creation of effective SDIs.

1.2   NSDI Phenomenon

Some of the main milestones in the emergence of the NSDI phenomenon are set out in Table 1.1. From this it can be seen that its origins date back almost 20 years to the establishment of the Australian Land Information Council (ALIC) in January 1986 (later the Australia New Zealand Land Information Council) as a result of an agreement between the Australian Prime Minister and the heads of the state governments to coordinate the collection and transfer of land-related information between the different levels of government and to promote the use of that information in decision making (ANZLIC, 1992, p. 1). The feature that distinguishes this body from similar bodies set up by other governments around this time in France (Conseil National de l'Information Geographique in 1985) and the Netherlands (National Council for Real Estate Information in 1984) is its emphasis on the need for coordination between the different levels of government. In

TABLE 1.1

Some NSDI Milestones

1986	Australian Land Information Council set up to coordinate the collection and transfer of land-related information between the different levels of government
1990	U.S. Federal Geographic Data Committee set up to coordinate the development, use, sharing, and dissemination of surveying, mapping, and related spatial data
1993	U.S. Mapping Science Committee report on "Toward a coordinated spatial data infrastructure for the nation"
1994	Executive Order 12906 "Coordinating geographic data acquisition and access: the national spatial data infrastructure"
1997	First global spatial data infrastructure conference in Bonn, Germany
1998	First generation of NSDIs paper identifies 11 initiatives from all parts of the world
1998–2000	54 countries respond to GSDI survey
2002	Crompvoets claims that 120 countries are considering NSDI development

this respect it anticipated the NSDI debate, which began in earnest 10 years later.

The second milestone occurred in 1990 when the U.S. Office of Management and Budget (OMB) established an interagency Federal Geographic Data Committee (FGDC) to coordinate the "development, use, sharing, and dissemination of surveying, mapping, and related spatial data." This set out the main objectives of an NSDI in the following terms:

- Promoting the development, maintenance, and management of distributed database systems that are national in scope for surveying, mapping, and other related spatial data
- Encouraging the development and implementation of standards, exchange formats, specifications, procedures, and guidelines
- Promoting technology development, transfer, and exchange
- Promoting interaction with other existing Federal coordinating mechanisms that have an interest in the generation, collection, use, and transfer of spatial data ... (OMB, 1990, pp. 6–7).

Until this time, the term "National Spatial Data Infrastructure" was not in general use, although a paper was presented by John McLaughlin at the 1991 Canadian Conference on Geographic Information Systems in Ottawa entitled "Toward national spatial data infrastructure." Many of the ideas contained in that paper were subsequently developed and extended by the U.S. National Research Council's Mapping Science Committee in their report on "Toward a coordinated spatial data infrastructure for the nation" which was published in 1993. This report recommended that effective national policies, strategies, and organizational structures need to be established at the federal level for the integration of national spatial data collection, use, and distribution. To realize this goal, it further proposed that the powers of the FGDC should be strengthened to define common standards for spatial data management and to create incentives to foster data sharing, particularly among federal agencies.

The turning point in the evolution of the NSDI phenomenon came in the following year in the United States with the publication of an Executive Order 12906 signed by President Bill Clinton on 11th April 1994 entitled "Coordinating geographic data acquisition and access: the national spatial data infrastructure." The Order set out in some detail the main tasks to be carried out and defined time limits for each of the initial stages of the NSDI. Apart from the core task of interagency coordination through the FGDC, these tasks included the establishment of a national geospatial data clearinghouse and the creation of a national digital geospatial data framework. The Order also gave the FGDC the task of coordinating the federal government's development of the NSDI and required that each member agency of the committee hold a policy level position in their organization. In this way it significantly raised the political visibility of geospatial data collection, management, and use not only among federal agencies but also nationally and internationally.

This document had an immediate impact on thinking in the European Union where a meeting of key people representing geographic information (GI) interests in each of the member states was organized by Directorate General XIII (now the DG Information Society and Media) in Luxembourg in February 1995. The main task of this meeting was to consider and further develop the ideas set out in the first of what became a series of draft documents on the subject of "GI 2000: toward a European geographic information infrastructure" (CEC, 1995). One of the outcomes of this debate in Europe was the decision to hold the first of what subsequently became a regular series of global spatial data infrastructure (GSDI) conferences at Bonn in Germany in September 1996 under the patronage of the EU commissioner-in-charge of DG XIII Martin Bangemann. The fledgling European Umbrella Organisation for Geographic Information (EUROGI) played an important role in planning this conference which brought together representatives from the public and private sectors and academia for the first time to discuss matters relating to NSDIs at the global level.

After the second GSDI conference in Chapel Hill, North Carolina in 1997, the author carried out a survey of the first generation NSDIs (Masser, 1999), which showed at least 11 NSDIs were in operation in various parts of the world by the end of 1996. What distinguished these NSDIs from other GI policy initiatives was that they were all explicitly national in scope and their titles all referred to geographic information, geospatial data, or land information and included the term "infrastructure," "system," or "framework." This first generation included relatively wealthy countries such as the United States and Australia as well as relatively poor countries such as Indonesia and Malaysia.

The rapid rate of NSDI diffusion after 1996 is highlighted by the findings of a survey carried out by Harlan Onsrud for the GSDI (www.gsdi.org). This shows that, between 1998 and 2000, 54 countries responded positively to his questionnaire: 21 of these came from the Americas, 14 from Europe, 13 from Asia and the Pacific, and 6 from Africa. The number of positive responses to this survey was nearly five times the number of first generation NSDI countries identified up to the end of 1996, whereas the data collected by Crompvoets (2002) suggested that as many as 120 countries might be considering projects of this kind. Given these circumstances it is felt that the term "NSDI phenomenon" is a reasonable description of what has happened in this field over the last 10 years.

1.3 Nature of an NSDI

The extent of the NSDI phenomenon is all the more surprising, as there is no clear consensus about what constitutes an NSDI. Many NSDIs have a strong project dimension which focuses on concrete goals such as the completion of the national topographical database. Others are much more process-oriented and focus mainly on strategic issues such as capacity building and the modernization of government. This is partly due to the different interpretations that can be given to the notion of infrastructure. To some people, infrastructure means tangible physical assets like roads and railway networks. To others, it is a strategic process of policy formulation and implementation carried out by governments to ensure that their GI assets are managed in the interests of the nation as a whole (Barr and Masser, 1997). This includes not only the tangible assets but also the individuals and institutions that are needed to make it a functional reality.

The definition put forward by the Global Spatial Data Infrastructure Association (www.gsdi.org) conveys some of the complexity of the issues involved. It defines a global (and implicitly a national) SDI as follows:

> A (National) Spatial Data Infrastructure supports *ready access to geographic information*. This is achieved through *the coordinated actions of nations and organisations* that promote awareness and implementation of complimentary policies, common standards and effective mechanisms for the development and availability of interoperable digital geographic data and technologies *to support decision making at all scales for multiple purposes*. These actions *encompass the policies, organisational remits, data, technologies, standards, delivery mechanisms, and financial and human resources* necessary to ensure that those working at the (national) and regional scale are not impeded in meeting their objectives. (emphasis added)

The italicized phrases of this comprehensive, but rather complex, definition show that there are four key concepts underlying the NSDIs. The first of these states their overriding objective is to promote ready access to the GI assets that are held by a wide range of stakeholders in both the public and private sectors with a view to maximize their overall usage. The second concerns the need for coordinated action on the part of governments to ensure that the overriding objective is achieved. The next part of this sentence gives some examples of the kind of actions that are required from governments. The third key element stresses the extent to which NSDIs must be user driven. Their primary purpose is to support decision making for many different purposes and it must be recognized that many potential users may be unaware of the original purposes for which the data was collected. Finally the last sentence illustrates the wide range of activities that must be undertaken to ensure the effective implementation of an NSDI. These include not only technical matters such as data, technologies, standards, and delivery mechanisms but also institutional matters related to organizational responsibilities and overall national information policies as well as questions relating to the availability of the financial and human resources needed for this task.

1.4   Evolution of NSDI Concept

It is useful to distinguish two stages in the evolution of the NSDI concept: a first generation consisting of a relatively small number of countries and a much larger second generation. The origins of the first generation go back to the 1980s, while the starting point for the second generation is around the year 2000. There are important differences in approach between the two generations, and there is also a growing emphasis on implementation in the latter. In addition, the second generation has to consider the role of NSDIs within the local to global hierarchy of SDIs.

1.4.1   Key Features of the First Generation NSDIs

The findings of the first generation study show that NSDIs come in all shapes and sizes (Masser, 1999). Table 1.2 indicates that it included initiatives from Australia, Canada, Indonesia, Japan, Korea, Malaysia, the Netherlands, Portugal, Qatar, the United Kingdom, and the United States. They included some recent initiatives which had as yet little to show other than good intentions (in 1996) as well as some more established initiatives which had already achieved a great deal. This generation constituted some very small countries with some very large ones, some relatively wealthy countries with relatively poor ones, and countries with and without federal systems of government.

The primary objectives of these NSDIs were to promote economic development, to stimulate better government, and to foster environmental sustainability. The notion of better government is interpreted in several different ways in them. In many countries, it means better strategic planning and resource development. This is particularly the case in developing countries such as Indonesia and Malaysia. Planning, in the sense of a better state of readiness to deal with emergencies brought about by natural hazards,

TABLE 1.2

The First Generation NSDIs

Australia	Australian Spatial Data Infrastructure
Canada	Canadian Geospatial Data Infrastructure
Indonesia	National Geographic Information Systems
Japan	National Spatial Data Infrastructure
Korea	National Geographic Information System
Malaysia	National Infrastructure for Land Information Systems
Netherlands	National Geographical Information Infrastructure
Portugal	National System for Geographic Information
Qatar	National Geographic Information System
United Kingdom	National Geospatial Data Framework
United States	National Spatial Data Infrastructure

was also an important driving force in the establishment of the Japanese National Spatial Data Infrastructure, while the National Geographic Information System in Portugal was also seen as an instrument for modernizing central, regional, and local administrations.

Most of these infrastructures contain the three main elements defined in the U.S. Executive Order:

1. Some form of mechanism for coordinating NSDI activities.
2. An acceptance that some core or framework datasets are common to a very wide range of applications and need to be made interoperable with one another during NSDI implementation.
3. A recognition of the need to create various types of metadata service to increase user awareness of what data is available in each country.

The first generation can be divided into two broad categories with respect to their status: those that are the result of a formal mandate from government and those that have largely grown out of existing GI coordination activities. The first category includes Portugal where the National Geographic Information System was created by the Decreto Lei of 53/90 and the United States where the NSDI was the subject of an Executive Order by the President in April 1994. The second category consists of countries such as Australia where current discussions regarding SDI are essentially an expansion of earlier discussions regarding the national land information strategies. The transformation of the Dutch Council for Real Estate Information into a National Council for Geographic Information in 1992 also marked a significant step toward the development of a national GI infrastructure for the Netherlands.

The scope of these infrastructures can also be looked at from two different standpoints: the range of substantive GI interests that is represented in the different coordinating bodies and the extent to which the main stakeholders are directly involved in the process. With respect to the former, the membership of the U.S. Federal Geographic Data Committee covers a very wide range of substantive interests at the federal level. In contrast, the Indonesian and Malaysian National Geographic Information Systems tend to be focused mainly on surveying and mapping activities associated with land management. With respect to the latter, there are important differences between the NSDIs in terms of the extent to which the main stakeholders are involved in the management of them. The majority of first generation NSDI initiatives were limited to the public sector and most were largely concerned with central or federal government activities. Although essentially public sector in scope, ANZLIC is unusual in this category because it is centrally concerned with the interface between different levels of government. A notable exception to this rule was the Canadian Geospatial Data Infrastructure (CGDI) that brings together representatives from all levels of government together with the private sector and academia.

1.5   From the First to the Second Generation of NSDIs

The development of the Internet and the World Wide Web has had a profound impact on the transition from the first to the second generation of NSDIs. This was recognized by the U.S. Mapping Science Committee in 1999 in their report on distributed geolibraries. In their view "the WWW has added a new and radically different dimension to its earlier conception of the NSDI, one that is much more user oriented, much more effective in maximizing the added value of the nation's geoinformation assets, and much more cost effective as a data dissemination mechanism." As a result they conclude that the notion of "distributed geolibraries reflect the same thinking about the future of geospatial data with its emphases on sharing, universal access and productivity but in the context of a technology that was not widely accessible in 1993" (Mapping Science Committee, 1999).

Given these and other developments, Rajabifard et al. (2003) argue that the development of the second generation of NSDIs began around the year 2000. The second generation consists of two distinct groups: first generation NSDIs that have evolved from a product to a process model, and a substantial number of new entrants from all parts of the world. These entrants have benefited substantially from the growing body of materials that is available on the Web and elsewhere on the experiences of the first generation. A major factor in the dissemination process has been the steady increase in GSDI conference presentations available on the Web and the creation of the NSDI cookbook through the combined efforts of its technical working group. The cookbook was launched at the Capetown GSDI 4 conference in 2000 and is regularly updated on the GSDI Web site (www.gsdi.org).

Notwithstanding the technological innovations that had taken place during the last decade, the distinctive feature of the second generation of NSDIs is the shift that has taken place from the product model that characterized most of the first generation to a process model of an NSDI. Rajabifard et al. (2003) argue that database creation was to a very large extent the key driver of the first generation and that, as a result, most of these initiatives tended to be led by data producers and national mapping agencies. The shift from the product to the process model is essentially a change in emphasis from the concerns of data producers to those of data users. The main driving forces behind the data process model are data sharing and reusability of data collected by a wide range of agencies for different purposes at various times. Also associated with this change in emphasis is a shift from the centralized structures that characterized most of the first generation

NSDIs to the decentralized and distributed networks that are a basic feature of the Web.

Rajabifard et al. (2003) concluded that developing a successful SDI initiative depends at least as much on issues such as political support within the community, clarifying the business objectives which the SDI is expected to achieve, securing sufficient project funding, and enlisting the cooperation of all members of the community as on technical issues relating to spatial data quality, standards, software, hardware, and networking. Consequently, the creation of successful NSDIs must be seen as a socio-technical rather than a purely technical exercise.

Some of the implications of this change in emphasis are highlighted in a recent study comparing Australian, Canadian, and the U.S. experiences with respect to NSDI implementation (Masser, 2004). The findings of this study suggest that leadership involves a great deal more than coordination. Whereas coordination implies to some extent a reactive mode of operation within well-established structures, leadership implies a more proactive mode in situations where it may be necessary to create new forms of organization. This is evident in some of the different partnership structures that have emerged in all three countries to facilitate NSDI implementation. Table 1.3 shows that at least five different types of partnerships are in operation. These range from the restructuring of existing government agencies to the establishment of many joint ventures involving different combinations of the key stakeholders.

The study also highlights the extent to which effective NSDI implementation involves the active participation of many different agencies at the

TABLE 1.3

Some New Organizational Structures Facilitating NSDI Implementation in Australia, Canada, and the United States

Type	Main Task	Example
Restructuring within government structures	Creation and maintenance of an integrated land information database	Land Victoria, Australia
External to government structures	Delivery of wide range of eGovernment services	Service New Brunswick, Canada
Joint ventures consortium of data producers	Integration of datasets held by state and commonwealth agencies	Public services mapping agencies consortium, Australia
Joint ventures by key data users	Maintenance and dissemination of core datasets	Alberta Spatial Data Warehouse, Canada
Joint ventures by a wide range of data producers and users	Creation and sharing of core datasets	US-I team initiative

subnational and national levels. In each of these three countries, the lead agency in NSDI formulation is the national or federal government. However, its effective implementation lies to a considerable extent in the hands of the state and local government agencies who act as lead agencies at the subnational level. The findings of the analysis suggest that there is both a top-down and a bottom-up dimension to the relationships between the different levels involved in the NSDI implementation process. NSDI strategies drive statewide SDI strategies, and statewide SDI strategies drive local-level SDI strategies. As most of the detailed database maintenance and updating tasks are carried out at the local level, the input of local government also has a considerable influence on the process of SDI implementation at the state and national levels. The outcomes of such processes from the standpoint of the NSDI are likely to be that the nature of SDI implementation will vary considerably from one subnational agency to another. Consequently the NSDI that emerges from this process will be a collage of similar, but often quite distinctive, components that reflect the commitments and aspirations of the different subnational governmental agencies.

Another feature that distinguishes the second generation NSDIs is the increasing attention that is now being given to the creation of commercial opportunities for private-sector companies in NSDI implementation process. This is particularly evident in Canada through the work of the Geomatics Industry Association of Canada. This body has played an important role in the formation of the CGDI. The desire to exploit the commercial opportunities that are being created by the implementation of NSDIs is also recognized in Australia through Ministry of Industry's spatial information industry action agenda (DITR, 2001). An Australian Spatial Industry Business Association has been set up and is likely to play a major role over the next few years in the ASDI as lead agency for the implementation of the Action Agenda. In both Canada and Australia, an important theme in NSDI development is the need to develop export opportunities and increase the international competitiveness of the national GI industry.

1.6 Toward a Hierarchy of SDIs

The experience of the second generation NSDIs illustrates the links that have to be developed with different types of subnational administration to facilitate NSDI implementation. The notion of a hierarchy of SDIs can also be extended to higher levels. The Global Spatial Data Infrastructure Association provides a global international forum for the exchange of ideas and experiences, while bodies such as the EUROGI and the United Nations Permanent Committees for Geographic Information in Asia and the Pacific and the Americas perform a similar function at the regional level (Masser et al., 2003).

TABLE 1.4

From Global to Local SDIs

Global and Regional SDIs	National SDIs	Local SDIs
Global and regional forums for collaboration and the exchange of ideas and experiences	Strategic initiatives concerned with the management of national information assets	Municipal and provincial initiatives concerned with the operational needs of day-to-day decision making

Table 1.4 describes the main tasks that are carried out at the different levels of the hierarchy. From this it can be seen that the global and regional SDI bodies have a strong interest in strategic issues and are actively engaged in capacity building among their members. An interesting exception to this rule which falls somewhere between the regional and national levels in the hierarchy is the European Union's infrastructure for spatial information in Europe (INSPIRE) initiative. This initiative uses the European Union's political machinery "to make available harmonized sources of geographical information in support of the formulation, implementation and evaluation of Community policies. It relates to the base information collected in the Member States in order to respond to a wide range of policy initiatives and obligations at local, regional, national, international level." The current initiative deals largely with the information that is required to support environmental policy, but it is envisaged that the framework will be extended over time to include information specific to other policy sectors (http://inspire.jrc.it/).

The table also shows that the national level occupies a central position in this hierarchy as the critical link between the higher and lower levels (Rajabifard et al., 2000). It is also the link at which strategic initiatives regarding the management of national GI assets are formulated and implemented. It should be noted, however, that the term "national" is used in this context in a relative, rather than an absolute, sense. For example, the NSDI of the United States is federally driven, and other public-sector stakeholders at the state, county, and city levels have only a subsidiary role in its development, as does the private sector. This led a panel set up by the U.S. National Academy of Public Administration (1998) to recommend the establishment of a broadly representative National Spatial Data Council to complement the FGDC in providing national leadership and coordination for the NSDI. The U.S. Federal Government has not acted upon this recommendation.

Table 1.4 also shows that the main tasks associated with SDI development at the subnational levels are closely linked to the operational needs of day-to-day decision making. The Property Information Project (PIP) developed by the Australian State of Victoria provides a good example of the kind of collaborative effort that is needed to create a statewide SDI. The basic objective of this project is "to establish a common geospatial infrastructure between local and state government based around the digital cadastral map base" (Jacoby et al., 2002). This reflects the need of the state for information about proposed property developments that is handled by its 78 local government agencies (LGAs). Although the state maintains the cadastral map base there is often little or no commonality between the LGA data and their database. Because of this reason, Land Victoria obtained funding to match or reconcile each LGA database with that of the state. It was agreed that each LGA would be allowed free use of the state's database and would be periodically supplied with updates. In turn they had to agree to adopt Land Victoria's version and advise them of all proposed plans and subdivisions in their areas. PIP provided a well-structured approach that was independent of vendors as well as a low-risk path to GIS implementation for the LGAs. Given these circumstances, it is not surprising to find that 75 out of 78 LGAs had signed up to the scheme by the end of 2001, and as a result Land Victoria has been able to drastically reduce the amount of duplicative maintenance work within the state.

1.7   Discussion

The nature of the NSDI phenomenon, the ways in which NSDIs have evolved over the last decade and a half, and their place with the emerging hierarchy of SDIs have been discussed in the previous sections of this chapter. Here we consider some of the broader strategic questions that are associated with the creation of effective SDIs. These include the length of time that is likely to be required to create an effective NSDI, the costs that this is likely to incur and who will pay them, the links between NSDIs and eGovernment, and the cultural barriers that will have to be overcome by both data producers and data users during NSDI implementation.

1.7.1   How Long Will It Take to Create an Effective NSDI?

The old adage that Rome was not built in a day is equally applicable to NSDIs. The creation of NSDIs is a long-term task that may take years or even decades before they are fully operational. This process is likely to be an evolving one that will also reflect the extent to which the involved organizations are changing themselves over time. To some extent, it is also dependent on the availability of main elements of the institutional context that are needed to facilitate the NSDI implementation. Because of the particular institutional context that has emerged in Australia, e.g., owing to the administrative duties that have been allocated to the states, the task should be much simpler and take less time than will be the case in the United States. However, the existing institutional context can create barriers to implementation that hinder effective implementation in some countries and that the rate of progress may be faster in some less-developed countries where there are fewer obstacles of this kind to overcome. The rate of progress is also likely to be strongly influenced by the need for substantial capacity-building efforts to ensure that the maximum use is made of NSDI efforts.

Major changes in the form and content of NSDIs can be expected over time. A good example of this is the changes that have taken place to Britain's National Geospatial Data Framework that was launched in 1996. Its original Web site is no longer operational and enquirers are transferred to the Gigateway Web site that is funded by the Government through its National Interest Mapping Services Agreement with Ordnance Survey and administered by the Association for Geographic Information. Alongside these developments has been the recent emergence of SDIs for some of the regions within the United Kingdom that has been stimulated by the devolution of some powers to elected regional assemblies in Northern Ireland, Scotland, and Wales. The best developed of these initiatives is the proposals that have been made for a GI strategy for Northern Ireland by its Department of Culture, Arts and Leisure (2002).

1.7.2 How Much Will NSDIs Cost and Who Is Going to Pay for Them?

The answers to these questions will vary considerably from country to country. In Australia, for example, the creation and maintenance of statewide SDIs is to some extent self-financing because of the close links between surveying and mapping and cadastral activities at the state level. In the United States, on the other hand, financial considerations have bedeviled the development of the NSDI and led to a search for new and innovative ways of funding such activities (Urban Logic, 2000). In other countries, key data producers and data users have been compelled to join forces to fund the creation of essential core datasets. This is the case, for example, in the Netherlands where the Dutch Cadastre and Dutch Telecom have created a series of consortia with the municipalities and the regional public utility companies to create and maintain a large-scale map of the Netherlands (Murre, 2002).

In responding to this question, it should also be stressed that the costs of funding important NSDI elements such as coordination and maintaining metadata services are relatively small in comparison with the very high costs of core database creation and maintenance. This means that some of the key components of an NSDI can be put into place without incurring high costs. To deal with the high costs of core database creation, it will also be necessary in most countries to exploit alternative information sources such as remotely sensed data in addition to conventional survey technology. Much work can be done in this way without incurring the delays that are inevitably associated with conventional database creation.

1.7.3   What Is the Connection between NSDI and eGovernment?

There is a close connection between NSDI and eGovernment strategies. The geographic information held in NSDIs is an important input to eGovernment at all levels. It is also being increasingly recognized that the economic potential of public-sector information is considerable for the development of knowledge-based economies. In the European Union, for example, it has been estimated that the content sector already has a market size of 433 billion Euros and employs 4 million people. Geographic information has also been singled out as one of the potentially most valuable components of public-sector information because it supports a wide range of economic activities (see, e.g., Pira International Ltd and others, 2000).

One consequence of the growing importance of eGovernment and the emergence of knowledge-based economies is that GI policy is being increasingly seen as part of broader national and international information policy. In the EU, for example, a directive on the reuse of public sector information was approved by the European and the Council of Ministers in 2003 (CEC, 2003). Its main objective is to overcome what is seen as one of the major barriers to realizing the economic potential of public-sector information by ensuring that "the same basic conditions apply to all players in the European information market, that more transparency is achieved on the conditions for re-use and that unjustified market distortions are removed" (CEC, 2003, p. 3). Given these circumstances it seems likely that the future success of the EU's INSPIRE initiative will depend to a large extent on the way in which this directive is implemented by the national member states.

1.7.4   What Cultural Barriers Must Be Overcome During NSDI Implementation?

The development of NSDIs will require some fundamental changes in the organizational and institutional cultures of both data producers and data users. Until recently large data producers such as the national mapping and cadastral agencies have been natural monopolies in most countries and this is strongly reflected in their organizational cultures. However, Groot (2001) argued that in future they will be increasingly operating in competitive markets because of the growing number of commercially available substitutes for their products and services. Given the demands arising out of eGovernment and the knowledge-based economy described above, it will be necessary to devise regulatory frameworks to ensure that there is a level playing field for both these public-sector agencies and their competitors with regard to the development of value-added products and services in a free market.

The most important cultural barriers facing data users in connection with NSDI implementation are linked to the need to operate collectively with other users at higher and lower levels in the administrative hierarchy as well as with similar bodies in the neighboring areas. The problems associated with this task have been explored in detail by Wehn de Montalvo (2003) in her research on spatial data sharing which examines perceptions and practices in South Africa. As a result of an extensive qualitative analysis based on the findings of a number of semi-structured interviews with practitioners, she has developed and empirically tested a conceptual framework based on the assumptions underlying the theory of planned behavior in social psychology. The findings of this quantitative analysis generally bear out the relationships postulated in the theory and also give considerable insights into the factors that determine the willingness to share spatial data. They also indicate the relatively limited commitment that exists among those involved to promote data sharing despite the existence of high-profile initiatives such as the South African National Spatial Information Network whose successful implementation is dependent on a high-level spatial data sharing.

References

Australia New Zealand Land Information Council (ANZLIC), 1992, *Land information management in Australasia 1990–1992* (Canberra: Australia Government Publishing Service).

Barr, R., and Masser, I., 1997, Geographic information: a resource, a commodity, an asset or an infrastructure? In *Innovations in GIS 4*, edited by Kemp, Z. (London: Taylor & Francis), pp. 234–248.

Commission of the European Communities, 1995, *GI 2000: towards a European geographic information infrastructure* (Luxembourg: Directorate General XIII).

Commission of the European Communities, 2003, The re-use of public sector information, Directive 2003/98/EC of the European Parliament and of the Council, *Official Journal of the European Union, L 345*, 90–96.

Crompvoets, J., 2002, Global developments of national spatial data clearinghouses, *Proceedings of the 6th Global Spatial Data Infrastructure Conference* (GSDI 6), 15–19 September, Budapest, Hungary.

Department of Culture, Arts and Leisure, 2002, *A geographic information strategy for Northern Ireland: a consultation document* (Belfast: Department of Culture, Arts and Leisure).

Department of Industry, Tourism and Resources (DITR), 2001, *Positioning for growth: spatial information industry action agenda*, http://www.industry.gov.au (Canberra: DITR).

Executive Office of the President, 1994, Coordinating geographic data acquisition and access, the National Spatial Data Infrastructure, Executive Order 12906, *Federal Register 59*, pp. 17671–17674.

Groot, R., 2001, Reform of government and the future performance of national surveys, *Computers Environment and Urban Systems*, 25, 367–387.

Jacoby, S.J., Smith, J., Ting, L., and Williamson, I., 2002, Developing a common spatial data infrastructure between state and local government, *Int. J. Geographical Information Science*, 16, 305–322.

McLaughlin, J., 1991, Towards a national spatial data infrastructure, *Paper presented at the Canadian Conference on GIS March 1991* (Ottawa: Canadian Institute of Surveying and Mapping).

Mapping Science Committee, 1993, *Toward a coordinated spatial data infrastructure for the nation*, National Research Council (Washington D.C.: National Academy Press).

Mapping Science Committee, 1999, *Distributed geolibraries: spatial information resources*, National Research Council (Washington D.C.: National Academy Press).

Masser, I., 1999, All shapes and sizes: the first generation of National Spatial Data Infrastructures, *Int. J. Geographical Information Science* 13, 67–84.

Masser, I., 2004, *GIS Worlds: creating spatial data infrastructures* (Redlands, CA: ESRI Press).

Masser, I., Holland, P., and Borrero, S., 2003, Regional SDIs. In *Development of Spatial Data Infrastructures: from Concept to Reality*, edited by Williamson, I., Rajabifard, A., and Feeney, M.E. (London: Taylor & Francis).

Murre, L., 2002, GBKN—The large scale base map of the Netherlands: a map that took 25 years to complete! *Proceedings of the 6th Global Spatial Data Infrastructure Conference*, Budapest, Hungary, http://www.gsdi.org.

National Academy of Public Administration (NAPA), 1998, *Geographic information in the 21st century: building a strategy for the nation* (Washington D.C.: National Academy of Public Administration).

Office of Management and Budget (OMB), 1990, *Coordination of surveying, mapping and related spatial data activities*, Revised Circular A-16 (Washington D.C.: Office of Management and Budget).

Pira International Ltd, University of East Anglia, and KnowledgeView Ltd, 2000, *Commercial Exploitation of Europe's Public Sector Information* (Luxembourg: Office for Official Publications of the European Communities).

Rajabifard, A., Williamson, I., Holland, P., and Johnstone, G., 2000, From local to global SDI initiatives: a pyramid building blocks. *Proceedings of the 4th Global Spatial Data Infrastructure Conference*, Cape Town, South Africa.

Rajabifard, A., Feeney, M.E., Williamson, I., and Masser, I., 2003, National spatial data infrastructures. In *Development of Spatial Data Infrastructures: from Concept to Reality*, edited by Williamson, I., Rajabifard, A., and Feeney, M.E. (London: Taylor & Francis).

Urban Logic, 2000, *Financing the NSDI: National Spatial Data Infrastructure*, http://www.fgdc.gov (Reston, VA: FGDC).

Wehn de Montalvo, U., 2003, In search of rigorous models for policy oriented research: a behavioural approach to spatial data sharing, *URISA Journal*, 15(2), 19–28.

2

Urban Neighborhood Pattern Recognition Using High Spatial Resolution Remotely Sensed Data and Point-Based GIS Data Sources

Victor Mesev and Paul McKenzie

CONTENTS

2.1 Introduction

The recent revitalized interest in satellite urban remote sensing is as much the result of strengthening links with GIS as it is of the breakthrough in super-high spatial resolution satellite sensor data (Mesev, 2003a). For the first time, the rationale for using satellite sensor data in urban planning can go beyond coarse approximations of land-use characterization and become more sympathetic to the needs for up-to-date precision maps of land-use delineation, which in turn are critical for government policy decisions on individual household behavior and interaction (Donnay, 1999). As such, there is now also a realistic possibility that urban remote sensing can begin to emulate environmental remote sensing and play an increasingly central role for supporting national and regional policies on the estimation of population change (Chen, 2002), the calculation of quality-of-life indices (Lo, 2003), and the evaluation of transport flow, as well as the precise delineation of urban features and measurement of size, density, and height of buildings (Barnsley et al., 2003).

2.1.1 Difficulties of Remotely Sensing Urban Land Use

Wider adoption of data from Earth-orbiting satellite sensors by policy makers for urban mapping and monitoring is restricted essentially by two groups of factors: scale and land-use inference (Forster, 1985). Scale is commonly associated with the spatial resolution of sensor instrumentation, which determines the smallest discernible unit of measurement (the pixel) and is the direct consequence of the instantaneous field-of-view capabilities of the satellite sensor. In other words, the spatial resolution of a sensor determines the degree of spatial disaggregation and therefore the level of spatial clarity in the satellite image. Traditionally, satellite sensor data have been recorded at spatial resolutions considered too coarse for successful and routine identification and categorization of urban features (Welch, 1982). Until 1999, the most widely available satellite sensor data for optical urban observation have been obtained from Landsat multispectral scanning system (MSS) sensors at a spatial resolution of 79 m, the Landsat thematic mapper (TM) series of sensors at 30 m, and SPOT high-resolution visible (HRV) at 20 m (multispectral) and 10 m (panchromatic) scales. Data at these scales were never designed to represent precisely the intricate spatial variations and heterogeneity of the physical layout of built structures, and as a consequence, pixels typically represent aggregations of disorganized mixtures of urban land cover types (both built and natural) at inconsistent proportions. They, instead, are more suitable for citywide applications, focusing, for instance, on a more macro differentiation of built-up structures from natural biophysical land covers, or the categorization of residential density suitable only as input data for urban growth and density profile models. The second limitation to the more widespread adoption of satellite sensor data for urban monitoring, and hence policy making, is more conceptual rather than technical (Geoghegan et al., 1998). This limitation refers to the subjective interpretation of socioeconomic constructs and policy-driven administrative mixtures of urban land use from the biophysical configuration of land cover properties. This is where the advantage of remote sensing over GIS as a more objective data collection source quickly diminishes when human occupation is inferred from discrete multispectral radiometric values that only represent the reflective and emittance properties of the physical landscape.

Various combinations of coarse scales and loose inference between land cover and land use have historically restricted both scope and accuracy in urban applications using satellite sensor data. Rising to the challenge, some methodologies recognized the inherent heterogeneity of images of urban areas and developed techniques that manipulated pixels not in isolation but within groups using textural (Möller-Jensen, 1990; Myint, 2003), contextual, and spatial properties and arrangements. When classifying, the difficulty of representing highly heterogeneous urban pixels was also acknowledged by methodologies designed to incorporate information from outside the spectral domain (e.g., fuzzy sets, neural nets, and Bayesian modifications by Mesev, 1998, 2001). Other work includes three-dimensional urban remote sensing employing laser-induced detection and ranging (LIDAR; Barnsley et al., 2003) and interferometric synthetic aperture radar (SAR; Gamba et al., 2000; Grey et al., 2003), as well as promising advances in hyperspectral image interpretation and multiple endmember spectral mixture models (Rashed et al., 2003). The intention in all cases is to increase accuracy by reducing uncertainty in both the discrimination of urban land cover pixels and the inference of land-use pixels. However, success in improving urban land-use accuracy has been small to negligible, usually qualified by local site-specific and time-specific conditions.

Instead, research on the spatial, rather than the spectral, characterization of urban land cover has gained momentum in recent years bringing about the development of a number of quantitative indices. Among the most widely applied have been the scale invariant properties of fractal geometry, which many proponents have argued are capable of measuring the structural complexity and fragmentation of highly heterogeneous urban land cover (De Cola, 1989). These same objectives were also tested using syntactic pattern recognition systems employing graph-based methodologies (Barnsley and Barr, 1997) and automated expert systems (Tullis and Jensen, 2003). More recently, landscape or spatial metrics have revitalized fractal geometry as part of a suite of indices, including the contagion and patch density measurements (Wu et al., 2000; Herold et al., 2002; Greenhill et al., 2003), as well as spatial metrics directly related to urban sprawl (Hasse and Lathrop, 2003). Central to the formulation of these metrics is the concept of photomorphic regions, or homogeneous urban patches, which are routinely extracted from aerial photographs but are more of a challenge from satellite imagery (Aplin et al., 1999). The two principal criticisms of spatial metrics are that their functionality is completely dependent on an initial spectral characterization of the satellite imagery, and that they are conspicuously absent from the actual process of the characterization of homogeneous classes. In this sense, they are merely measuring the outcome of the classification regardless of the accuracy.

2.1.2 Aggregated Urban Spatial Patterns

The common factor in most of the methodologies outlined above that restricts improvements in classification and pattern recognition accuracy is undoubtedly the inability to measure urban land use at a scale fine enough to identify individual building characteristics and hence infer human behavioral processes. If the objective is to delineate the maximum extent of human settlement then traditional approaches using coarse spatial resolution imagery and aggregated government statistics may suffice. However, such citywide measures have yet to convince planners and decision makers of their importance and, as a consequence, play only peripheral roles within local government policies (Donnay, 1999). If proponents of remote sensing and GIS want to rebuild the reputation of their data, they need to seriously tackle the limitations of aggregated urban models and begin to embrace the challenges of disaggregated alternatives. Although such models may be more demanding theoretically and technically, they are nonetheless essential pragmatically.

Moreover, aggregate models are the standard vehicles for extraneous information commonly used in the augmentation of remotely sensed images representing urban land use (Chen, 2002). Most of these extraneous data are extracted from national population censuses. This is understandable given their ease of access, wide range of socioeconomic indicators, almost complete coverage, and reasonably reliable representation. However, census information is considered too confidential by most governments to report at the individual household level. Instead, households are aggregated by areal tracts of assumed uniformity. Standard dasymetric techniques (Langford, 2003), those using surface models (Martin et al., 2000), and statistical relationships (Chen, 2002) can help redress this homogeneity, especially at the urban fringe, but the majority of tracts within the city remain summative. The aggregated scale used in reporting census records may not be considered a major hindrance to integration with satellite imagery if the areal resolution is of comparable size, typically 900 and 400 m^2 for the Landsat TM and SPOT HRV sensors, respectively. Besides, these scales are adequate for many citywide applications, in particular monitoring urban sprawl and modeling density attenuation (Longley and Mesev, 2002). Moreover, accuracy in image interpretation using these relatively coarse sensor data with aggregated census tracts is highly variable, due in large part to site-specific inconsistencies in the imagery and the effects of the ecological fallacy and modifiable areal unit problem in the aggregated census tracts.

2.1.3   Disaggregated Urban Spatial Patterns

Instead of relying on aggregated census tracts, we will consider an alternative scale of measurement; disaggregated GIS data at the point level of abstraction (Harris and Longley, 2000). Point data representing individual buildings are grouped into localized two-dimensional spatial patterns characterizing various types of both residential and commercial developments (Mesev, 2003b). Suffice to say, the level of characterization available from point patterns is highly limited with only two properties distinctively and consistently measurable. These are density and linearity, both calculated by standard nearest neighbor and linear-adjusted nearest neighbor indices, respectively. Although point-based spatial patterns correspond with the location of buildings, they are nevertheless dimensionless and as such bear an incomplete relation with the actual physical size of the buildings (Harris and Chen, 2004). For a more complete representation, the reliance is on high spatial resolution imagery to delineate the physical extent of building size and shape. Point-based patterns can then be used to identify the configuration of various types of residential and commercial land use from second-order, classified satellite imagery. Their spatial properties will further be explored within an image pattern recognition system designed to identify and characterize urban structural patterns. The spatial resolution of the satellite sensor data has to match the disaggregated scale of these point-based patterns. As such traditional Earth-observation imagery are far too coarse to identify individual buildings. Instead, the advent of the latest generation of high spatial resolution imagery at unprecedented scales, such as 4 m (multispectral) and 1 m (panchromatic) from the IKONOS sensor (Space Imaging), has restored enthusiasm in more precise urban mapping by satellite imagery. Rather than interpret coarse approximations of residential or commercial groups of buildings, it is now possible to identify individual properties with a reasonable degree of spatial delineation. Of course, aerial photography has a long history in exact urban mapping, but unlike satellite imagery dynamic urban monitoring is hampered by the high cost of photographic acquisition, limited spatial coverage, and a lack of spectral information.

The long-term goal of the research outlined in this chapter is to reaffirm the inextricable links between satellite imagery and GIS data by identifying a role for point-based building patterns within image processing that would improve both the accuracy and precision of urban land-use interpretation from this very fine scale satellite imagery.

2.2 Point-Based Geographies

Point-based data are the ultimate in disaggregation. In Euclidean geometry terms, points are dimensionless; instead of size and shape, they represent the precise locations of individual entities. This chapter will introduce databases that hold point-based information representing the location of every postal delivery address in the United Kingdom. Created by the Ordnance Surveys of the United Kingdom and known as ADDRESS-POINT in Great Britain and COMPAS (COMputerized Point Address Service) in Northern Ireland both databases hold spatial and attribute information on the location of delivery addresses as well as whether they are residential or commercial properties. Tables 2.1 and 2.2 list the complete array of information from both databases.

The planimetric coordinates of the point data representing postal delivery buildings are claimed to be precise to within 0.1 m (50 m in some rural

TABLE 2.1

COMPAS Attribute Table

Field	Format	Description
Area	I7	Plan area of building (m^2)
Deletion date	I9	Date of building deletion from dataset
Date of survey	I9	Date building entered dataset
Eastings (1 m)	I6	Irish grid (1 m resolution)
Geocode	I12	Unique building reference
Irish grid sheet	A3	OSNI 1:10,000 map sheet
Level number	I1	0, dwelling house; 1, other use
Locality	A30	Townland name
Northings (1 m)	I6	Irish grid (1 m resolution)
Plan	A5	OSNI 1:1250/2500 plan
Postcode	A7	Approx. 14 addresses
Primary addressable object	A60	Premise number, name, organization
Street	A100	Street name
Town	A30	City name
Nondwelling buildings		
Level number	I10	Category of nondwelling building
Type	A3	Type building: commercial, communal

Source: Adapted from OSNI.

areas) of the actual location of the building. Both databases were created primarily using the Royal Mail's postcode address file (PAF) along with ground survey measurements and are updated on a frequent interval (usually in 3 months). Note that the COMPAS dataset includes an additional

TABLE 2.2

ADDRESS-POINT Attribute Table

Field	Format	Description
Address key	I8	Key identifying addresses
Building name	A50	If number not available
Building number	I4	Range 0–9999
Change type	A1	Insert, change, or delete
Change date	D6	Date of last change to record
Department name	A60	For organizations
Dependent locality	A35	Subdivision of post town
Eastings (0.1 m)	I8	National grid (0.1 m resolution)
Northings (0.1 m)	I7	National grid (0.1 m resolution)
OSAPR	A18	OS unique identifier
Physical status	I1	E.g., planned
PO box number	A6	PO box number
Postcode	A7	Approx. 14 addresses
Positional quality	I1	Accuracy of seed addresses
Post town	A30	Name of post town
Royal Mail version	I8	Date of last PAF update

Source: Adapted from OSGB.

attribute relating to the area of the point representing the location of a building. This areal feature is an approximation of the actual two-dimensional size but not the shape of the building.

Together these national datasets represent a major conceptual advance toward the deconstruction of the geography of urban areas and away from aggregated census tracts. In disaggregating the geographical distributions of individual households, these point data provide unique opportunities to view the urban landscape as a surface of discrete entities rather than the traditional and administratively convenient patchwork of aggregated and uniform surfaces partitioned only by artificial zones. Disaggregated surfaces also offer, for the first time, the possibility of analyzing the spatial configuration and density of individual addresses within neighborhoods, which are typically hidden by zonal representations (for a sensitivity analysis of density, refer to the work by Harris and Chen, 2004). The geographic coordinates of each point are positioned to correspond with the center of the building, and as such the only measurable parameters that can be derived are those relating to the distribution of points indicating density (compactness or sparseness) and arrangement (linearity or randomness). All in all, the creation of ADDRESS-POINT and COMPAS is a major step forward in the pursuit of "framework" data that encapsulate the desire for higher quality urban information not just for image-pattern recognition and classification improvement but also for all urban-based spatial data analyses.

2.2.1 Point-Based Geographies of Bristol and Belfast

Preliminary work using ADDRESS-POINT and COMPAS is centered on the cities of Bristol in South-Western England and Belfast in Northern Ireland (Figure 2.1).

Both cities have large populations, approximately 375,000 for Bristol and 278,000 for Belfast, at the time of the 2001 census; and both are relatively dense, occupying built areas of around 200 and 160 km^2, respectively. Similarities in the two cities are further evident in high traffic congestion and complex morphological structures, representing archetypal postindustrial economies. The mix of residential and commercial land-use patterns are highly interlaced and only physically discernible by remote sensing in some parts of the city centers and the more recent peripheral commercial estates. The spatial arrangement of residential street patterns are both highly variable, ranging from dense interwar linear design to modern curved geometrical layouts.

Almost every postal address is known by the post office to be in current use, although the databases also contain other minor categories, such as properties under construction. Frequent maintenance updates by the Ordnance Surveys of Great Britain and Northern Ireland ensure that the databases are much more contemporary than the census, although they do not contain any of the socioeconomic variables associated with the census. At first this may seem to be a major disadvantage but not when it is considered

FIGURE 2.1
Location of Bristol and Belfast. (© Crown Copyright Ordnance Survey. All rights reserved.)

that the only variables from the census to have had any impact on image classification are population and household categories (Mesev, 1998). Moreover, even these standard census variables are difficult to accommodate into standard image classifications given the aggregate nature of census tract representations. Instead, here we explore the possibility of calculating indices that characterize the spatial distributions of postal points and how these indices can be used to infer land use from classified land cover.

2.3 Nearest Neighbor Indices of Point Distributions

The nearest neighbor technique is statistically straightforward to calculate and ideal for expressing spatial distributions (Pinder and Witherick, 1973). It compares the *observed* average distance connecting neighboring points (D_{OBS}) and the *expected* distance among neighbors in a random distribution (D_{RAN}). The statistic is an uncomplicated ratio, where randomness is represented by parity; a clustering tendency has values approaching zero, and perfect uniformity toward a theoretical value of 2.15. The nearest neighbor statistic R is expressed as:

$$R = D_{OBS}/D_{RAN} \qquad (2.1)$$

where D_{OBS} is the total measured Euclidean distance between neighboring points divided by the total number of points (n), and D_{RAN} is calculated as

$$D_{RAN} = \frac{1}{2\sqrt{n/A}} \tag{2.2}$$

where (n/A) is the density of points within area A, in other words the number of postal points per unit of area, where area is constant throughout. One of the many strengths of the nearest neighbor statistic is the facility to compare spatial distributions on a continuous scale, especially when area (A) is constant.

2.3.1 Belfast and Bristol Nearest Neighbor Indices

The samples of nearest neighbor indices (R) documented in Tables 2.3 and 2.4 match the eight residential and six commercial postal point arrangements displayed in Figures 2.2 through 2.5 for the cities of Belfast and Bristol (derived from work by McKenzie, 2003).

Of the combined 16 different residential neighborhoods, 10 are successfully identified as having strong clustering patterns (residential types 1, 2, 4, and 7 for Belfast in Table 2.3 and Figure 2.2; and residential types 1, 2, 4, 5, 7, and 8 for Bristol in Table 2.4 and Figure 2.3). As expected, given the compact architecture of terraced (back-to-back row) housing in U.K. cities, the most clustered neighborhoods are demonstrated by the linear patterns of residential types 1 and 2 in Belfast and residential types 1, 2, 4, and 5 for Bristol. However, although their nearest neighbor values may be

TABLE 2.3

Density and Nearest Neighbor Statistics for Belfast [N = density (area constant); R = nearest neighbor; LN = linear density (area constant); LR = linear nearest neighbor]

Sample	N	R	LN	LR
Residential_1	581	0.547	30	0.470
Residential_2	514	0.584	12	0.543
Residential_3	184	1.306	26	0.761
Residential_4	340	0.584	24	0.609
Residential_5	571	1.040	51	1.237
Residential_6	400	0.976	27	1.115
Residential_7	230	0.642	15	0.619
Residential_8	247	0.936	21	0.868
Commercial_1	23	0.959	9	1.097
Commercial_2	65	0.508	24	0.623
Commercial_3	20	0.948	6	1.327
Commercial_4	378	0.577	15	0.414
Commercial_5	323	0.689	21	0.536
Commercial_6	35	0.759	16	0.538

TABLE 2.4

Density and Nearest Neighbor Statistics for Bristol [N = density (area constant); R = nearest neighbor; LN = linear density (area constant); LR = linear nearest neighbor]

Sample	N	R	LN	LR
Residential_1	1479	0.466	53	0.452
Residential_2	898	0.614	20	0.503
Residential_3	365	0.942	11	1.595
Residential_4	906	0.563	38	0.538
Residential_5	640	0.585	19	0.589
Residential_6	81	1.564	14	0.995
Residential_7	673	0.697	11	0.660
Residential_8	1117	0.555	26	0.681
Commercial_1	54	0.243	8	0.422
Commercial_2	204	0.364	27	0.396
Commercial_3	71	0.409	22	0.433
Commercial_4	41	0.724	15	1.104
Commercial_5	673	0.301	41	0.526
Commercial_6	242	0.478	21	0.450

very similar, it is plainly apparent that Residential type 1 in Figure 2.3 exhibits a far denser concentration of postal points (N). The same situation applies between residential types 2 and 4 for Bristol (both inner-city local government–owned estates) and Residential type 5 (a more affluent 1980s peripheral estate) where, although all three have similar nearest neighbor values, only Residential type 5 has a considerably lower density profile. For Belfast (Figure 2.2) nearest neighbor similarities between residential types 1, 2, 4, and 7 are again punctured by marked differences in density, this time higher N values for types 1 and 2. Conversely, some types have very similar densities yet somewhat dissimilar nearest neighbor values. For example in Belfast, residential types 1 and 2 have considerably more clustered patterns than Residential type 5 but all three share similar densities. The same applies to the residential types 1 and 2 in Bristol over Residential type 5. Samples demonstrating more uniform nearest neighbor tendencies are characteristic of lower density, more sporadic residential patterns. Generally speaking, changes in R values are inversely in proportion to changes in D. For example, samples with the highest nearest neighbor values (residential types 3 and 5 for Belfast and Residential type 6 for Bristol) also have the lowest densities. On the ground, these residential types are examples of highly affluent low-density areas of Belfast and Bristol with large dwellings (as demonstrated by area-proportioned COMPAS points labeled as **D** in Figure 2.9). Overall, what is clear is that if nearest neighbor and postal point densities are taken together they are valuable measures for identifying and characterizing different residential types.

The same measurements can also be applied to commercial postal points (Figures 2.4 and 2.5). Strong clustering patterns (low N values) are

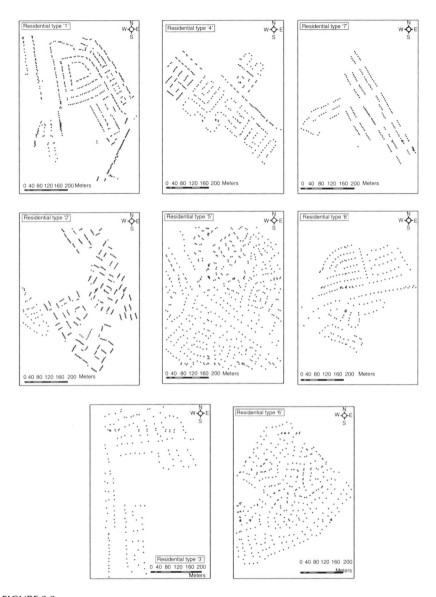

FIGURE 2.2
Residential COMPAS arrangements for Belfast. (Courtesy of COMPAS. Reproduced under permit 70109 from the Ordnance Survey of Northern Ireland map with the permission of the Controller of HMSO. © Crown Copyright 2007.)

indicative of linear developments in Commercial type 5, the city center represented by Type 4, and peripheral estates characterized by types 2 and 6, all in Belfast (Figure 2.4), and types 2 and 6, 5, and 3, respectively, for Bristol (Figure 2.5). However, given the nature of commercial development,

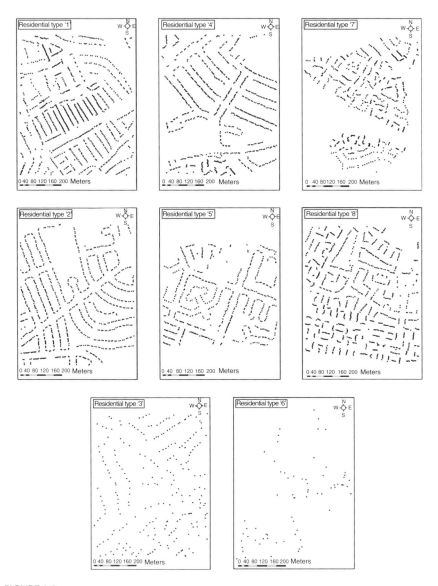

FIGURE 2.3
Residential ADDRESS-POINT arrangements for Bristol. (Courtesy of ADDRESS-POINT.
© Crown Copyright Ordnance Survey. All rights reserved.)

density values are highly variable, highest in linear developments and city centers, less so in industrial estates, and lowest in predominantly residential areas (the latter exemplified by commercial types 1 and 3 in Belfast and Type 4 in Bristol). As with residential types, if both nearest neighbor and density values are taken in combination then unequivocal differences can be revealed.

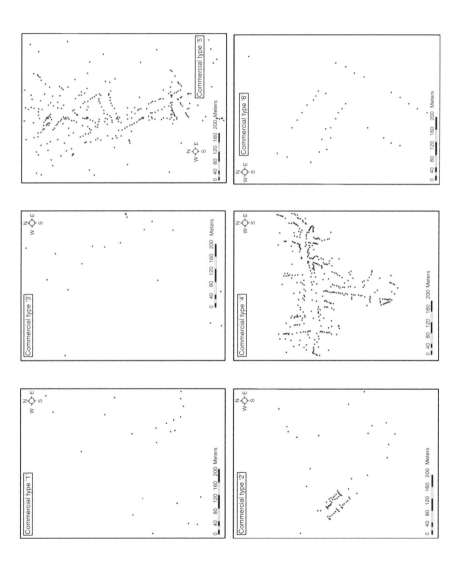

FIGURE 2.4

Commercial COMPAS arrangements for Belfast. (Courtesy of COMPAS. Reproduced under permit 70109 from the Ordnance Survey of Northern Ireland map with the permission of the Controller of HMSO. © Crown Copyright 2007.)

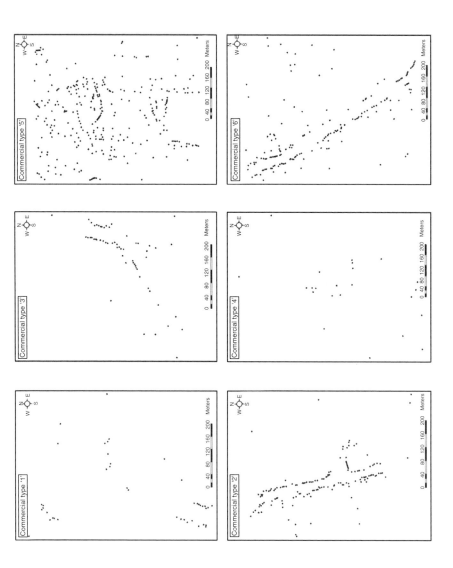

FIGURE 2.5
Commercial ADDRESS-POINT arrangements for Bristol.

2.3.2 Linear Nearest Neighbor Indices of Point Distributions

The conventional nearest neighbor statistic is effective for measuring clustering patterns, but as it stands, it lacks the ability to detect spatial arrangement.
A variant of two-dimensional nearest neighbor analysis is the linear adjustment (LR) devised by Pinder and Witherick (1975). Instead of measuring all
observed distances between neighboring points, D_{OBS} is determined from a
linear sequence (L) of consecutive points (LN) in all directions (essentially
linear subsets demonstrated by Figure 2.6), while LD_{RAN} is

$$LD_{RAN} = 0.5\left(\frac{L}{n-1}\right) \tag{2.3}$$

What constitutes a linear arrangement is of course dependent on scale. Any
grouping of buildings regardless of configuration can be considered linear if
the scale is fine enough. As a consequence, a minimum of 20 buildings was
considered representative of a linear formation of both residential and
commercial postal points. In a sense, the figure is arbitrary but was determined by a series of sensitivity tests designed to establish representation
within each sample types.

On the whole LR values are similar to the original R, and LN is similar in
relative terms to N (Tables 2.3 and 2.4). The exceptions are those samples
with either a lack of any definitive linear arrangements or loose linear
structures. For instance, although Residential type 8 for Bristol (Figure 2.3)
exhibits high-clustering and high-density properties, similar to types 1, 2,
and 5, a lack of consistent linearity is revealed by an LR value of 0.881

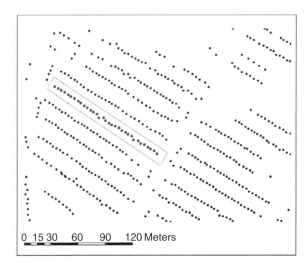

FIGURE 2.6
Selecting linear nearest neighbor arrangements. The highlighted linear sequence of postal
points satisfies the minimum 20-point threshold. (Reproduced under permit 70109 from
the Ordnance Survey of Northern Ireland map with the permission of the Controller of
HMSO. © Crown Copyright 2007.)

compared to 0.565 for *R*. Other, less noticeable cases of slight increases in LR are evident in residential types 5 and 6 in the Belfast example. On the other hand, Residential type 3 for Belfast is typical of a sample with a highly uniform or random nearest neighbor (1.306) but a more clustered linear nearest neighbor (0.761). Similar cases can also be observed from the commercial samples in Figures 2.4 and 2.5. The most striking is Belfast Commercial type 6, which exhibits uniform tendencies (a high *R*-value of 0.759) but higher linearity (LR = 0.538). Finally, the sparse commercial distributions of types 1 and 2 from Belfast and 4 from Bristol shows marked increases in LR values over *R*, clearly demonstrating a lack of linear behavior.

2.4 Using Postal Point Data to Infer Land Use from IKONOS Imagery

Summaries of residential and commercial postal point patterns using the nearest neighbor and linear nearest neighbor techniques can be used to identify similar spatial patterns in classified imagery. In addition, postal points can also be used to identify two types of misclassifications: confusion between built and nonbuilt land covers (especially the spectral similarity between built structures and bare rock or bare soil), and confusion between residential and commercial built land uses. These typical misclassifications are illustrated in Figure 2.7 within a methodology and flow of operations that begins with the spatial association of classified pan-sharpened IKONOS sensor data with postal points.

However, before the methodology is examined in more detail, it is important to first emphasize the notorious difficulty of spectrally classifying urban land use, mostly because of the overwhelmingly large proportion of pixels that represent mixed land cover types. As a rule, those pixels which almost entirely represent built surfaces with impervious, highly specular characteristics are straightforward to classify. The problem arises with pixels that represent not only impervious surfaces but also considerable proportions of biophysical properties (Ridd, 1995). As a result, most misclassifications occur in areas of low building density within urban areas as well as along urban fringes. Various approaches for alleviating this problem have tried to piece together "missing" portions of the misclassified urban morphology. Most have resorted to using extraneous GIS data, usually housing information from the census, and most, as this chapter has repeatedly alluded, have used these data at the aggregated level of representation. Notable advances in research on census disaggregation have centered on the use of a trend surface model to inform contextual labeling at the post-classification stage, along with the ability to inform training samples and modify the prior probabilities of the classical maximum likelihood

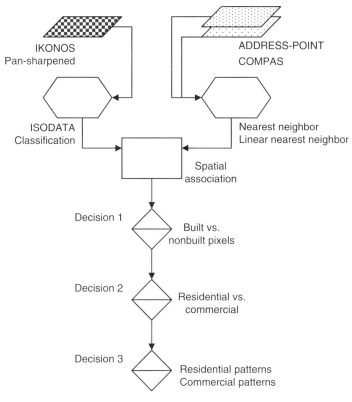

FIGURE 2.7
Methodology and flow of operations.

discriminant function (Mesev, 1998). The surface model, which was built on a distance decay interpolation procedure, is essentially a dasymetric technique that seeks to both identify the built geography and at the same time eliminate the nonresidential urban areas (Martin et al., 2000). The disaggregation effect of the surface model is noticeable and it serves as a useful vehicle for importing census ancillary information into the classification process by producing sharper estimates of the spatial distribution of property types (terraced, detached, semidetached, and apartments) than could be obtained from the image alone. Although reasonable improvements in classification accuracy are reported, the disaggregation is still at a relatively coarse scale and the information within the surface cell is an average number of households, not individual dwellings.

Postal points, on the other hand, represent the entire distribution of individual dwelling units within a city, and as such are the ultimate in disaggregated surfaces. They convey valuable information on local spatial association—density and arrangement—information that is surprisingly overlooked in research on urban image classification, especially given the spatial nature of class distributions and the inherent limitations of spectral data. Postal points represent the location of all individual buildings and as such are an invaluable source of information for identifying misclassified pixels that represent not only omitted buildings (errors of omission) but have also erroneously included the location of buildings (errors of commission). The identification of both types of errors should inevitably lead to a better understanding of the reasons behind misclassified urban pixels.

In the Bristol and Belfast examples, the use of postal points has identified a number of misclassified urban pixels from IKONOS imagery. Most of these are at the urban fringe and include the spectral similarities between pixels representing built land cover and those representing bare rock or bare soil. In addition, some misclassifications have been identified between pixels representing residential and commercial land use. In both instances, the identification of misclassified pixels uses the categorization of IKONOS imagery using the ISODATA algorithm available from the ERDAS Imagine 8.6 proprietary software. Spatial masks and iterative spectral clustering using panchromatic-sharpening of all four available multispectral IKONOS bands (blue, green, red, and near infrared) then yielded a reasonably accurate classified image of 88% for the built land cover, and 72% and 67% for the residential and commercial land uses, respectively. The ground truth for these accuracy assessments was provided by the interpretation of digital aerial photographs collected in September 2001 by GeoInformation International in Cambridge as part of its Cities Revealed series. The photographs were captured at spatial resolutions of 12.5 cm, with a 15.25 cm camera focal length, and at a height of 3200 m. Figure 2.8 illustrates the close correspondence between COMPAS points and the aerial photograph for Belfast.

In addition to identifying misclassified pixels, the spatial distribution of postal points, characterized in terms of density and arrangement, can be used to infer types of residential and commercial developments identifiable in the IKONOS imagery classified as built. Figure 2.9 illustrates the spatial correspondence between a subset of the classified IKONOS image of Belfast and the location of postal points from the COMPAS dataset. Note that the postal points (open squares for residential and shaded squares for commercial) have been scaled in proportion to the area of the buildings, based on the "area" attribute unique only to COMPAS. White areas of the image have been classified as buildings, light gray as roads, and dark gray as nonbuilt biophysical land covers. Four distinctive residential building types are evident in Figure 2.9 and are labeled as follows:

A. This is an area of very high-density residential buildings, showing a highly linear arrangement, similar to the patterns represented by residential types 1, 2, and 4 in the Belfast samples (Figure 2.2), and residential types 1, 4, and 8 in Bristol (Figure 2.3).

FIGURE 2.8
Spatial correspondence between Cities Revealed aerial photograph of Belfast (at spatial resolution of 15 cm) and COMPAS points (squares proportioned by area attributes). (Courtesy of COMPAS. Based on the Ordnance Survey of Northern Ireland map with the permission of the Controller of HMSO © Crown Copyright 2007. Permit 70109.)

B. In contrast to **A**, these are lower density areas with less of a linear arrangement, and demonstrating properties displayed by residential types 4 and 8 in Belfast, and residential types 2, 4, and 5 in Bristol.

C. Here, medium density and a linear arrangement are recognizable and associated with residential types 7 and 8 in Belfast, and type 8 in Bristol.

D. Lastly, a very low-density pattern of residential buildings, indicative of a combination of residential types 3 and 8 in Belfast, and type 3 in Bristol.

As this subset is a residential suburb of Belfast, only one commercial pattern is observable; at **E** which is similar to commercial types 1 and 2 in Belfast, and types 1 and 4 in Bristol. Note the large area of classified built pixels to the west of **X**. These correspond to nonresidential buildings, in this case, a school. This is because the highly irregular form of nonresidential buildings can be seldom

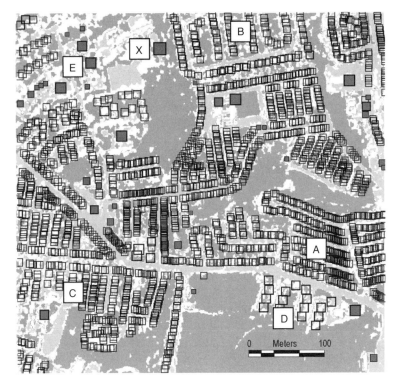

FIGURE 2.9
Classified IKONOS image of Belfast subset: white areas represent pixels classified as residential, dark gray areas classified as nonbuilt, and light gray as roads. Residential (open squares) and commercial (shaded squares) buildings from the COMPAS dataset are represented in proportion with area attributes. Note the very high density, linear arrangement of residential buildings at **A**; lower density, curvilinear distribution at **B**; medium density, linear arrangement at **C**; and very low density, more uniform configuration at **D**. As this subset is a residential suburb of Belfast, only one commercial pattern is observable at **E**. The large area of classified built pixels to the west of **X** corresponds to nonresidential buildings, in this case, a school. (Courtesy of COMPAS. Based on the Ordnance Survey of Northern Ireland map with the permission of the Controller of HMSO © Crown Copyright 2007. Permit 70109.)

represented by convenient square point patterns, regardless of the area attribute. Finally, a quick comparison with Figure 2.8 reveals many misclassified pixels throughout the subset. Most of these are a result of shadows, a problem rarely confronted in conventional spectral or even spatial classifications.

All five building types have been identified visually and are therefore highly subjective. What is now needed is a more automated approach that can improve precision and objectively implement the comparison of sample nearest neighbor indices of both residential and commercial land uses with classified imagery. But at the same time be able to also accommodate these same nearest neighbor indices within an image classification methodology. Research is currently examining the feasibility of an automated image pattern recognition system to facilitate both of these objectives.

2.5 Conclusions and Further Research

The proprietary postal products outlined in this chapter represent two disaggregated urban datasets with tremendous potential for inferring land use from remote sensing. They can be used to help discern at least three geographies: the built environment, and the structural configuration of residential and commercial land use, all of which are spatially exhaustive, regularly updated, and highly precise. Ongoing research has been presented on the implementation of the nearest neighbor and linear nearest neighbor indices for calculating unique summaries of the various structural patterns of postal points representing density and linear arrangements of residential and commercial buildings. These summaries have an immense future for inferring land use from land cover patterns classified from high spatial resolution remotely sensed data. Results so far, for the cities of Bristol and Belfast, are most encouraging but further testing is crucial if an even closer relationship between imagery and postal information can be statistically established. In particular, research breakthroughs are needed in linking classified land cover with land use by means of not only nearest neighbor techniques but also entropy maximization (Harvey, 2002); as well as the resolution of nonresidential from residential land-use patterns using targeted training sample selections; neighborhood differentiation using invariant fractal dimensions (Longley and Mesev, 2002); and urban growth using spatial metrics (Pesaresi and Bianchin, 2001).

More testing and refinement is necessary before a fully automated pattern recognition system can be fully implemented and more results evaluated. The importance of such work is especially relevant given the recent proliferation of very high spatial resolution imagery at $4/1$ m and $2.4/0.6$ m from the IKONOS and QUICKBIRD sensors, respectively. Urban land cover information at such spatial resolutions is discrete and highly identifiable. Similar discrete subsidiary information, for example in the form of postal points, is critical for converting image-derived urban land cover into urban land use within a reliable degree of consistency and accuracy. This will address the common restricting factor in most traditional classification and pattern recognition methodologies, namely the undoubted inability to measure urban land use at a scale fine enough to identify individual building characteristics and hence infer human behavioral processes.

The methodology of characterizing postal points for use in image pattern recognition is also an effective means for integrating GIS data with remote sensing where the benefits of both are harmonized in the pursuit of greater accuracy. A future research direction would be to establish direct relationships between socioeconomic information within census tracts and the spatial distribution of postal points. In this way, both attribute and spatial indicators would be readily available to inform multispectral classifications of urban areas. However, for the time being, research is focused on the spatial utility of postal points and how spatial indices can be used to identify and infer land use from high spatial resolution land cover data. Once residential and commercial characteristics from postal points are established and comprehensively tested, a situation is envisaged where land cover patterns can be routinely categorized into a variety of types. A fully automated procedure, with many more postal point configurations, will allow consistent pattern recognitions both across settlements and through time. The advent of disaggregated models, built by postal points, represents a major step forward in precision urban mapping, which is intuitively more realistic than the uniformity of traditional areal representations. The resulting high-precision maps may at long last become central to many government planning and policy-making decisions.

Acknowledgments

Cities Revealed is the copyright of the GeoInformation Group (www. crworld.co.uk).

References

Aplin, P., Atkinson, P., and Curran, P., 1999, Per-field classification of land use using the forthcoming very fine spatial resolution satellite sensors: problems and potentials, in *Advances in Remote Sensing and GIS Analysis*, edited by Atkinson, P. and Tate, N., pp. 219–239 (Chichester: John Wiley & Sons).

Barnsley, M.J. and Barr, S.L., 1997, A graph-based structural pattern recognition system to infer land use from fine spatial resolution land cover data, *Computers, Environment and Urban Systems*, 21, 209–225.

Barnsley, M.J., Steel, A.M., and Barr, S.L., 2003, Determining urban land use through an analysis of the spatial composition of buildings identified in LIDAR and multispectral image data, in *Remotely Sensed Cities*, edited by Mesev, V., pp. 47–82 (London: Taylor & Francis).

Chen, K., 2002, An approach to linking remotely sensed data and areal census data, *International Journal of Remote Sensing*, 23, 37–48.

De Cola, L., 1989, Fractal analysis of a classified satellite image, *Photogrammetric Engineering & Remote Sensing*, 55, 601–610.

Donnay, J.-P., 1999, Use of remote sensing information in planning, in *Geographical Information and Planning*, edited by Geertman, S. and Openshaw, S., pp. 242–260 (Berlin: Springer).

Forster, B.C., 1985, An examination of some problems and solutions in monitoring urban areas from satellite platforms, *International Journal of Remote Sensing*, 6, 139–151.

Gamba, P., Houshmand, B., and Saccani, M., 2000, Detection and extraction of buildings from interferometric SAR data, *IEEE Transactions on Geoscience and Remote Sensing*, 38, 611–618.

Geoghegan, J., Pritchard, L., Jr., Ogneva-Himmelberger, Y., Chowdbury, R.R., Sanderson, S., and Turner, B.L., 1998, "Socializing the pixel" and "pixelizing the social" in land use and land cover change, in *People and Places: Linking Remote Sensing and Social Science*, edited by Liverman, D., Moran, E.F., Rindfuss, R.R., and Stern, P.C., pp. 51–69 (Washington, DC: National Academy Press).

Greenhill, D.R., Ripke, L.T., Hitchman, A.P., Jones, G.A., and Williamson, G.G., 2003, Characterization of suburban areas for land use planning using landscape ecological indicators derived from Ikonos-2 multispectral imagery, *IEEE Transactions on Geoscience and Remote Sensing*, 41, 2015–2021.

Grey, W.M.F., Luckman, A.J., and Holland, D., 2003, Mapping urban change in the UK using radar interferometry, *Remote Sensing of Environment*, 87, 16–22.

Harris, R.J. and Chen, Z., 2004, Giving dimension to point locations: urban density profiling using population surface models, *Computers, Environment and Urban Systems*, 29, 115–132.

Harris, R.J. and Longley, P.A., 2000, New data and approaches for urban analysis: modelling residential densities, *Transactions in GIS*, 4, 217–234.

Harvey, J.T., 2002, Estimating district census populations from satellite imagery: some approaches and limitations, *International Journal of Remote Sensing*, 23, 2071–2095.

Hasse, J. and Lathrop, R.G., 2003, A housing-level approach to characterizing residential sprawl, *Photogrammetric Engineering and Remote Sensing*, 69, 1021–1030.

Herold, M., Scepan, J., and Clarke, K.C., 2002, The use of remote sensing and landscape metrics to describe structures and changes in urban land uses, *Environment and Planning A*, 34, 1443–1458.

Langford, M., 2003, Refining methods for dasymetric mapping using satellite remote sensing, in *Remotely Sensed Cities*, edited by Mesev, V., pp. 137–156 (London: Taylor & Francis).

Lo, C.P., 2003, Zone-based estimation of population and housing units from satellite-generated land use/land cover maps, in *Remotely Sensed Cities*, edited by Mesev, V., pp. 157–180 (London: Taylor & Francis).

Longley, P.A. and Mesev, V., 2002, Measurement of density gradients and space-filling in urban systems, *Papers in Regional Science*, 81, 1–28.

Martin, D.J., Tate, N.J., and Langford, M., 2000, Refining population surface models: experiments with Northern Ireland census data, *Transactions in GIS*, 4, 343–360.

McKenzie, P., 2003, Urban neighbourhood patterns: dasymetric links between IKONOS imagery and new point-based GIS data sets, *Unpublished MSc Thesis*, School of Environmental Sciences, University of Ulster.

Mesev, V., 1998, The use of census data in urban image classification, *Photogrammetric Engineering and Remote Sensing*, 64, 431–438.

Mesev, V., 2001, Modified maximum likelihood classifications of urban land use: spatial segmentation of prior probabilities, *Geocarto International*, 16, 39–46.

Mesev, V., 2003a, *Remotely Sensed Cities* (London: Taylor & Francis).

Mesev, V., 2003b, Neighborhood pattern recognition from mailing information: links with satellite imagery, *Online Journal of Space Communication*, 3 (http://satjournal. tcom.ohiou.edu/pdf/Mesev.pdf).

Möller-Jensen, L., 1990, Knowledge-based classification of an urban area using texture and context information in Landsat-TM imagery, *Photogrammetric Engineering and Remote Sensing*, 56, 899–904.

Myint, S.W., 2003, The use of wavelet for feature extraction of cities from satellite sensor images, in *Remotely Sensed Cities*, edited by Mesev, V., pp. 109–134 (London: Taylor & Francis).

Pesaresi, M. and Bianchin, A., 2001, Recognizing settlement structure using mathematical morphology and image texture, in *Remote Sensing and Urban Analysis*, edited by Donnay, J.-P., Barnsley, M.J., and Longley, P.A., pp. 55–67 (London: Taylor & Francis).

Pinder, D.A. and Witherick, M.E., 1973, Nearest neighbour analysis of linear point patterns, *Tijdschift voor Economische an Sociale Geographie*, 64, 160–163.

Pinder, D.A. and Witherick, M.E., 1975, A modification of the nearest-neighbour analysis for use in linear situations, *Geography*, 60, 16–23.

Rashed, T., Weeks, J.R., Roberts, D., Rogan, J., and Powell, R., 2003, Measuring the physical composition of urban morphology using multiple endmember spectral mixture models, *Photogrammetric Engineering and Remote Sensing*, 69, 1011–1020.

Ridd, M.K., 1995, Exploring A V-I-S (vegetation–impervious surface–soil) model for urban ecosystem analysis through remote-sensing—comparative anatomy for cities, *International Journal of Remote Sensing*, 16, 2165–2185.

Tullis, J.A. and Jensen, J.R., 2003, Expert system house detection in high spatial resolution imagery using size, shape, and context, *Geocarto International*, 18, 5–15.

Welch, R., 1982, Spatial resolution requirements for urban studies, *International Journal of Remote Sensing*, 3, 139–146.

Wu, J., Jelinski, E.J., Luck, M., and Tueller, P.T., 2000, Multiscale analysis of landscape heterogeneity: scale variance and pattern metric, *Geographic Information Sciences*, 6, 6–16.

3

Geodemographics

Richard Webber

CONTENTS

3.1   Context

Geodemographics is a term used to define an increasingly important field of research that involves the classification of consumers according to the type of residential area in which they live. The practice was pioneered in the early 1970s to assist governments with the identification of inner-city communities for which different policy interventions were appropriate (Webber, 1975; Webber and Craig, 1978). Since the early 1980s, the application has subsequently spread to commercial organizations who have sought to tailor their investments in facilities and in communications to the specific interests of the local communities that they service (Weiss, 1988; Sleight, 2004). Today most of the large consumer-facing international brands use geodemographic classification to improve their business performance in applications such as retail-site location, the setting of local sales targets, the distribution of promotional material, customer relationship management, and risk management. As governments seek to adopt proven techniques from the private sector, recent years have witnessed a renewed interest in the application of geodemographic classifications in sectors such as policing, health and education, and areas of public sector service provision which absorb high levels of funding but for which responsibility is devolved to local delivery units because of the wide variations in service need at a local level. During recent years the use of geodemographics has extended beyond the United States and the United Kingdom to cover most of continental Europe and much of East Asia.

Because of the geographical nature of the application, most users of geodemographics recognize the need for the investment in some form of information system for manipulating the geographical information they hold regarding the home locations of their customers, the postal, administrative, media, and sales geographies used in their business, and the locational information they hold about their outlets and those of their competitors. However, many geographers have found it more difficult to recognize the differences between conventional GIS and geodemographic information systems than their similarities. This has often led to a failure to recognize the bespoke investments that are needed in software solutions as well as in data and visualization tools in order to sustain effective returns from this form of analysis.

Many successful commercial applications of GIS to the analysis of human behavior involve common elements structured in familiar ways but in a bespoke development. The developer is likely to work to a brief which will list the most critical applications to which the system will be put. An assessment is made from various datasets needed to support the application. These will be referenced to each other and configured within an established set of software tools. Query opportunities will be made available to users via some form of network. Operators will then be trained in the use of the system to support the set of applications agreed at the outset of the project.

Typically the system will then be capable of supporting additional queries. However, in practice, the modifications needed to support extra functions will often need to be handled by information specialists. Such a model, to which real-life applications only approximate, typically proves highly effective in applications which are predictable, involve use by operators on a routine rather than an occasional basis, support operational rather than strategic queries, and where operational savings are easy to quantify and demonstrate. Elsewhere, and commonly in academic and research environments, users make use of powerful GIS packages to undertake a series of bespoke analyses.

The key difference between geodemographic information systems and mainstream GIS is that whereas conventional GIS tools and datasets are application independent, geodemographics involves the structuring of GIS software and geographical databases in a generic form which is designed to support a general class of users thought to have similar application requirements. The customer of a geodemographic system therefore purchases, or more often leases, an application which is largely prebuilt, and in which different types of data are preconfigured both in relation to each other and standard GIS tools. Such systems are then supported with an ongoing training, consultancy, and updating service which is of a standard level of service and supplied at prices based on a standard rate-card.

Such an approach necessarily reduces the specificity of each application because the product itself is generic. However, the approach does assure users of access to standard industry methods of tackling particular applications. The other principal benefit is the lower cost of access to these applications and, in a commercial environment, the security of knowing that one is no longer at a competitive disadvantage to rivals who may have the resources to design and commission their own systems.

3.2   Origins of Geodemographics: The Classification of Residential Neighborhoods

Geodemographics originated as a distinct concept in 1974. During that year geographers in the United States and in the United Kingdom independently experimented with the concept of a nationwide classification of residential neighborhoods using the finest level of geography for which census statistics were published in these two countries. These were the "block group" in the United States and the "census enumeration district" in the United Kingdom. Using cluster analysis techniques, researchers identified that whereas every census output area was unique, there were nevertheless significant numbers of census output areas whose demographic patterns were broadly similar. By using the computer to search census output areas whose demographics were broadly similar across all the different topics covered by the census, it was possible to identify a limited number of neighborhood types to which every census output area to a varying degree approximated. By examining the key features which differentiated each of these clusters from their respective national averages, it was possible to create statistical profiles to help researchers understand the function that each type of neighborhood played in a complex urban residential system.

What transformed a basic urban research tool into a concept of relevance to a much wider audience was the emergence of tools which could relate residential addresses to these neighborhood clusters on a national basis. This was made possible in the United States by the development of geocoding systems. These allowed researchers to take a list of names and addresses and append block group identifiers to them. Using the correspondence table listing the classification assigned to each block group, this made it possible to code each individual address by a type of neighborhood. Finally,

TABLE 3.1

Variations in Victimization Rates for Different Types of Crimes
in North and East Devon [Rates as a Percentage of the Average
Rate for the Study Area]

Mosaic Groups	Incidents per 1000 Households	Same Postcode as Offender	Offender Detected
High-income families	69	58	60
Suburban semis	70	58	64
Blue-collar owners	112	95	143
Low rise council	146	145	193
Council flats	318	414	383
Victorian low status	193	216	227
Town houses and flats	118	117	112
Stylish singles	198	225	177
Independent elders	56	58	43
Mortgaged families	98	96	115
Country dwellers	72	72	60

by comparing the proportions of these names and addresses falling within
each class of neighborhood with the corresponding proportions for the
country as a whole, it became possible to profile an address file, in other
words to identify whether the persons' addresses one was analyzing were
predominantly from high- or low-income areas, from areas of young people
or old, from urban or rural, and from ethnic or white neighborhoods. In the
United Kingdom, exactly the same method of analysis could be used provided that the address files contained postcodes, using a correspondence
table between postcodes and census output areas.

Table 3.1 provides a good example of how police use information from
operational databases to identify variations in the level of victimization
experienced in different types of neighborhood within a force area, variations in the success of the police in clearing up the crime, and variations in
the concentrations of offenders between neighborhood types. Although the
areas of "Independent Elders" generate very few offenders and victimization rates are low, these prosperous retirees have legitimate complaints that
the police are relatively ineffectual at apprehending offenders in these areas
compared, for example, with areas of low-rise local authority housing.

3.3   Applications of Neighborhood Classification Systems

Although both in the United States and the United Kingdom the principal
intended use of neighborhood classifications was for public policy applications, the tool rapidly "escaped" into the private sector. In 1978, by a curious
accident of history, Ken Baker, who was the head of statistics at the United
Kingdom's largest consumer research company, the British Market

Research Bureau (BMRB), and who was troubled by the possible bias in the location of the respondents to the Target Group Index survey for which he was responsible, attended a seminar on social deprivation in order to evaluate the possible role that a neighborhood classification could play in analyzing respondent bias. Taking away from the seminar a copy of the classification, Ken coded up a 12-month sample of survey respondents with the classification in order to check its representativeness.

As an afterthought Ken decided it might be interesting to examine various consumer behaviors by type of neighborhood and began to realize that neighborhood classifiers linked to market research data provided very interesting and highly actionable insights to consumer marketers (Baker et al., 1979).

Unlike their public sector counterparts, consumer marketers are unable to have questions of interest to them included as questions on a decennial census. Whereas educational administrators can and do require the census to carry a question on educational qualifications and housing policy experts can successfully argue for the inclusion of questions on tenure, number of rooms, accommodation, and in some countries, age of dwelling, consumer marketers have to make do with asking questions on market research surveys whose coverage is typically restricted to a set of 40,000 respondents in any 1 year.

The significance of this is that the data relevant to most public policy issues are available for geographical areas of very great detail, whereas the data relevant to consumer marketers are unlikely to be statistically reliable below the level of the standard region or regional media area. In order to develop advertising campaigns at a local area level, consumer marketers need some method which a geodemographic classification can provide for interpolating reliable estimates of product and service needs for individual streets and communities at local level from random sample data collected at national level. In contrast, public sector professionals do not.

The applications that neighborhood classifications were first used to support were the recruitment of new customers. Businesses, through their advertising agencies, were continuously on the lookout for media which were particularly cost effective in reaching specific target audiences. High levels of sophistication were applied by agencies to the selection and purchase of TV spots. On the contrary, owners of more localized media channels, such as radio, door-to-door distribution, poster sites, and direct mail, were unable to provide the same level of detail about the audiences they could reach, which was ironic since by being more local in their coverage they were potentially much more attractive to advertisers who were interested in reaching tightly defined consumer groups. Table 3.2, which was derived from Testologen, a market research survey in Sweden, would be helpful to businesses in the leisure market with their media targeting as well as with distribution.

In 1979, the U.S. market research organizations Simmons and the United Kingdom's BMRB initiated a service whereby clients could access tabulations of consumer behavior analyzed by a residential neighborhood

TABLE 3.2

Variations in Leisure Activities by Type of Neighborhood [Index Values: 100 Represents the National Average, Sweden]

Mosaic Groups	Very Interested in Hunting	Very Interested in Playing Golf	Own a Caravan	Own a Summer House
A. Well-educated metropolitans	40	149	18	112
B. Low–middle income earners	53	121	35	99
C. Pensioner areas	70	72	66	107
D. Low educated in villages	92	81	84	87
E. Younger low income	58	60	51	51
F. High-income villas	58	175	65	148
G. Terraced houses and villas	74	124	136	116
H. Middle incomes, detached houses	109	92	144	95
I. Countryside	222	50	161	81

Source: Testologen.

(Weiss, 1988). This enabled media owners to link national research profiles to local demographic data so as to create credible statistics showing the goodness-of-fit between the target audiences that were reached by their titles and the target audiences of the most heavily advertised brands. Local newspapers began to equip their sales forces with neighborhood profile of their circulation areas as the centerpiece of their sales proposition. Door-to-door distributors enabled do-it-yourself (DIY) retailers to avoid the waste of having their catalogs dropped through letter boxes in high-rise flats. The post office vigorously promoted the rental of names and addresses selected geodemographically. Poster companies began to break up their stock by the geodemographics of passing drivers and pedestrians.

In parallel the people responsible for the siting of new stores began to recognize the merit of using statistical information as well as experience and judgment in the evaluation of new sites. In businesses whose boards of directors did not subscribe to "seat of the pants" methods, it was often a requirement that a geodemographic profile of a proposed new store's trade area should be included in any investment appraisal.

While the link between neighborhood classifications and market research allowed the targeting of new customers to be undertaken in a more scientific manner, the facility for geocoding (or postcoding) excited the interest of those responsible for the targeting of communications to existing customers. The link allowed a bank or a mail-order company for the first time to examine whether there were differences between the types of neighborhood in which their profitable and unprofitable customers lived; whether their new customers were similar or different to their old customers, in which sorts of neighborhoods they had good payers, and in which bad payers. Likewise the link allowed a mail-order company to compare the demographics of customers between who bought shoes and who bought washing machines.

Using geodemographic classifications, they were able to profile customers at a much more detailed level than was possible using national market research surveys where the list of questions that could be asked of their own customers was limited and the number of respondents too small to support answers other than to very broad-brush questions (Sleight, 2004).

3.4 Methods of Accessing Geodemographic Information

When confronted with these possibilities most marketers realized that it would probably be inappropriate to use traditional GIS to undertake the particular forms of analysis that they required. The systems were too complex in their functionality, too expensive to install, and often unsuited for use other than by specialists. Marketers in any case found it difficult to articulate what they needed, since their needs followed necessarily unpredictable changes in the competitive landscape. For an application which was not operational and therefore had no definable cost savings, it was difficult to prepare a financial justification.

Finally, it was evident that some of business applications, such as the profiling of customers, were not standard features in any GIS software and, while extra functions could be added, this would involve further expense, delay, and uncertainty.

In the United States and the United Kingdom, it therefore became evident to vendors of the classification systems that the majority of client needs could seldom be cost effectively met without the development of a set of integrated software tools for delivery with the classification system itself (Longley and Batty, 2003). While obviously there is an overlap between their functions and those of standard GIS tools, there are a number of notable differences based on a need for the following requirements:

- The requirement to geocode a file, to display the distribution of records by type of neighborhood, and to index this against an external distribution which might, for example, be the national distribution or the distribution of an entire customer file.

- The requirement to accumulate statistics for ad hoc areas and to compare their distribution with that of, for example, the country as a whole or a set of comparator store catchments.

- The requirement to accumulate from a set of base zones a reusable set of user-defined zones, such as store catchments or media areas, to which other data may be accumulated for analysis at a later time. In many instances these areas will be overlapping and not mutually exhaustive in coverage.

- The requirement to rank order these user-defined areas by concentrations of particular demographic segments and to rank order the zones within them.
- The ability to use the neighborhood profiles of a consumer behavior and of a user-defined zone to interpolate relative levels of consumer spend by zone, and to report and rank either behaviors (according to their relative frequency by zone) or zones (by frequency of behavior).

In addition to specific functional features these systems also needed to meet user needs by incorporating a number of databases other than the census, as follows:

- The requirement that tabulations, rankings, and comparisons be available for standard media area definitions
- The requirement to define isochrones around points of interest
- The requirement to define the location and characteristics of key shopping centers
- The requirement that census-based population and household counts be updated on an annual basis with best current estimates

Table 3.3 illustrates how developers and retailers would typically describe differences in the population make up of two different regional shopping centers on the south coast of England, their catchment areas having been defined in terms of 45 min drive times.

TABLE 3.3

Population Characteristics of Two Shopping Catchment Areas (Defined as 45 min Drive Times) in Plymouth and Bournemouth

Mosaic U.K. Groups	Plymouth %	Bournemouth %	Plymouth as % of Bournemouth %
A. Symbols of success	3.75	10.19	37
B. Happy families	15.16	11.55	131
C. Suburban comfort	15.78	16.93	93
D. Ties of community	18.23	11.24	162
E. Urban intelligence	6.91	5.76	120
F. Welfare borderline	5.38	2.82	191
G. Municipal dependency	7.97	1.64	485
H. Blue-collar enterprise	13.73	9.57	144
I. Twilight subsistence	2.29	2.59	89
J. Grey perspectives	7.12	24.15	29
K. Rural isolation	3.63	3.54	103

These analysis tools, as is so often the case, not only incorporated popular methods of analyzing consumer data, with their standard report formats and specialized terminology, but, in due course, met the industry's needs for standards for negotiation. For example, when advertisers buy from media owners or retailers negotiate with developers, both parties want to use a common currency, whether in terms of mutually understood neighborhood classifications, industry agreed methods of updating population estimates, and commonly understood formats for displaying area statistics.

The synergy from the use of common standards is enhanced to the extent that third parties begin to make use of them. For example in some national markets, such as Japan, the adoption of geodemographic classifications is often inhibited until the most widely used market research surveys are coded up or until geocoding companies offer geodemographic coding as part of their address recognition systems. Only as a result of user demand will vendors of mailing lists set up arrangements whereby the names they rent can be selected on the basis of geodemographic category. If is often only when key suppliers to a particular vertical market support these classification systems that clients within it adopt the classifications as a standard.

3.5 Relation between Suppliers and Users

Insofar as the tools incorporate forms of analysis that are standard within particular industries there has been a recognition on the part of vendors and users that standard fixed annual fee licensing is a more appropriate charging mechanism than a once off fee linked to a smaller annual support charge.

The annual fee arrangement enables the user who is unsure of the financial payback of the system to enter into an agreement without long-term commitment. The user is assured not just of advice on how to use the system but some measure of strategic consultancy. The annual fee will typically incorporate updates of all the input datasets and automatic reconstruction of their linkages, and automatic upgrades to the software. Such an arrangement is, therefore, not unlike that between a client and a professional services organization.

Large organizations such as Marks & Spencer and McDonalds operate a culture that requires standard solutions to standard requirements throughout their international operations. It is for this reason that the strength of the relationship between a client and a supplier causes a supplier to invest in the development of geodemographic services in new international markets. For example, while evaluating the markets of Hong Kong and Japan for international expansion, Marks & Spencer made it a requirement of its relationship with Experian that the company should use its best endeavors to build geodemographic classifications.

3.6 Internationalization of Geodemographics

Since the first geodemographic systems were launched in the United States and United Kingdom in 1974 equivalent classifications have been introduced to 19 markets around the world as shown in Table 3.4, a process which has been extensively documented by Weiss (2000).

The take up of geodemographics in different countries has depended on a number of factors. In general, it is easy to introduce geodemographics in countries which make available census statistics at a fine level of geographic detail. It is for this reason that systems work more effectively in countries such as Canada, Italy, Peru, Finland, and Sweden (as well as in the United States and the United Kingdom) than in Spain and Germany. On the other hand, in markets where the cost of accessing the data is very high or where there is a virtual monopoly on the linkage between address and census geography, such as Italy, it is difficult to commercialize the service.

In a number of countries where census statistics are unavailable (the Netherlands) or available only at a coarse level of geography (Spain and Germany), geodemographic classifications have been built for finer levels of geography using statistics from sources other than the census. For example, the electoral roll is used extensively as a data input in both Spain and the United Kingdom, whereas in the Netherlands market research interviews and mail-order data are key alternative input sources. In New Zealand, public data on buildings allow geodemographic classifications to be taken right down to the building level.

There is no doubt that in some countries neighborhood differences operate at a much finer scale than in other countries. The United States, China, and Hong Kong are examples of countries where differences tend to be more evident at a coarse than a fine level. In Hong Kong, most blocks of flats are more populous than a U.S. census block. In Italy, France, and the United Kingdom, by contrast, the mesh of social differentiation operates at a much finer scale—in other words to identify the type of neighborhood a consumer lives in, you need to use the demographics for quite a small geographical area around which that consumer lives. In general, it seems that geodemographic differentiation operates at a coarser scale in those countries where population growth is faster. Indeed, within the United Kingdom, it is evident that in cities that grew very rapidly over a limited period of time, such as Glasgow, Liverpool, and Middlesbrough, one can see particular types of neighborhood extending over much larger contiguous residential areas than in those cities that have grown at a slower or more consistent rate, such as London and Bristol.

In each country the classifications are unconstrained. That is to say there is no a priori determination of what the clusters should be. This is left to the computer algorithm which was used to construct these classifications to determine, which it does according to its own optimization criteria

TABLE 3.4

Countries Covered by the Mosaic International Network

Country	Level of Geography	No. of Units	No. of Types	Households per Geographic Unit	Total Household Estimate
Australia	Mosaic microsegment	314,078	41	22	7,002,346
Belgium	Street segment	160,000	30	25	4,000,000
China (part)	Street	17,366	34	2,265	39,338,000
Denmark	Mosaic area	32,594	34	75	2,444,550
Finland	Mosaic unit	192,000	30	20	3,840,000
France	Ilot	246,109	52	50	22,300,000
Germany	Building	15,966,793	38	2	31,933,586
Great Britain	Postcode	1,632,261	52	14	22,851,654
Greece	Postcode/census block	49,143	33	85	4,100,000
Hong Kong	Street block group	2,757	30	570	1,571,490
Japan	Grid cell	184,684	39	400	73,873,600
Netherlands	Six-digit postcode	389,756	41	17	6,625,852
New Zealand	Mesh block	38,365	38	50	1,918,250
Northern Ireland	Postcode	44,219	36	13	574,847
Norway	Electoral unit	13,648	30	180	2,200,000
Republic of Ireland	Mosaic area	17,016	32	40	1,800,000
Spain	Mosaic area	621,408	48	27	16,662,339
Sweden	Mosaic Områden	68,461	34	56	3,833,816
United States	ZIP + 4	20,000,000	60	8	137,750,000

(Webber, 2004). For this reason, the types of neighborhood that are created do differ significantly from one country to another.

In general, in countries with large populations and fine geographic detail more residential differentiation will be apparent. The geodemographic systems in these countries will therefore have more categories. This is why the U.S. systems have more different clusters than the ones in Hong Kong and Ireland. Factors that cause the number of categories to vary include the extent of social housing (none in Brazil or Peru), young singles live away from their parents (which they do more in Canada and Australia than in Spain and Italy), the level of ethnic diversity (high in the United States), and the propensity of old people to retire to geriatric neighborhoods, often by the coast (high in the United Kingdom and Australia).

Notwithstanding the differences in the sources of input data, in the questions covered by the census, in the level of granularity of the census output areas, and the size of each country, there are significant similarities as well as differences among the geodemographic classifications that are created in different countries. Almost every country (except Hong Kong) has a set of clusters characterized by high levels of education, late marriage, employment in service occupations, young age profile, high mobility, and location close to the centers of very large cities. Such clusters, such as the "Elite Urbanas" segment in the Spanish Mosaic (Figure 3.1), almost invariably have high proportions of people working for international companies,

FIGURE 3.1
"Elitas Urbanas" classification of Spanish Mosaic. This is typical of similar categories found in classifications around the world.

TABLE 3.5

"Global Mosaic" Categories

Code	Label
A	Agrarian heartlands
B	Blue-collar self-sufficiency
C	Career-focused materialists
D	Deindustrial legacy
E	Educated cosmopolitans
F	Farming town communities
G	Grays, blue sea, and mountain
H	Hardened dependency
K	Inner-city melting pot
L	Low-income elderly
M	Midscale metro office workers
O	Old wealth
S	Shack and shanty

high levels of consumption of print media, a predilection for eating in quality restaurants, and good knowledge of international trends. This is the most cosmopolitan type of neighborhood that occurs within each country and many of its members would be more at home with their counterparts in similar types of neighborhood in other countries than they would with residents from blue-collar neighborhoods in their own homeland.

In contrast, most geodemographic classifications contain one or more types characterized by blue-collar employees, with low levels of education, high levels of home ownership, and elderly age profiles. These tend to exhibit high levels of community involvement, low levels of crime, and as good places to experience what is quintessential national about a country's cuisine. People from such neighborhoods, if they do travel abroad, want to meet others of a similar background and are least interested in absorbing alien cultures and other countries' food in particular.

Table 3.5 shows an attempt by Experian International to organize the different geodemographic clusters into common global categories. In each country, each type is classified into a single one of 13 different groups from "Old wealth" to "Dependency hardened." Not every one of these 13 categories is found in every market.

3.7 Limitations of Geodemographic Analysis

How well do these classifications work? To answer this question satisfactorily we need to consider the expectations that users may have of the systems. Clearly not everyone who lives in a single census output area will share the same age, housing characteristics, and socioeconomic profile. To the degree that census output areas are themselves heterogeneous, no geodemographic classification can be as good as a classification built at the person level, at least in relation to predicting the demographics of its members. To this degree a neighborhood classification will always be a second best to person-level demographics in instances for which this is the required basis for targeting.

A different way of looking at their effectiveness is to consider how similar to each other are the different census output areas (or other zones) that are grouped together into a common geodemographic cluster. On this criterion, the systems can be said to be very efficient. Typically the loss of variability of the original dataset that would be lost by grouping output areas into clusters is around 50%. For many key variables, the loss of variance is very much less. Geodemographic clusters are particularly uniform in relation to housing type, tenure, dwelling age, and size. They are also particularly uniform in relation to car ownership, income, and travel to work, in relation to population density and to employment in agriculture. The proportions of the population married or single and the proportions with children are typically very similar across zones in any given category. On the other hand, zones tend to be less uniform in relation to the industrial sector in which employees work, in terms of age structure and, in most countries, ethnic origin. They are least uniform in relation to the proportion of women who work. In general, therefore, they are more uniform and predictive of behaviors that are related to building type and to status than to life stage and industrial structure.

An interesting measure of the efficiency of the U.K. postcode-level classification is that less variability in the input data is lost by grouping residents into 61 geodemographic categories than into 9000 postcode sectors. Knowing that a person lives in a cluster such as "New urban colonists" is more predictive of his or her demographics than knowing that he or she lives in postcode sector N6 4.

Returning to the issue of within-cluster uniformity it is not uncommon for ethnic minorities to be in the minority even in a cluster labeled "Asian enterprise" or for pensioners to be in a minority in a cluster "Sepia memories." One has to be careful not to suppose that labels necessarily apply to everyone in the cluster. However, if one's objective is to target specific groups, whether these be relevant to a public service, such as people burgled in the last year, on to a private organization, such as people who have purchased a new car, the key issue is whether a geodemographic classification is more effective in locating such groups as a targeting system based on any other single criterion, such as age, educational qualification, income, or whatever. Tests have established that over a wider range of consumer behaviors multivariate geodemographic classifications are typically as good as but not better than univariate demographic classifications such as age, gender, income, and so on. These tests have shown that geodemographics is seldom the best discriminator and seldom the worst among comparator demographics.

Whether the systems work or not may also be considered in terms of the nature of error. Some users of geodemographics, such as mail-order companies, use geodemographics as one of a set of criteria for selecting or deselecting customers for communications. In such instances whether or not there is a systematic pattern of bias at a regional or local level is unimportant. On the other hand, for a retailer using the system as an input to an estimate of market potential on a local basis, it is important that neighborhoods of a similar category should behave in a uniform manner throughout the country. In this context such systems would not work very well for predicting the local demand for snow chains in a country with such diverse weather as the United States. Nor would it work well for predicting the level of demand for whiskey or porridge in various supermarket catchment areas in the United Kingdom. These are examples of food products whose consumption varies on a regional level with Scots having particular predilections for both. On the other hand, for the estimation of the proportion of local children likely to apply to university in a particular town, geodemographics should work well because there is unlikely to be systematic error at the cluster level for such behavior. We say unlikely because such an assumption may well not have been tested. In summary, for interpolation from national to local levels, the systems probably work well in the majority of cases but more research would be useful to improve our understanding of the contexts in which the method is likely to be least effective.

3.8 Geodemographics and Government

As mentioned earlier, geodemographic classifications were developed initially to help with the spatial allocation of government programs for inner-city regeneration. Prior to this application, and much of the time since, deprivation has been viewed as a one-dimensional characteristic, much like temperature, and the objective of much statistical analysis has been to identify the associated measures which, once combined, could position each census output area on an appropriate ordinal scale.

One of the uses of geodemographics in this regard was to provide a formal evaluation of the input variables used to measure deprivation. Two examples will suffice. In the 1991 U.K. census, there are two alternative measures of overcrowding available from the census. One measure is the proportion of households living at over one person per room, and the other measure is the proportion living at over 1.5 persons per room. Left to decide which indicator is the more appropriate, housing specialists will debate which level of overcrowding is today considered acceptable. Left to decide which indicator is more appropriate, geodemographers will note that the sorts of neighborhood with high levels of persons living at over 1.5 persons per room are areas of large old divided houses (with very spacious rooms) in very wealthy areas of inner London. Here there are significant numbers of dwellings with two persons in one room.

This behavior is associated with foreign visitors and young singles adapting to very high rents. The alternative measure, in contrast, is associated with very poor peripheral council estates where councils tend to place families with four or more children in three-bedroom houses. The two measures are not measuring differences in level of overcrowding as much as the reason why households may be overcrowded.

A second nice example is taken from a Shanghai regional classification. Table 3.6 illustrates the contrasting demographics of the best and worst educated clusters in the city region. The cluster with the lowest level of unemployment (intensive farming) is also the cluster which seems to have the lowest income, namely the cluster with the most own account farmers. This cluster, despite (or because of) its low incomes has the highest incidence of residents who work 7 days a week, who are not retired, and are not unemployed. Unemployment in this context is not a meaningful indicator of deprivation.

These two examples show the potential use of classifications as methods of evaluating the appropriateness of potential social indicators. Such analysis could be undertaken using regression. However, the use of geodemographics is quicker and simpler. Often patterns which one was not looking

TABLE 3.6

Contrasting Demographies of Top Areas and Intensive Farming in Shanghai [Values Show Proportions of Different Demographic Groups Expressed as a Percentage of the Regional Average]

	Mosaic A1	Mosaic G32
	Comfortable Living/Top Areas	Rural/Intensive Farming
Aged 20–24	79	53
Aged 55–59	144	148
Two children	33	206
Percentage females 25–44 who are single	219	15
No education	38	252
University general and postgraduate	242	5
Farming	1	977
Manufacturing	74	43
Finance and real estate	193	3
Government occupations	199	21
Housewife	46	133
Retired	134	26
Unemployed	86	10
Work 7 days a week	37	199
Self-built house	2	428
House bought from public sector	292	1
Space 40 plus square meters	145	121
Use firewood for heating	0	1529

for emerge unsought—in the manner of true data mining. More often the contextual background provides an explanation of the relationship as well as evidence of it.

The other intended use of geodemographics in deprivation studies was to provide a clearer context for the differentiation of programs according to the type not just the level of deprivation. At the rather crude level of electoral ward that is used to assemble neighborhood statistics in the United Kingdom, this distinction is less evident than it is at postcode level. The original Liverpool Inner Area Study, which spawned the first U.K. neighborhood classification, was able to differentiate three quite distinct types of environment for which quite different policy interventions were appropriate. One of these types consists mostly of dockside council estates with very high levels of unemployment and low levels of skill. Despite their hardship these were the areas with strong community spirit which exhibited some social controls over the most blatant examples of deviant behavior. The problems of these areas was primarily rooted in the decline of the local source of employment, the docks, and the poor level of skills needed for winning jobs elsewhere.

A second type for deprivation occurred among areas of older terraced housing. In days when lack of an inside toilet and incidence of damp and rodents was a cause of major concern, these neighborhoods would score high on a complex deprivation measure. In practice these were the thriving communities, where unemployment was lower than might be expected given the social class profile, but ones where outworn infrastructure and a dilapidated environment were the principal source of stress.

The third type of neighborhood was characterized by large old houses split up into tiny flats occupied by single parents, students, young professionals, dropouts, drug addicts, and prostitutes. Lack of social cohesion and support was the key source of stress in these areas which, in terms of income and educational attainment, fared no worse than the average for the city.

Figure 3.2 illustrates the architectural and environmental characteristics of these three different sorts of neighborhood.

The United Kingdom has, in recent years, seen a renewed interest in public sector applications of geodemographics, particularly in education, health, and crime. In a more evidence-based policy environment more interest is shown in what works where, and it is felt that public services could be more effectively delivered if they were supported by the targeting systems that are routinely used in the commercial sector. The speed at which such innovations can be made depends upon the linking of geodemographic classifications to datasets managed by the public sector. Examples of such databases are the Hospital Episode Statistics (HES), the Driver and Vehicle Licensing Office database, the Land Registry database, the Pupil Level Annual School Census (PLASC), and the British Crime Survey. While some of these generate revenue from an enhanced level of information service to their commercial clients, the last two have provided

(a)

(b)

(c)

FIGURE 3.2
Different types of disadvantaged neighborhoods in Liverpool. (a) Large Victorian villas, young, mobile, rootless singles, (b) Older terraces, strong community network, and (c) Peripheral council estates, high levels of antisocial behavior.

evidence which has been of practical value to universities and to police forces wanting to apply greater selectivity to the manner in which they handle individual clients.

Table 3.7 provide a generic overview of the key different neighborhoods in Great Britain, relating the level of social capital and the degree of trust to the types of crime these neighborhoods are particularly prone to suffer, and the crime prevention and detection strategies that are most likely to work within them.

3.9 Neighborhood Classification Systems in China

Mosaic China was constructed using statistics from the 2000 census for three provinces: Beijing, Guangzhou, and Shanghai. This can be attributed in part to the difficulty of obtaining statistics from other regions, there being no single source of statistics for the whole country, in part to the interest of overseas marketers which at present is concentrated in three large city regions. The total enumerated population living in these three provinces in 2001 was 39,405,000 divided between 17,366 enumeration areas.

The statistics available for these three regions are uniform in definition. For each of their enumeration districts there are some 396 published counts which describe the population in terms of their demographics, housing, and employment characteristics. For example, it is possible to access counts of people by age, marital status, ethnic origin, and number of years' residence within the locality. Features of the housing that are covered in the statistics are the decade in which it was built, the number of rooms in each dwelling, and their floor space. Information is provided on the ownership of the dwellings, on toilet and sanitation facilities, and means of heating. The Chinese authorities also provide information on whether a building is used solely for residential or for mixed residential and commercial purposes. Published information on employment include whether the population is employed, sick, unemployed or studying, the occupations and the industries that people work in, and their level of schooling and/or university qualifications. Information is also provided on number of days worked per week.

From this set of 396 data items, a set of 76 statistically reliable variables were created for use in the classification. Among these are a number of complex variables, such as for example the average number of rooms per person and the average number of square meters per dwelling.

To build the classification, a positive weight was given to only 66 of these variables, the other 10 being used only for analyzing differences between the clusters and not for creating them. Each of these variables was given a weight, reflecting to the level of influence it was thought appropriate to assign to that variable in determining the cluster each enumeration area should be allocated to. This weight was set for each variable on a scale from

TABLE 3.7

Social Capital, Crime, and Policing Methods in Different Neighborhood Types

Neighborhood Type (Mosaic)	Social Capital				Crime Profile					Appropriate Options
	Level of Trust	Informal Contacts	Formal Association	Social Capital	Fear of Crime	Crime Level	Clear-Up Rate	Source of Offenders	Type of Crime	
A. High-income families	High	Low	Medium	Weak community involvement	Fairly low	Low	Low	Imported	High-value burglaries/fraud	Alarms; high-level home security systems; private patrols; car trackers
B. Suburban semis	High	High	High	Strong support for community action	Medium	Low	Medium	Imported	Burglary	Neighborhood watch; geographic policing; target hardening
C. Blue-collar owners	High	Medium	Medium	Low	High levels of self-reliance	Medium	Medium	Medium	Vehicles/domestic violence/drinking	Neighborhood watch
D. Low rise council	Low	Medium	Low	Strong local knowledge	High	High	High		Petty theft/domestic violence/teenage vandalism	Community development; victim support
E. Council flats	Low	Low	Low	Self-policing gangs	Very high	Very high	High	Indigenous	Gangs/domestic violence/drugs	Community development; victim support; zero tolerance; high-intensity assurance

F. Victorian low income	High	High	Medium	Informal community networks	Moderate	High	Moderate	Indigenous	Burglary/attack/racial harassment	Development of community links; victim support groups; youth initiatives
G. Stylish singles	Low	Low	Low	Low levels of community involvement/students	Moderate	High	Moderate	Indigenous	Equipment theft	Engagement of ethnic leaders; campaign for reporting
H. Town houses and flats	Low	Medium	Low	Centered around local shops	Medium	Moderate	Medium		Burglary	Development of community links; postcode marking
I. Independent elders	High	High	High	Strong networks/people at home during day	Low	Low	Low	Imported	Fraudsters/identity issues	Physical evidence of police, reassurance; personal alarms; cameras/CCTV
J. Mortgaged families		Medium	Medium	High	Moderate networks/people at work during day	Medium	Medium	Medium	Burglary/vehicles	Child security; crocodile
K. Country dwellers	High	High	High	Strong networks/few perpetrators	Low	Low	Low	Imported	Theft of antiques	Reassurance on response times; dogs; alarms; cameras; postcode marking

one to seven based on experience and judgment rather than formal rules. The variable given the highest weight in determining the cluster allocation was the variable "average square meters per dwelling." Its weight was seven times that of the variable with the lowest weight.

After various experiments with different selections of variable weights and different numbers of clusters, the preferred solution was the one which contained 33 separate clusters. The loss of variance resulting from the grouping of the 17,366 enumeration areas into 33 clusters was only 39.5%. As would be expected this loss of variance was much greater for some of the input variables than for others. For example, the loss of variance of the variable "% dwellings built prior to 1949" was only 10.4% while the loss of variance of the variable "% miscellaneous working status" was as high as 90.2%.

The 33 clusters were then grouped up into a coarser nine-level categorization to help with interpretation and mapping. This grouping contributed a further loss of only 16.3% of the original variance. In other words, by grouping 17,366 areas into only nine separate categories, as much as 44% of the initial variability in the data was still retained. This statistic is significantly higher than in other countries, reflecting the high level of within-cluster uniformity and high level of between-cluster variability in China

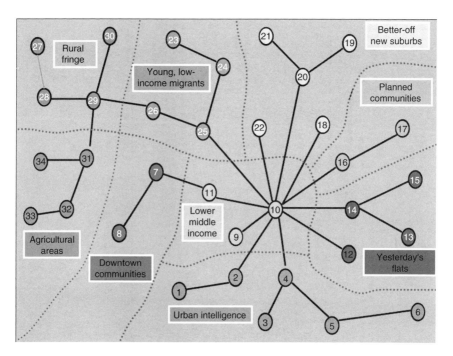

FIGURE 3.3
Minimum spanning tree of the clusters in the Mosaic China. This illustrates the different dimensions whereby Chinese residential neighborhoods are differentiated from each other.

compared with other countries. One reason why this statistic is so high in China is because of the comparatively large average population size of the enumeration areas, over 2200 people. On the other hand, it might be expected that a classification covering three cities in very different climatic zones and with very different industrial bases would not achieve such a high level of within-cluster uniformity.

In general, one wants to achieve a classification in which the population size of the clusters varies within as narrow bands as possible. In the case of China, this variation is larger than in other countries, with the smallest cluster, a set of rural enumeration districts containing concentrations of ethnic minorities, containing only 0.10% of the population. In the largest cluster, areas of older people renting collective flats accounts for 6.77% of the population. This variation in population size is justified by the fact that the smaller of these two clusters contains the largest amount of within-cluster variance of any of the 33 clusters, whereas the "collective flats" cluster, although very large in terms of population, consists of areas which are more similar to each other than is the case in any other cluster.

The minimum spanning tree of the clusters shown in Figure 3.3 provides clear evidence of the large number of different dimensions whereby Chinese residential neighborhoods are differentiated from each other.

3.10   Using Multilevel Geography to Improve Discrimination in the United Kingdom

For many years, U.K. users of neighborhood classification systems had supposed that the finer the resolution of the zoning system used to build a neighborhood classification, the higher would be its performance in discriminating between areas of different types of human behavior. It was this assumption that led developers of neighborhood classification to supplement the statistics that are published by the census offices with additional data items collected from other sources and summarized at the level of the unit postcode, a level of geography one-tenth the average size of a geographical census output area.

In recent years, this assumption has been questioned and it has been suggested that in certain domains, such as the level and type of crime, voting preferences, and children's performance in school, characteristics of the broader community in which a resident lives may have significant incremental predictive power over and behind the characteristics of the immediate microenvironment (Webber, 2004).

To investigate these possible effects when rebuilding its U.K. neighborhood classification based on the results of the 2001 census, Experian decided to create for each of the United Kingdom's 1.5 million full postcodes a series of 1, 2, and 5 km circles. For each of these 4.5 million zones, Experian then created average values on a number of variables derived from the census, including the percentage of economically active persons engaged in agriculture, in manufacturing and mining, and in services, and the proportion in professional and managerial occupations. In addition, Experian created for each area an average level of population density and, for each postcode centroid, a series of metrics indicating relative accessibility to retail shopping centers, other populated areas, and coastal retirement resorts. Each of these variables were included in the set of variables used to construct the classification and were given weights considered appropriate.

On the completion of the classification, tests were undertaken to establish how effectively the new classification discriminated across a set of 120 customer files and lifestyle respondent databases. These databases were not used to build the classification but were used, in this way, to compare its discriminative power with that on the equivalent classification system, built largely based on 1991 census statistics, that it was designed to replace. Having made this comparison, the analysts responsible for the classification undertook further tests to establish whether the discriminatory performance of the classification could be improved either by up-weighting or by down-weighting the weights given to the "higher level area" variables relative to the weights given to the variables for micro areas.

Tests demonstrated that noticeable improvements could be achieved by significantly up-weighting the weights initially given to the higher level area variables until collectively they accounted for 1.4% of the total weight. By a process of "hill climbing," it was proved possible to generate an improvement in the discriminatory performance of the new classification of 1.8%.

During the process of interpreting the different clusters, it became evident that the weight given to employment data for higher level areas had been particularly effective in isolating council housing in what historically had been important mining towns. Subsequent analysis of information from the British Crime Survey showed that respondents in these clusters experienced very much lower levels of crime than did residents in otherwise similar housing estates in larger cities with an economy more dependent on services. Likewise the use of higher level data was influential in causing inner London neighborhoods to coalesce into a distinctive set of clusters clearly separate from those in outer London and in provincial cities. Accessibility data also seem to have the effect of differentiating villages from which people commuted to nearby towns from other upmarket villages in more remote areas of high landscape value. The use of the accessibility variables was also particularly effective also in causing neighborhoods of the Shetlands and Western Isles which are not dependent on agriculture to be assigned to clusters generally characterized by "rural isolation" rather than by "municipal dependency."

As a result of these tests, Experian have arrived at the opinion that the most effective neighborhood classification is built by using information on employment for multiple levels of geography within a single classification system, while using information on housing and household characteristics only at the very finest level of geography for which they can be obtained.

3.11 Conclusions

Geodemographic classifications have established themselves with consumer marketers as a very useful tool for segmenting consumers. They have been used as a research tool to better understand the profile of the market for specific products and services, media, and brands. They have demonstrated an ability to discriminate both the developed and developing markets. The development of geodemographics has been facilitated by an increased willingness of government organizations to release census statistics in aggregated form at the level of geographical resolution which meets the needs of commercial users.

However, it is evident that the extent to which these new tools are adopted does depend significantly on the level of local skills in the application of geographical information to marketing and communications as well as on the granularity of the postcode system.

By leaving the commercialization of these systems to a small number of international consulting organizations, problems exist for academic users in knowing how to access this information. In the United Kingdom, Experian have come to an arrangement with the Economic and Social Research Council whereby the Mosaic classification can be accessed free of charge by bona fide academic researchers. Notwithstanding this agreement, academic users often have difficulty in learning the applications of the classification systems as well as the meaning of the categories. Because of their confidential nature and the different priorities of commercial clients, it is often difficult to gain access to the results of demonstration projects. Outside the United States, the United Kingdom, and Australia, there is little evidence of access by the academic community to geodemographic classifications.

Nevertheless the information resource does represent a potentially very useful opportunity for academics to improve their understanding of the processes whereby particular groups segregate themselves within urban systems and hence of the contemporary patterns of migration and gentrification which cause neighborhoods to change (Butler and Robson, 2003). These resources should also facilitate a better understanding of the significance of area effects within multilevel modeling and improve the accuracy with which models can be used to interpolate local estimates from national surveys.

However, for these research opportunities to properly inform policy interventions, it is necessary that government organizations should recognize the value to researchers of the geodemographic coding of many of the large operational datasets which they maintain. Datasets in the United Kingdom, such as the HES, the Pupil Level Annual and School Census, and the British Crime Survey, are examples of files which have now been coded by type of residential neighborhood, making it possible to provide robust benchmarks for service delivery standards within highly localized areas. Where, how, and by whom such geodemographically coded datasets are best analyzed is an important issue. Is this best undertaken within government, by commercial contractors, or by academic research institutes?

While academics and commercial organization can be relied upon to find productive applications for these research assets, there is little evidence of successful application in environments where operations are highly devolved such as policing, health, and education where, on account of the fragmented structure of delivery, there are few mechanisms for improving performance by reference to evidence from national datasets. On the other hand, one should not ignore the power of consumer-oriented Web sites such as "Up My Street" (www.upmystreet.com) to raise awareness of the value of these classifications among intelligent lay audiences or indeed even to school children. The intrinsic appeal of the characterization of streets and addresses to the man in the street may ultimately be an important reason why this approach to organizing geographical information achieves widespread recognition in the long term.

References

Baker, K., Bermingham, J., and McDonald, C., 1979, The utility to market research of the classification of residential neighbourhoods, Paper presented at the Market Research Society Conference, Brighton.

Butler, T. and Robson, G., 2003, *London Calling—The Middle Classes and the Remaking of Inner London* (Oxford: Berg).

Longley, P. and Batty, M. (editors), 2003, *Advanced Spatial Analysis: the CASA Book of GIS*, pp. 233–266 (Redlands, CA: ESRI Press).

Sleight, P., 2004, *Targeting Customers, Third Edition, How to Use Geodemographic and Lifestyle Data in Your Business* (Oxford: NTC Publication).

Webber, R., 1975, Liverpool Social Area Study, 1971 Data, *PRAG Technical Paper No 14* (London: Centre for Environmental Studies).

Webber, R., 2004, Designing geodemographic classifications to meet contemporary business needs, *Journal of Interactive Marketing*, 5(3), 219–237.

Webber, R. and Craig, J., 1978, Socioeconomic classifications of local authority areas, *OPCS Studies on Medical and Population Subjects*, No. 35 (London: HMSO).

Weiss, M., 1988, *The Clustering of America* (New York: Harper and Row).

Weiss, M., 2000, *The Clustered World* (Boston, MA: Little Brown & Co).

Section II   *Methodological Advances*

4

Routing out the Hot Spots: Toward Using GIS and Crime-Place Principles to Examine Criminal Damage to Bus Shelters

Andrew Newton

CONTENTS

4.1　Introduction

This chapter describes initial efforts to utilize GIS technology to cross-reference crime data on one aspect of the public transport journey, bus shelter damage, with information on socio-demographic conditions, land use, and infrastructure, covering the county of Merseyside in the North West of England. GIS are used in conjunction with spatial statistical analysis to explore the nature, manifestation, and patterns of damage to bus shelters. Evidence of clustering is found, and one-fifth of all damage for a year is shown to occur at 2.5% of all bus shelters. The findings also suggest that particular neighborhood types, as well as certain characteristics of socio-demographic and physical environments, are more likely to experience shelter damage than others. This implies that bus shelter damage is related in a systematic and predictable way to known attributes of a shelter's location. This prompts a discussion of the use of a combination of GIS and other crime-mapping techniques developing our knowledge of the nature and extent of, and the theoretical reasons underlying, crime and disorder on public transport.

Public transport crime: what is it, and why does it exist? The police in the United Kingdom do not record incidents of crime and disorder on public transport systems as a separate category. This might imply that it is an area not worthy of research and further attention. However, recent findings by the then Department of the Environment, Transport and the Regions (DETR, 1998) suggest that patronage on public transport could be increased by 3% at peak and 10% at off-peak times if fear of crime and disorder on public transport journeys were to be reduced. These findings also highlight the importance of public transport availability as a means of gaining access to health, leisure, and other facilities, and thus in making a contribution to minimize social exclusion. Any attempt to reduce fear of crime on public transport requires a fuller understanding of both the nature and extent of crime and disorder on public transport, and environmental characteristics that may help to explain this crime. These environmental features are likely to include land use, socio-demographic influences, and features of the physical infrastructure, such as the layout of buildings and the spaces between them. The techniques used in this chapter have been applied to other areas of crime research (Johnson et al., 1997; Bowers and Hirschfield, 1999). Here, GIS are used in conjunction with spatial statistical analysis to explore the nature, manifestation, and patterns of crime and disorder on public transport, and, in particular, criminal damage to bus shelters. In an attempt to offer some explanation for the spatial patterns identified, it is necessary to draw upon theoretical perspectives that relate crime in general to its environment. Some relevant theories are now highlighted, before the methodology and findings of this research are discussed in more detail.

4.2 Theories Relating Crime to Its Environment

Environmental criminology is concerned with describing and explaining the place and space of crime. Place of crime refers to the location of crimes, and space of crime refers to spatial factors that may help to explain the location of crime. The two core concerns of environmental criminology are to describe and explain the distribution of criminal offences, and to describe and explain the distribution of crime offenders (Bowers, 1999). This research concentrates on the former concern, where crimes happen. The spatial distribution of many offences (crime events) has been shown to be nonrandom (Eck and Weisburd, 1995), and attention has focused on analyzing when and where these crime events occur and the environmental factors that may help to explain the occurrence of these incidents.

The three major theories of environmental criminology concerned with the distribution of crime events are routine activities theory (Cohen and Felson, 1979), the rational choice perspective (Cornish and Clarke, 1986), and crime pattern theory (Brantingham and Brantingham, 1993). Routine activities theory states that, for a criminal event to occur there must be a convergence in time and space of three factors: (a) the presence of a motivated offender, (b) the absence of a capable guardian, and (c) the presence of a suitable target. Whether or not these elements converge or coincide is a product of the routine activities (day-to-day movements) of potential victims and offenders.

A rational choice perspective suggests that offenders will choose their targets and achieve their goals in a manner that can be explained. This has its roots in economic theory and seeks to explain the way in which crimes are distributed spatially by weighing up the potential cost of a crime (chance of apprehension and cost of journey) against its possible benefits (potential reward and ease to commit). The offender rationally chooses the situation with the highest net outcome. The development of these two theories led to a growing recognition that they were not necessarily mutually exclusive, and a combination of both theories may help to explain crime events. A significant development in this was the development of crime pattern theory. This argues that "crime is an event that occurs when an individual with some criminal readiness level encounters a suitable target in a situation sufficient to activate that readiness potential" (Brantingham and Brantingham, 1993, p. 266).

This multidisciplinary approach to understanding crime contends that crimes are patterned, but these patterns are only discernible when crimes are viewed as etiologically complex, occurring within, and as a result of a complex environment. Places are linked with desirable targets and the situation or environment within which they are found, by focusing on how places come to the attention of particular offenders.

Eck and Weisburd (1995) further emphasize the importance of place as essential to crime pattern theory. They discuss how theories of place and crime have merged, in order to develop a crime event theory. Here, crime is examined at the microscale (individual or the smallest levels of aggregation). Crime and its environment can be analyzed at different levels of aggregation, from the individual (micro) to subpopulation (meso) to population (macro) analysis. Given a set of high crime locations, a crime pattern theorist may focus upon why and how offenders converge at these locations, whereas a routine activity theorist would be concerned with explaining the movement of targets and the absence of possible guardians. Both theorists may produce valid explanations, yet these may be supportive or differ substantially, and even a combination of both may be useful in explaining the crime.

One final important concept is that of *crime attractors* and *crime generators* (Brantingham and Brantingham, 1995). A crime generator is an area that attracts large numbers of people for reasons other than to commit a crime. At particular times and places, the concentration of victims and offenders in these locations produces an "unexpected" opportunity for the offender to commit a crime. Shopping centers, sports stadiums, and public transport interchanges are examples of this. Crime attractors are places that offenders visit owing to knowledge of the area's criminal opportunities, such as bars and prostitution areas.

4.2.1 Crime on Public Transport

Applications resulting from the above theories include situational crime prevention (Clarke, 1992), hot spot analysis (Buerger et al., 1995), opportunity theory (Barlow, 1993), and targeted policing (McEwen and Taxman, 1995). Although these have been applied to analyze crime and disorder in a number of areas, including domestic and commercial burglary, assault, theft, and robbery (Brown et al., 1998; Ratcliffe and McCullagh, 1998; Jupp et al., 2000), there has been only a limited amount of research into crime and disorder on public transport. Pearlstein and Wachs (1982) provide evidence that crime on public buses is concentrated both in time and space. Levine et al. (1986) use results from survey and observational data to demonstrate that bus crime incidents tend to be high on routes passing through high crime areas. Block and Davis (1996) examined street robbery data in Chicago and found that, in low crime rate areas, crime was concentrated near rapid transit rail stations. LaVigne (1997) demonstrates how unusually low crime rates on the Metro, subway system of Washington, D.C., can be explained by reference to some aspect of its environment. A recent paper by Loukaitou-Sideris (1999) uses empirical observations, mapping, and survey research to examine the connection between criminal activity at bus stops and environmental factors. Ten high crime bus stops were analyzed along with four low crime "control" stops. This empirical research indicates that environmental attributes and site conditions at bus stops do have an impact on crime levels, and further research is required to better understand and measure this effect. It has been demonstrated that the environment plays an important role in the location of crime events on public transport systems. There does not seem to have been any attempts to produce a systematic evaluation of the nature, extent, and causes of crime and disorder on public transport.

4.2.2 Crime Events

Central to the understanding of environmental criminological theories and their applications is the concept of a crime event. An event is something that occurs (Barlow, 1993) and the theories discussed above all depict this event as a nonmoving event at a particular time and location (a static event). When considering the public transport system, a "whole journey approach" is needed (DETR, 1999). This incorporates all parts of the bus journey, including walking from destination point to a bus stop, waiting at a bus stop, traveling on a bus, transferring between stops, and traveling from bus stop to arrival point. In terms of the bus journey, there are three possible scenarios in which a crime event can occur:

- Waiting at a bus, train, or tram stop (the waiting environment)
- On board a mode of public transport (bus, train, and tram)
- Transferring between stops on foot (departure point to stop, between stops, stop to destination point)

The first and third situations both describe a static crime event. The middle possible scenario, however, implies the crime to be moving (nonstatic). Here the fundamental question arises: Can the existing theories of environmental criminology be applied or adapted to explain crime and disorder on public transport? The growth of new technologies has allowed increased sophistication in the mapping and analysis of crime data, particularly with the evolution of GIS. The challenge is to map the location of a crime event that occurs on a moving public transport vehicle. Ideally, a global positioning system would be used, but, at present, this is likely to prove expensive. If a crime were reported along a section of a route, this would demarcate where the crime event occurred (although not necessarily the movement of the crime offender). This could then be captured in a GIS as a static event, at a unique time period, together with information about crime events at stops and stations, alongside information about the physical infrastructure, land use, socio-demographic and other associated environmental features. This would allow existing theories of crime and place to be tested and either applied or adapted. The location of crime events could be represented as points (at stops) and lines (sections of a route).

One major advantage of a GIS is its ability to combine data from different sources, and for the spatial relations between these to be investigated. The use of a GIS as a framework for analysis opens up the possibility of carrying out a systematic evaluation of the nature and extent of crime and disorder on public transport and its juxtaposition with associated environmental characteristics. It is believed that this could lead to the development of an evidence base that would enable management to make informed decisions about resource targeting and policy formulation, and to monitor and evaluate strategies that have been implemented. This research represents an initial attempt to develop a systematic approach capable of evaluating the nature, extent, and causes of crime on public transport. It was noted earlier that the police in the United Kingdom do not record incidents of crime and disorder on public transport as a separate category. Indeed, the lack of available data that exists on the location of crime on buses restricts the spatial analysis that can be performed, since crime is reported specific to an entire route and not pinpointed to a precise location. Bus shelter damage is recorded to individual stops with $X–Y$ coordinates, and hence this research examines data on bus shelter damage to pilot whether further research in this area is deemed appropriate or not.

This study uses data obtained by Merseytravel, the Public Transport Executive Group (PTEG) for Merseyside. It relates to bus shelter damage on Merseyside for the year 2000. There were 3116 incidents of shelter damage recorded, costing approximately £400,000 in repairing the damage. In comparison, police records of shelter damage for this period consist of only eight incidents. This highlights both the problem of underreporting and the lack of available data on crime and disorder on public transport.

This study will address the following questions:

- Is bus shelter damage concentrated at particular stops and areas?
- Do particular neighborhoods suffer from raised levels of shelter damage?
- Do bus stops act as crime generators?

4.3   Characteristics of the Study Area

Merseyside is a metropolitan county in the North West of England and is an area where public transport is particularly important as it is estimated that over 40% of the population do not have access to a car (1991 Census of Population). Merseytravel is responsible for coordinating public transport services on Merseyside and acts in partnership with bus and rail operators to provide local services. The deregulation of bus services in 1986 resulted in bus services being operated by a number of commercial companies. This adds difficulties in acquiring reliable and consistent data concerning crime and disorder on buses, since operators report information in a nonstandardized fashion. Maritime and Aviation Security Services (MASS) also operate on a private contract as a rapid response service dedicated to buses in Merseyside. There are also two rail operators (First North West and Arriva) who are responsible for local rail services, with security provided by the British Transport Police (BTP) who police the rail network nationally.

4.4 Data

The following section describes the data utilized in this research, highlighting its advantages and limitations.

4.4.1 Bus Shelter Damage

Data on the number of incidents and cost of damage to bus shelters, for a 12-month period (January–December 2000) were obtained from Merseytravel. Data fields indicated the date of an incident, the cost of an incident, and the type of incident. Incident types have been assigned to classification groups to include smashed panels, graffiti, and other incidents of vandalism. Each bus stop is uniquely referenced with an X and Y coordinate with an accuracy of 1 m. Bus stop type is also categorized to distinguish between bus posts (concrete posts), conventional displays (CDs which are two metal posts holding a single glass or plastic panels displaying timetable information), and bus shelters.

The major disadvantage of this data set is that it only indicates when an incident is reported, not when it occurred. It is assumed that events are reported up to 24 h during weekdays and up to 62 h at weekends after the event occurred. No indication of the time of day is given.

4.4.2 Census Variables and Geodemographics

From the 1991 Census of Population, 35 selected variables were extracted at enumeration district (ED) level. The ED is the smallest unit of the census for England and Wales for which data are available. Geodemographics is a term used to describe the construction of residential units or neighborhoods from the Population Census. Geodemographic classifications are based on the use of cluster analysis to assign each ED to a district cluster or area type based on variables reflecting their demography, social and economic composition, and housing type (Brown, 1991). This research uses the SuperProfile lifestyle classification, based on data from the 1991 census and other descriptive information from other sources such as the electoral roll and consumer surveys (for further information, refer to the work by Brown and Batey, 1994). Britain's 146,000 EDs were broken down into 160 SuperProfile neighborhood types, a broader 40 target markets, and the most general classification of 10 SuperProfile lifestyles (see Appendix 4.1 for selected pen pictures of lifestyles). Caution should be exercised in the interpretation of these descriptions which seek to highlight distinctive features of the lifestyles based on an index table comparing the cluster means of selected indicators with the corresponding national mean value. Further, caution is required in comparing data from 1999 with 2000 shelter damage data although no comparable contemporary imformation on social, demographic, economic and housing types existed at the time of writing. It is important to offset the limitations of such a classification with the insights they may provide for the analysis of crime and its relationship with the environment.

4.4.3   Index of Local Conditions

This area-based index of deprivation was produced at ED level using six indicators of deprivation from the 1991 Population Census (Department of the Environment, 1995). For the purposes of this research, the 2925 Merseyside EDs were ranked by their index of local conditions (ILC) score and then grouped into 10 groups (deciles), each containing 10% of the EDs. Other indexes that could be utilized are the 1998 Index of Local Deprivation (ILD) and the 2000 Index of Multiple Deprivation (IMD). The former of these at ED level is also based on 1991 census variables, and the latter is only available at ward level (http://www.ndad.nationalarchives.gov.uk/CRDA/24/DS/1998/1/4/quickref.html).

4.4.4   Recorded Crime Data

Data on a number of crime types for the period January–December 2000 were obtained from the Merseyside Police's Integrated Criminal Justice System (ICJS). This data is known to be subject to a degree of underreporting (British Crime Survey, 2000). The categories obtained include criminal damage, drugs-related, robbery, other violence, and all recorded crime. Data were also acquired for the same period for calls to the police from command-and-control records. These are service calls to the police, not recorded levels of crime, and are subject to overreporting. They have been used as an indication of demand from the public for police intervention or "formal social control" (Bowers and Hirschfield, 1999). The categories of incident for which call records were provided are "disorder" and "juvenile disturbance." All these data sets were supplied aggregated to ward level, of which there were 118 covering Merseyside in 1991.

4.5   Methodology

All the data were compiled in a GIS. Stop references were captured using their X and Y coordinates, while all other data were transferred using the point centroids of their respective census ED or ward level coverage. The GIS intersect command was used to join bus stops to the ED in which they were situated. This method enables a profile to be constructed of damage at each shelter with environmental variables (SuperProfile lifestyles, selected census variables, % open space and % built areas, the ILC decile, and selected recorded crime and command-and-control data). The GIS program used was ArcView v3.1. This data was then exported into a statistical package (SPSSv10.0) to enable the further statistical analysis of the spatial data.

Analysis was undertaken to establish whether the point data relating to damage to bus shelters displayed evidence of clustering. CrimeStat v1.1 was the package used for this (http://www.ojp.usdoj.gov/nij/maps/). Both the nearest neighbor index (NNI) and Ripley's K-statistic were calculated. The first of these measures tests if the distance to the average nearest neighbor is significantly different from what would be expected by chance. If the NNI is 1, then the data is randomly distributed. If the NNI is less than 1, the data shows evidence of clustering. An NNI result greater than 1 reveals evidence of a uniform pattern in the data. A test statistic (the Z-score) was also produced; the more negative the Z-score, the more confidence that can be placed in the NNI result. It is not a test for complete spatial randomness and only examines first-order or global distributions. The Ripley's K-statistic compares the number of points within any distance to an expected number for a spatially random distribution. It provides derivative indices for spatial autocorrelation and enables the morphology of points and their relationship with neighboring points to be examined at the second, third, fourth, and nth orders, thus enabling the identification of subregional patterns. In Crime-Stat, these values are transformed into a square-root function, $L(t)$, at 100 different distance bins. To reduce possible error, rectangular border correction for 10 simulation runs was applied.

ArcView was used for visual analysis, producing proportional circles of hot spot damage and comparing these with choropleth maps displaying related environmental characteristics aggregated to ED and ward levels. The "hot spot" function in CrimeStat produced statistical ellipses of hot spot clusters that were also displayed using ArcView. An important consideration is that the production of these visualizations is subject to user input, and modification of the classification ranges and inputs used produces different visualizations. In CrimeStat, three parameters, the probability a cluster was obtained by chance, the minimum number of points per cluster, and the number of standard deviations for the ellipse, can all be altered, resulting in different visualizations. The benefit of this type of analysis is that possible relationships can be visualized and demonstrated without, or prior to, employing statistical analysis.

Resource target tables (RTTs) compare the number of stops damaged with the total number of stops. Bus stop incidents are ranked in descending order of incident frequency at each stop. Cumulative counts of incidents as a percentage of all incidents are constructed, and cumulative percentages are calculated. These are compared with the corresponding cumulative counts and percentages of bus stops. This gives an indication of the extent to which the incidents are concentrated at particular bus stops or groups of bus stops. An initial assumption in undertaking this analysis was that only certain types of stop (shelters and conventional displays) would be damaged. Thus, a separate RTT was constructed from which other stop types were excluded (notably, concrete poles).

All bus stops were assigned to a particular ED using a GIS-based operation, and from this, the number and cost of incidents of shelter damage could be cross-referenced with SuperProfile lifestyle, ILC decile, and selected 1991 census variables. In addition to this, the bus stops were also cross-referenced with a number of police-recorded crime, and police command-and-control variables aggregated to ward level. This data was exported from ArcView into a statistical package (SPSSv10.0), which enabled statistical analysis of the relationships between bus shelter damage and selected environmental factors. Two possible errors arise here. Using aggregated data (at ED and, especially, at ward level) increases the possibility of error related to the ecological fallacy (Martin and Longley, 1995). The ability of a GIS to adjust the levels of aggregation of data can result in further error attributed to the modifiable areal unit problem, whereby different aggregations can yield differing interpretations of the same data (Openshaw and Taylor, 1981). The Spearman's rank correlation was chosen as an appropriate nonparametric method for two-tailed bivariate correlation of non-normally distributed data. In addition to this, the number of bus stops that suffered shelter damage in each SuperProfile lifestyle were calculated and compared with the frequencies of what damage would be expected on the basis of the number of stops in each lifestyle using Chi-square (χ^2) analysis. This technique has previously been applied to burglary data (Bowers and Hirschfield, 1999).

To examine the temporal patterns of shelter damage, variations in cost were produced on a monthly basis for the whole of Merseyside. At present no information exists on hourly variations, and daily variation would be biased as incidents reported on the weekend (Friday p.m. through Monday a.m.) are reported as Monday. The data was split into the five districts of Merseyside, but to account for the disproportionate number of shelters in each district the rate of shelter damage per 100 shelters per month for each district was calculated. This was also compared with the rate for shelter damage per month per 100 shelters for Merseyside.

4.6   Findings and Discussion

Nearest neighbor analysis (NNA) and Ripley's *K*-statistics were produced using CrimeStat to derive for evidence of clustering in the data. The NNI calculated was 0.1346 and the test statistic (*Z*) value was −102.2862. This implies a very strong likelihood that the average nearest neighbor is significantly nearer than would be expected by chance, and the global distribution of damaged bus shelters displays evidence of clustering. An important consideration is whether the distribution of shelters themselves is clustered. The NNI of all the shelters is 0.2278 implying that the location of shelters themselves is clustered. However, the larger NNI value of all shelters compared to the damaged shelters implies the clustering of damaged shelters is over and above the clustered distribution of all shelters themselves. The *L*(*t*) values produced for the Ripley's *K*-statistic using the CrimeStat

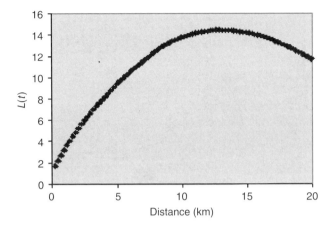

FIGURE 4.1
$L(t)$ values using Ripley's K-statistic compared with the distance between points.

software are plotted against the distance bins between points (Figure 4.1). This demonstrates that the $L(t)$ increases up to a distance of about 13 km before starting to decrease again. This also provides evidence for clustering at some higher orders than first-order clustering.

A GIS was used to visualize the outcome of the hot spot analysis of the shelter damage. Figure 4.2 shows proportional circles of hot spots, and compares them with first- and second-order nearest neighbor hierarchical (NNH) ellipses produced in CrimeStat. The advantage of NNH clusters is that they can be applied to an entire data set, but may still indicate small areas of clusters. Only those points closer than expected by chance are clustered at the first level, before these clusters are reclustered. Linkages between several small clusters and higher ordered clusters can be readily observed. The resulting images provide a method of portraying hot spots, depicting patterns that can be combined with other data within the framework provided by the GIS. The clustered distribution of shelter damage on Merseyside can be readily observed from this image.

Figure 4.3 shows a choropleth map of the SuperProfile lifestyles in which the shading is restricted to the built-up areas with proportional circles of hot spot damage overlaid. This provides a visual representation of the possible relationship between bus shelter damage and lifestyle, and suggests a very strong correlation between bus shelter damage and the areas of highest deprivation (the least affluent lifestyle Have-nots). It also demonstrates the ability of GIS to cross-reference multiple data sets.

A number of methods of hot spot analysis exist (e.g., Crime Mapping Research Centre, 1998; Chainey and Reid, 2002). These include different methods of visual interpretation, choropleth mapping, grid cell analysis, point pattern analysis, and spatial autocorrelation. Techniques that could be applied to this data in the future include kernel density interpolation and

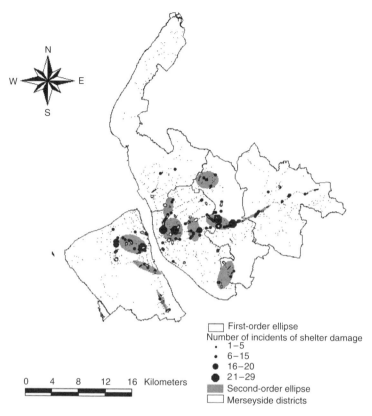

FIGURE 4.2
Proportional circles depicting incidents of bus shelter damage during Jan–Dec 2000, with first-
and second-order nearest neighbor hierarchical ellipses overlaid. (From 1991 Census: Digitised
Boundary Data (England and Wales).)

methods utilizing local indicators of spatial association (LISA). An example
of this is provided by Ratcliffe and McCullagh (1998). These allow for local
influences such as passenger flow numbers to be incorporated into the hot
spot analysis.

Thus far the clustered distribution of bus shelter damage has been demonstrated, but the techniques applied provide no indication as to the extent
to which incidents are concentrated at particular stops or in particular areas.
RTTs were produced to address this issue. An RTT was produced for all the
stops on Merseyside (Appendix 4.2). Over the year, 20% of all shelter
damage incidents occurred at 1% of all stops, 50% of all incidents at 5% of
all stops, and 100% of incidents at 25% of all stops. In terms of targeting
resources, this implies that all of the damage occurred at one-quarter of all
the stops. However, this includes all stop types including concrete poles, a
type where it is assumed that little or no damage can take place.

To allow for this, a further RTT was constructed for shelters and conventional displays only, with the stop type "concrete posts" excluded

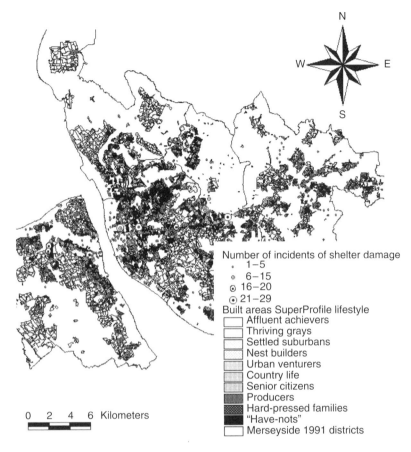

Number of incidents of shelter damage
- 1–5
- 6–15
- 16–20
- 21–29

Built areas SuperProfile lifestyle
- Affluent achievers
- Thriving grays
- Settled suburbans
- Nest builders
- Urban venturers
- Country life
- Senior citizens
- Producers
- Hard-pressed families
- "Have-nots"
- Merseyside 1991 districts

0 2 4 6 Kilometers

FIGURE 4.3
Bus shelter damage during Jan–Dec 2000 and SuperProfile lifestyles for a section of Merseyside.
(From 1991 Census: Digitised Boundary Data (England and Wales).)

(Table 4.1). A concentration of damage is evident, with 20% of the damage occurring at 2.5% of all shelters, 50% of damage at 10% of all shelters, and 100% of the damage at 58% of all shelters. Therefore, one-fifth of all damage occurred at 2.5% of all bus shelters, which in terms of volume equates to only 63 out of the 2556 bus shelters and CDs in Merseyside. The RTTs demonstrate that a concentration of shelter damage exists at particular stops and in certain areas and, when combined with a GIS, RTTs are a powerful tool in the identification and targeting of highly victimized stops.

The visual analysis suggests apparent relationships between criminal damage to bus shelters and its local environment, and further statistical analysis using bivariate correlations was deemed appropriate. This was to ascertain whether particular neighborhoods or environmental factors display a degree of correlation with bus shelter damage. Appendix 4.3 shows a detailed table of some selected results. It is evident from this that a positive correlation with the number of incidents of shelter damage is found for the

TABLE 4.1

Resource Target Table for the Bus Shelter Damage on Merseyside, Jan–Dec 2000

Incidents per Bus Shelter	Number of Affected Bus Shelters	Cumulative Number of Bus Shelters	Cumulative Number of Incidents	Cumulative Percentage of Bus Shelters	Cumulative Percentage of Incidents
29	1	1	29	0.04	0.76
27	1	2	56	0.08	1.47
25	1	3	81	0.12	2.12
24	1	4	105	0.16	2.75
23	1	5	128	0.20	3.35
20	1	6	148	0.23	3.88
17	1	7	165	0.27	4.32
16	3	10	213	0.39	5.58
15	4	14	273	0.55	7.15
14	5	19	343	0.74	8.99
13	2	21	369	0.82	9.67
12	5	26	429	1.02	11.24
11	13	39	572	1.53	14.99
10	14	53	712	2.07	18.66
9	10	63	802	2.46	21.02
8	22	85	978	3.33	25.63
7	29	114	1181	4.46	30.95
6	33	147	1379	5.75	36.14
5	60	207	1679	8.10	44.00
4	89	296	2035	11.58	53.33
3	151	447	2488	17.49	65.20
2	290	737	3068	28.83	80.40
1	748	1485	3816	58.10	100.00
0	1071	2556	n/a	100.00	n/a

percentage household lone parents, the percentage of an area open space, the percentage of youth unemployment, and the percentage of youths (age 15–25 years) in the area. All are significant at the 99% confidence level. These are possible indictors of a lack of capable guardianship and the presence of youths, and suggest they are important contributory factors to bus shelter damage. Interestingly, the percentage of male unemployment showed a negative correlation with incidents of bus shelter damage. This is possibly associated with high unemployment as an indicator of low mobility. Clearly further analysis of these patterns is appropriate when attempting to implement crime-reduction measures that design out crime. Examples of these include crime prevention through environmental design (CPTED) techniques (Pease, 1997).

Variables that provide information on passenger flows suggest there is a positive relationship between passenger numbers and bus shelter damage. Such a relationship is evident at the 99% confidence level for the following variables: the volume of passengers, percentage of households without a car, number of persons who travel to work on foot, and those who travel to work by car. Negative correlations are found between shelter damage and the following: the percentage households with one car, percentage home workers, percentage travel to work by car, and interestingly percentage travel to work by train, all significant at the 0.001 level. This adds weight to the claim that bus stops are crime generators. However, it is difficult to infer any causal relationships because data on other crime levels in the area would be required. The negative relationship with passengers using trains raises a number of questions. Does public transport facilitate, or displace crimes, for example? It is evident that information on damage to bus routes, train stations, train journeys, and other mode of transport needs to be assembled and built into this system so that such issues can be explored completely.

The police crime data aggregated to ward level shows positive correlation with shelter damage, although this is a very generalized measure. Youths causing annoyance and recorded criminal damage displayed the most significant correlations with shelter damage. To understand this relationship further, crime would need to be analyzed at finer levels of aggregation (at ED or using disaggregate data, for example). This could be coupled with information about land use in the vicinity of individual bus stops, and local population levels as this may also vary by time of day. This could then provide further insight into whether bus stops act as crime generators, and, if so, for what types of crime and at what times of day?

The SuperProfile lifestyle classification and the ILC both exhibit a positive relationship between levels of deprivation and levels of shelter damage (significant at the 99% confidence level). To examine this further, the number of damaged shelters located within each lifestyle area were compared with the amount of damage that would be expected based on the number of shelters in each lifestyle. Chi-square analysis was used for this and the results are shown in Table 4.2. The high positive relationship with "have-not" areas is evident. "Hard-pressed" and "producers" also experience greater than expected levels of shelter damage. In most affluent areas there is an underrepresentation of bus shelter damage. This suggests that there is a clear

TABLE 4.2

Correlation Coefficients for the Four Domains

Lifestyle	Number of Damaged Stops	χ^2-Value	Significance Level
Affluent achievers	518	50.74 (−)	0.001
Thriving greys	617	34.71 (−)	0.001
Settled suburban	825	31.03 (−)	0.001
Nest builders	683	0.8 (−)	n.s.[a]
Urban venturers	185	0	n.s.
Country life	28	1.57 (−)	n.s.
Senior citizens	445	0.02	n.s.
Producers	769	9.09	0.001
Hard-pressed	546	5.93	0.005
Have-nots	1366	92.66	0.001

[a] n.s., not significant.

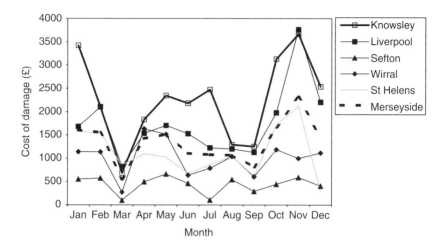

FIGURE 4.4
Merseyside shelter damage 2000: costs per 100 shelters by district.

social gradient in the degree to which neighborhoods are prone to shelter damage.

Figure 4.4 shows the cost of shelter damage per 100 shelters by month for the year 2000 in five Merseyside districts. Although the district Liverpool, which contains the city center, experiences a higher volume of incidents of shelter damage (Appendix 4.4). The rate of damage per shelter is highest in Knowsley. A distinct peak in the damage occurs in October and November. This is probably attributable to Halloween, Mischief Night, and Bonfire Night. In March and in the summer months a trough exists. One possibility is during school holiday's youths use buses and hence shelters less frequently, adding weight to the idea of shelters as crime generators. This data is only for 1 year, and hourly or daily variation plus comparisons with other years is desirable for future analysis.

4.7   Conclusions

This research has demonstrated the importance of the use of GIS, in combination with other techniques, to increase the knowledge of the nature and extent of criminal damage to bus shelters. It represents an initial attempt to develop a framework that should enable the identification of the levels and causes of crime and disorder on public transport. Such a framework should allow the testing of general theories of crime and disorder to see whether they can be applied or adapted to explain crime on public transport.

This task could be improved by extending the range of data sets utilized in this research. For example information on crime on individual bus routes, distinguished by category and with information about time of day could usefully be added in the future. It is contended that this could then be combined with data relating to crime on other modes of transport. Data on land use at the individual stop level should also be associated. The understanding of crime on public transport systems could be further enhanced by adding more disaggregate contextual data on other crimes in the surrounding areas and of local socio-demographic characteristics. Aspects of the physical infrastructure could be incorporated using OS landline data or aerial photographs.

This paper has presented preliminary evidence that damage at bus shelters is concentrated at particular stops and areas. Hot spot analysis, RTTs, and GIS have been used to identify and target these "high risk" stops and areas. There is evidence to suggest that particular neighborhoods, socio-demographic influences, and physical characteristics are more susceptible to shelter damage than others. Such areas include those in which high levels of deprivation are recorded, areas with large amounts of open space, and those with concentrations of youth populations. It is argued that this has implications for route planning and in tackling crime and disorder on public transport, which warrants further research.

There is some evidence in support of the notion of bus stops as crime generators. It is possible that bus stops act as generators of crime at certain times of the day and as crime attractors at other times. This may also vary for different types of crime, for example, criminal damage and robbery. Evidently further information on this is required. In summary, this paper has demonstrated the importance of further research into crime and disorder on public transport. It suggests that bus shelter damage is related to its environment, and discusses how GIS and other crime-mapping techniques can be combined to develop the knowledge of the extent of, and the theoretical reasons underlying, crime and disorder on public transport.

Acknowledgments

This work is based on data provided with the support of the ESRC and JISC and uses boundary material which is copyright of the Crown and the ED-LINE consortium. Datasets used: Bus stops—Merseytravel; Crime data—Mersey Police; Superprofile—Liverpool University.

References

Barlow, H., 1993, *Introduction to Criminology* (New York: Harper Collins).
Block, R. and Davis, S., 1996, The environs of rapid transit stations: a focus for street crime or just a risky place? In *Preventing Mass Transit Crime. Crime Prevention Studies, Volume 6*, edited by R.V. Clarke (Monsey, NY: Criminal Justice Press).

Bowers, K., 1999, Crimes against non-residential properties: pattern of victimization, impact upon urban areas and crime prevention strategies. Ph.D. Thesis, University of Liverpool.

Bowers, K. and Hirschfield, A., 1999, Exploring links between crime and disadvantage in north-west England: an analysis using geographical information systems. *International Journal of Geographical Information Science* 13(2), 159–184.

Brantingham, P. and Brantingham, P., 1993, Environment, routine and situation: toward a pattern theory of crime, In *Routine Activity and Rational Choice. Advances in Criminological Theory, Volume 5*, edited by R. Clarke and M. Felson, pp. 259–294 (New Brunswick, NJ: Transaction Publishers).

Brantingham, P. and Brantingham, P., 1995, Criminality of place: crime generators and crime attractors. *European Journal on Criminal Policy and Research: Crime Environment and Situational Prevention* 3(3), 5–26.

British Crime Survey, 2000, *The 2000 British Crime Survey (England and Wales)*. Home Office Statistical Bulletin 18/00 (London: Home Office).

Brown, P., 1991, Exploring geodemographics, In *Handling Geographic Information: Methodology and Potential Applications*, edited by I. Masser and M. Blakemore, pp. 221–258 (London: Longman).

Brown, P. and Batey, P., 1994, Characteristics of Super Profile Lifestyle and Target Markets: Index Tables, Pen Pictures and Geographical Distribution. *Super Profile Technical Notes 2* (Liverpool: URPERRL, Department of Civic Design, University of Liverpool).

Brown, S., Colo, A., Lawless, D., Lu, X., and Rogers, D., 1998, Interdicting a burglary pattern: GIS and crime analysis in the Aurora police department, In *Crime Mapping: Case Studies*, edited by N. LaVigne and J. Wartell, pp. 99–108 (Washington, DC: Police Executive Research Forum).

Buerger, M., Cohn, E., and Petrosino, A., 1995, Defining the "hot spots of crime": operationalising theoretical concepts for field research, In *Crime and Place: Crime Prevention Studies, Volume 4*, edited by J. Eck and D. Weisburd, pp. 237–258 (New York: Willow Tree Press).

Chainey, S., Reid, S., and Stuart, N., 2002, When is a hotspot a hotspot? A procedure for creating statistically robust hotspot maps of crime, In *Innovations in GIS 9: Socio-Economic Applications in Geographical Information Science*, edited by D.B. Kidner, G. Higgs, and S.D. White, pp. 21–36 (London: Taylor & Francis).

Clarke, R. (editor), 1992, *Situational Crime Prevention: Successful Case Studies* (New York: Harrow and Heston).

Cohen, L. and Felson, M., 1979, Social change and crime rate trends: a routine activity approach. *American Sociological Review* 44, 588–608.

Cornish, D. and Clarke, R., 1986, *The Reasoning Criminal: Rational Choice Perspectives on Offending* (New York: Springer).

Crime Mapping Research Centre, 1998, *Hot Spots: An Exploration of Methods* (http://www.ojp.usdoj.gov/nij/maps/).

Department of the Environment, 1995, *1991 Deprivation Index. A Review of Approaches and a Matrix of Results* (London: HMSO).

Department of the Environment, Transport and the Regions, 1998, *White Paper on the Future of Transport* (London: HMSO).

Department of the Environment, Transport and the Regions, 1999, *Young People and Crime on Public Transport* (London: HMSO).

Eck, J. and Weisburd, D., 1995, Crime places in crime theory, In *Crime and Place: Crime Prevention Studies, Volume 4*, edited by J. Eck, and D. Weisburd, pp. 1–34 (New York: Willow Tree Press).

Johnson, S., Bowers, K., and Hirschfield, A., 1997, New insights into the spatial and temporal distribution of repeat victimization. *British Journal of Criminology* 37(2), 224–241.

Jupp, V., Davies, P., and Francis, P. (editors), 2000, *Doing Criminological Research* (London: Sage).

LaVigne, N., 1997, *Visibility and Vigilance: Metro's Situational Approach to Preventing Subway Crime* (Washington, DC: National Institute of Justice).

Levine, N., Wachs, W., and Shirazi, E., 1986, Crime at bus stops. A study of environmental factors. *Journal of Architectural and Planning Research* 3(4), 339–361.

Loukaitou-Sideris, A., 1999, Hot spots of bus stop crime. The importance of environmental attributes. *Journal of the American Planning Association* 65(4), 395–411.

Martin, D. and Longley, P., 1995, Data sources and their geographic integration, In *GIS for Business Service Planning*, edited by P. Longley and G. Clarke, pp. 15–32 (New York: Wiley).

McEwen, T. and Taxman, S., 1995, Applications of computerised mapping to police operations, In *Crime and Place: Crime Prevention Studies, Volume 4*, edited by J. Eck and D. Weisburd, pp. 259–284 (New York: Willow Tree Press).

Openshaw, S. and Taylor, P., 1981, The modifiable unit areal problem, In *Quantitative Geography: A British View*, edited by N. Wrigley and R. Bennett, pp. 60–69 (London: Routledge and Kegan).

Pearlstein, A. and Wachs, M., 1982, Crime in public transit systems. An environmental design perspective. *Transportation* 11, 277–297.

Pease, P., 1997, Crime prevention, In *The Oxford Handbook of Criminology*, edited by M. Maguire, R. Morgan, and R. Reiner (Oxford: Clarenden Press).

Ratcliffe, J. and McCullagh M.J., 1998, Identifying repeat victimization with GIS. *British Journal of Criminology* 38(4), 651–662.

Appendices

Appendix 4.1   SuperProfile Lifestyle Pen Pictures

A short description of each lifestyle provides some idea of the distinguishing characteristics of these geodemographic groups based on the interpretation of an index table comparing the mean value of a selection of variables for each cluster with the corresponding mean value for the country as a whole, which is taken from Brown and Batey (1994). Lifestyles are alternatively numbered from 1 to 10.

Lifestyle A: Affluent Achievers

High-income families, living predominantly in detached houses. The affluent achiever typically lives in the stockbroker belts of the major cities and is likely to own two or more cars, which are top of the range, recent purchase and relied on for pursuit of an active social and family life. This type of person has sophisticated tastes. They eat out regularly, go to the theater and opera, and take an active interest in sports (e.g., cricket, rugby union, and golf). In addition they can afford several expensive holidays every year. Financially aware, with a high disposable income, affluent achievers often invest in company shares and specialized accounts. They use credit and charge cards frequently, and are likely to opt for private health insurance. Investments are followed closely in broadsheets such as the *Financial Times*, the *Times*, and the *Telegraph*. Other magazines bought may include *Hello*, *Harpers & Queen*, and *Vogue*.

Lifestyle B: Thriving Greys

Generally older than affluent achievers, possibly taking early retirement, the thriving greys are also prosperous. Their detached or semidetached homes have been completely paid for, and children have grown up and left home. Therefore, the greys have money to spare for investments or spending, on items such as a superior car. They eat out regularly, take one or two holidays a year, and are likely to play and enjoy going to the theater. This group is also financially aware and may invest in the stock exchange and opt for health insurance. The thriving greys read the broadsheets as well as more traditional magazines, such as *Women's Realm*, and *Woman and Home*.

Lifestyle C: Settled Suburbans

Well-established families in generally semidetached suburban homes. Settled suburbans are employed in white-collar and middle management positions, while in addition many wives work part-time. The lifestyle is fairly affluent, in that one or two package holidays a year may be taken, and the family can afford to purchase newer cars. They have taken advantage of government share offers in the past and often use credit cards. Many are mail-order agents. Typical publications read include the *Daily Mail*, the *Express*, *Ideal Home*, and *Family Circle*.

Lifestyle H: Producers

These more affluent blue-collar workers live in terraced or semidetached housing. Many are middle aged or older, and their children have left home. They work in traditional occupations and manufacturing industries, where unemployment has risen to a significant level. Most are well settled in their homes, which are either purchased or rented from the council. Leisure pursuits include going to the pub and betting on horse races. On TV, football and rugby league are the preferred sports. They do not spend money on cars and there is little planning for the future by way of financial investments. The *Sun*, the *Mirror*, and the *News of the World* are the most popular newspapers.

Lifestyle I: Hard-Pressed Families

Living in council estates, in reasonably good accommodation, unemployment is a key issue for these families. Most work is found in unskilled manufacturing jobs, if available, or on government schemes. The parochial nature of this group is emphasized by an unwillingness or inability to either move home or go on holiday. The most popular leisure activities are betting and going to pubs and clubs. On TV, sports such as football and rugby league are watched. Tabloids, particularly the *Sun*, the *Mirror*, and the *Daily Record* are the chosen daily papers.

Lifestyle J: Have-Nots

Single parent families composed of young adults and large numbers of young children, living in cramped flats. These are the underprivileged group who move frequently in search of a break. However, with 2.5 times the national rate of unemployment and with low qualifications, there seems little hope for the future. Most are on income support, and those who can find work are in low-paid, unskilled jobs. There are very few cars and little chance of getting away on holidays. Recreation comes mainly from the television and the take up of satellite and cable TV is high. Betting is also popular, particularly greyhound racing. The *Sun* and the *Mirror* are the most popular newspapers.

Appendix 4.2   Resource Target Table for All Shelter Types

Incidents per Bus Stop	Number of Affected Bus Stops	Cumulative Number of Bus Stops	Cumulative Number of Incidents	Cumulative Percentage of Bus Stops	Cumulative Percentage of Incidents
29	1	1	29	0.02	0.76
27	1	2	56	0.03	1.47
25	1	3	81	0.05	2.12
24	1	4	105	0.07	2.75
23	1	5	128	0.08	3.35
20	1	6	148	0.10	3.88
17	1	7	165	0.12	4.32
16	3	8	181	0.13	4.74
15	4	11	228	0.18	5.97
14	5	15	287	0.25	7.52
13	2	21	369	0.35	9.67
12	5	26	429	0.43	11.24
11	13	39	572	0.64	14.99
10	14	53	712	0.88	18.66
9	10	63	802	1.04	21.02
8	22	85	978	1.41	25.63
7	29	114	1181	1.88	30.95
6	33	147	1379	2.43	36.14
5	60	207	1679	3.42	44.00
4	89	296	2035	4.89	53.33
3	151	447	2488	7.39	65.20
2	290	737	3068	12.19	80.40
1	748	1485	3816	24.55	100.00
0	4563	6048	n/a	100.00	n/a

Appendix 4.3 Bivariate Correlation Results

Potential Indicators of Deprivation and Lack of Guardianship

		SuperProfile Lifestyles	ILC Decile	Male Unemployment	Youth (16–19 yr) Unemployed	% Open Space	% Lone Parents	% Youths (15–24 yr)	% Young Adults (25–44 yr)
Number of incidents of bus shelter damage	Spearman's ρ Significance (two-tailed)	0.228*	0.219*	−0.07*	0.145*	0.242*	0.165*	0.077*	−0.044**
		0.000	0.000	0.001	0.000	0.000	0.000	0.000	0.038
	N	2925	2925	2925	2925	2925	2925	2925	2925

Indicators of Passenger Volumes

		% Household with		% Home Workers	% Travel to Work			
	Passengers	No Car	1 Car		On Foot	By Car	By Bus	By Train
Number of incidents of bus shelter damage — Spearman's ρ	0.342*	0.231*	−0.207*	−0.075*	0.071*	−1.54*	0.177*	−0.083*
Significance (two-tailed)	0.000	0.000	0.000	0.000	−.001	0.000	0.000	0.000
N	2925	2925	2925	2925	2925	2925	2925	2925

Indicators of Other Crime Levels

	Command-and-Control				Recorded Crime		
	Youths Causing Annoyance	Disorder	Criminal Damage	Drugs	Other Violence	Robbery	All Crime
Number of incidents of bus shelter damage — Spearman's ρ	0.542*	0.526*	0.505*	0.428*	0.499*	0.485*	0.468*
Significance (two-tailed)	0.000	0.000	0.000	0.000	0.000	0.000	0.000
N	118	118	118	118	118	118	118

* Correlation is significant at the 0.01 level (two-tailed).

** n.s., not significant (p > 0.05)

Appendix 4.4A Merseyside Shelter Damage Jan–Dec 2000 (Cost per Month)

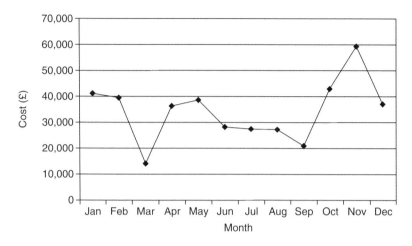

Appendix 4.4B Merseyside Shelter Damage 2000 (Cost per District per Month)

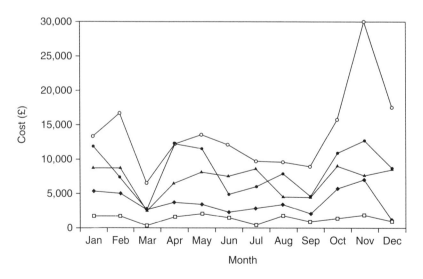

5

Policy Implications of Pockets of Deprivation in Scotland

Daniel Exeter, Robin Flowerdew, and Paul Boyle

CONTENTS

5.1 Introduction

There are many examples of research that make use of composite measures of material and/or social deprivation calculated at the area level to depict the socioeconomic status of an area (Townsend, 1987; Carstairs and Morris, 1991; Salmond et al., 1998; Department of the Environment Transport and the Regions, 2000; Kearns et al., 2000; Senior, 2002). Much of this research has focused on populations at the ward level in England and Wales, or the pseudo postcode sector level in Scotland. The ward and pseudo postcode sector level are typically used to preserve confidentiality of residents and minimize problems associated with statistics carried out on areas with a small number of observations.

Pockets of deprivation could be defined as small areas that are significantly more deprived than their neighboring areas. When deprivation indices are calculated at the ward and pseudo postcode sector levels, there is a possibility that the assigned value to an area might hide pockets of deprivation: enumeration districts in England and Wales or output areas in Scotland that are notably less affluent than the ward/pseudo postcode sector in which they are located. A recent paper by Haynes and Gale (2000) suggests that the large geographical space that is covered by a Census ward in rural areas might be too large to be homogeneous and that pockets of deprivation *do* exist. In rural areas, a "neighbor" might be a number of miles, rather than a matter of meters, away. Nevertheless, there is every possibility that the ward contains residents of less affluent status than is assigned to the ward by way of a socioeconomic indicator or deprivation index.

At present, many public sector organizations use both deprivation indices and/or health data to determine their catchment area's demographic profiles and resource allocation, typically at the pseudo postcode sector level (Mc Laren and Bain, 1998). Health boards, for example, are funded based on the Arbuthnott deprivation index, which accounts for remoteness (Scottish Executive Health Department, 1999). However, the use of the pseudo postcode sector for this type of analysis in rural areas potentially fails to identify small neighborhoods that are most in need of funding. In more remote parts of Scotland, the pseudo postcode sectors cover a large area and in some cases only include one or two output area(s), so the possibility of obscuring deprived households is significant.

This chapter investigates whether pockets of deprivation exist in Scotland using pseudo postcode sectors and output areas, and it is expected that such areas are particularly likely to exist in rural areas. We first use the Carstairs index of deprivation (Carstairs and Morris, 1991) to identify the most deprived output areas that are not located in the most deprived pseudo postcode sectors. Second, we consider the difference between output area Carstairs and the Carstairs score of the pseudo postcode sector in which the output area is located. A high positive value will identify situations in which an output area is significantly more deprived than the pseudo postcode sector in which it is located.

We also recalculate the Carstairs index excluding the car ownership variable, henceforth known as the carless Carstairs index, because access to a motor vehicle in rural areas is a necessity rather than a measure of affluence, as is often the case in urban areas (Higgs and White, 2000). We repeat the ranking and difference calculation on the carless Carstairs index.

The implications our results have on policy are then assessed in relation to health boards in Scotland. It is assumed that since deprivation indices are commonly used in resource allocation, areas that are more homogeneous are likely to gain more support from funding bodies. The more heterogeneous the output areas within a pseudo postcode sector, the more likely areas of significant deprivation will be ignored. Using the mean standard deviation of the difference between output area Carstairs and the Carstairs score of the pseudo postcode sector, we calculate the heterogeneity of deprivation in health boards across Scotland. This measure should give high values for health boards where pseudo postcode sectors are heterogeneous and low values where they are homogeneous.

5.2 Data and Methods

In this chapter, we use population density and the Carstairs index of deprivation for 1991 as surrogate measures for rurality and area-level deprivation and two approaches to identify pockets of deprivation. First, we simply identify deprived output areas that are not located within the most deprived pseudo postcode sectors and investigate these locations further by ranking the pseudo postcode sectors and output areas by deprivation. Secondly, we locate output areas that are notably more deprived than the pseudo postcode sectors in which they are located, identified by the difference between the Carstairs score at both levels of aggregation. We use the first measure as a spatial indicator of absolute inequality, whereas the second indicator is a measure of relative deprivation.

5.2.1 Population Density as a Measure of Rurality

Although the Office of National Statistics (ONS) provides an urban/rural index for England and Wales at the ward level, there is no similar index for Scotland that is commonly used. The General Register Office for Scotland (GROS) has produced a six-class scale that is based on population density within continuous areas. A value of one indicates a very urban settlement, such as Glasgow and Edinburgh, and a value of six indicates a very remote settlement, defined officially as "not in a locality" (GRO Scotland, 2001). However, the GROS index was difficult to replicate and the areas are not comparable to the census geography used here. In this work, a surrogate urban/rural measure has been calculated based on population density per square kilometer at the pseudo postcode sector level, which is shown in Figure 5.1. We categorized areas that had a population density of less than 100 as rural. Areas with a population density between 100 and 1000 were categorized as suburban, and above 1000 as urban.

As one might expect, population density is greatest in the urban areas of Aberdeen, Dundee, Edinburgh, and Glasgow. The population density decreases from the city centers out to the suburbs and diminishes as one moves further from the major cities. A very low population density in the Highlands will exist due to the topography of the mountain ranges making the area uninhabitable in many places. Population density is fairly high

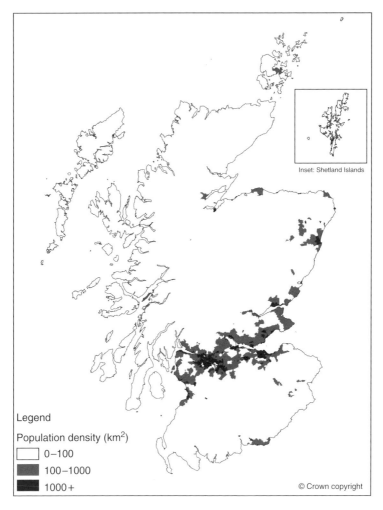

FIGURE 5.1
Population density in Scotland. (From 1991 Census: Digitised Boundary Data (Scotland), 1991 Census: Small Area Statistics (Scotland).)

along the east coast of Scotland between Aberdeen and Dundee, and also between Glasgow and Edinburgh, where settlements flank the main highways between these cities.

Pseudo postcode sector population density ranges continuously from 0.196 in one pseudo postcode sector in Orkney to 11704.07 persons per square kilometer in one pseudo postcode sector in Edinburgh, and there are four pseudo postcode sectors with extremely high population densities in Glasgow, where the densities range from 55609.48 to 19223.24 persons per square kilometer. These extreme outliers are possibly due to a high proportion of tower block housing in a small area.

5.2.2 Carstairs Index of Deprivation at the Pseudo Postcode Sector Level

There have been many definitions and interpretations of deprivation in the literature over the past two decades (Townsend, 1987; Carstairs and Morris, 1991; Salmond et al., 1998; Department of the Environment, Transport and the Regions, 2000; Kearns et al., 2000). In Scotland, the more commonly accepted measure of deprivation is the index developed by Carstairs and Morris (1991). The index is comprised of four standardized variables indicating levels of male unemployment, lack of car ownership, low social class, and living in overcrowded conditions. A negative Carstairs score indicates areas of relatively low deprivation and a positive score represents more relatively deprived neighborhoods. The index ranges from -4.07 to 15.63, with a mean of 0.081 and a standard deviation of 2.97 in output areas; and for pseudo postcode sectors the range is -7.49 to 12.87 with a mean of -0.58 and a standard deviation of 3.61. Some areas recording zero values for one or more components of the index affected the calculation of the Carstairs index at the output area level, however this does not influence our results, as we are interested in the extreme level of deprivation.

Figure 5.2 shows the distribution of the Carstairs Index of deprivation for pseudo postcode sectors in Scotland. The most deprived areas are located in urban areas of Glasgow, Edinburgh, and Dundee, with a few areas in the Western Isles also appearing more deprived. In general, those areas that were identified as having low population densities in Figure 5.1 appear less deprived in Figure 5.2.

While Figure 5.2 shows the spatial distribution of the Carstairs index of deprivation for the pseudo postcode sector level, the index has also been calculated at the output area level for analysis in this work, which produces a similar, denser pattern than the pseudo postcode sector level.

There was a slight positive relationship between output area Carstairs score and pseudo postcode sector population density, with a correlation coefficient of 0.206 ($p < 0.01$). This suggests that in areas of low population density, such as rural areas, material deprivation observed at the output area level is the lowest. Nevertheless, there were a number of urban areas that have low population densities, located in the central business district, or on the outskirts of the urban center.

5.2.3 Ranking the Carstairs Index

A preliminary investigation to identify pockets of deprivation was made by simply ranking all pseudo postcode sectors and output areas in Scotland by the Carstairs score, from the most deprived to the most affluent. We extracted the most deprived 10% pseudo postcode sectors and the most deprived 1% output areas for further analysis. This resulted in a sample of 101 pseudo postcode sectors from Dundee, Edinburgh, Glasgow, Inverness, Motherwell, Paisley, and Perthshire and another containing 380 output areas from postcodes within Aberdeen, Dundee, Edinburgh, Fife, Glasgow,

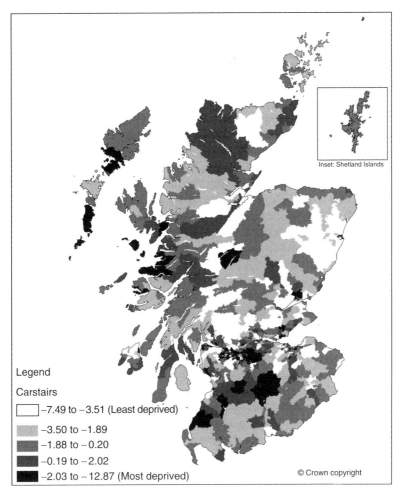

FIGURE 5.2
Carstairs index of deprivation. (From 1991 Census: Digitised Boundary Data (Scotland), 1991
Census: Small Area Statistics (Scotland).)

Inverness, Kilmarnock, Motherwell, Paisley, and Perthshire. Fourteen of the
101 pseudo postcode sectors had a population density less than 100 per
square kilometer.

   One would expect that the most deprived pseudo postcode sectors should
include the majority of the most deprived output areas. To test this hypothesis, a spatial overlay identified 314 of 380 output areas that were located
in 61 of the 101 pseudo postcode sectors sampled. All but one of these
output areas was located in pseudo postcode sectors with a population
density above 500 per square kilometer. The other output area was located
in a pseudo postcode sector with a population density of 12.16 per square
kilometer, located on the outskirts of Glasgow city.

The remaining 66 output areas that were not within the 10% pseudo postcode sector sample were located within 51 other pseudo postcode sectors across Scotland. The population density for these pseudo postcode sectors ranged from 11.16 to 9288.11 per square kilometer.

There were 40 pseudo postcode sectors in the 10% sample that did not contain any of the most deprived output areas sampled. These were located in rural, suburban, and urban areas, with a population density ranging from 0.5 to 82591.7. Thirteen pseudo postcode sectors had a population density less than 100 and are located in the Outer Hebrides, suburban Glasgow, and Motherwell. These areas of low population density were, however, not the most deprived pseudo postcode sectors in the sample with Carstairs scores ranging from 3.97 to 9.75.

5.2.4 Difference between Output Area and Pseudo Postcode Sector Carstairs Scores

One way in which pockets of deprivation might be identified is by subtracting the Carstairs score for the pseudo postcode sectors from the Carstairs score of an output area. When the difference is negative, the output areas are less deprived than is reported at the pseudo postcode sector level and a positive difference indicates the output areas that are more deprived than their pseudo postcode sector. The results of this analysis ranged from –13.68 to 15.75, with a mean of 0.08 and a standard deviation of 2.86. Figure 5.3 shows the spatial distribution of these results. On the whole, the distribution is relatively random, although many of the positive values were located in rural areas.

A closer examination of this distribution shows that 54.7% of the output areas have lower Carstairs scores than their pseudo postcode sector. In addition, of these output areas, 26.4% are located in pseudo postcode sectors with a population density less than 100 per square kilometer.

We extracted output areas that had a population density less than 100; a difference between the output area and pseudo postcode sector Carstairs score of more than 7; and an output area Carstairs score greater than 7. We aimed to identify deprived output areas in rural areas that were significantly more deprived than the pseudo postcode sector in which they were located.

These criteria identified a sample of 22 output areas, which can be categorized into three types: (1) small output areas within rural settlements; (2) small output areas in suburban areas; and (3) large output areas in rural areas, which are not part of a settlement. Examples of these three categories are shown in Figure 5.4. The Dalbeattie example (Figure 5.4a) identifies a small area on the outskirts of a rural settlement, which is notably more deprived than its pseudo postcode sector. The South Kessock example (Figure 5.4b) is an example of output areas located in an urban or suburban pseudo postcode sector. South Kessock is one of the three areas identified for having pockets of deprivation in which two or more relatively deprived output areas are located very closely to one another. However, South Kessock is the only example where this occurs within urban areas—the other two

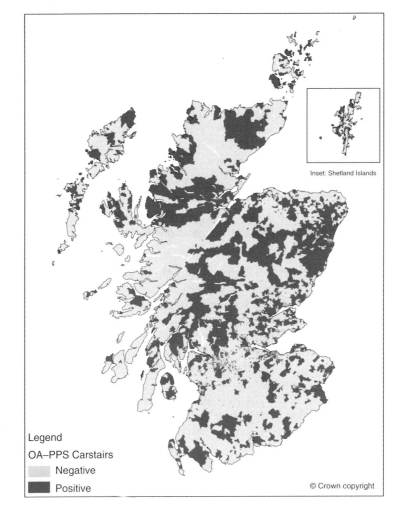

FIGURE 5.3
The difference between output area and pseudo postcode sector Carstairs scores. (From 1991
Census: Digitised Boundary Data (Scotland), 1991 Census: Small Area Statistics (Scotland).)

examples are located in rural settlements. The final example in Figure 5.4
shows one output area covering a comparatively large area and encompasses
a number of small communities that contain only one or two houses in
the village (Figure 5.4c).

Table 5.1 lists the five most relatively deprived rural output areas, taken
from the sample discussed above, and based on the difference between the
Carstairs score at the output area and pseudo postcode sector levels. It is
interesting that the Carstairs scores for these output areas are not near
the extreme end of the deprivation scale. In addition, it is interesting that
three of the five pockets of deprivation are located in the same pseudo
postcode sector and have a range of Carstairs scores for the output area.

Pockets of deprivation: Dalbeattie*

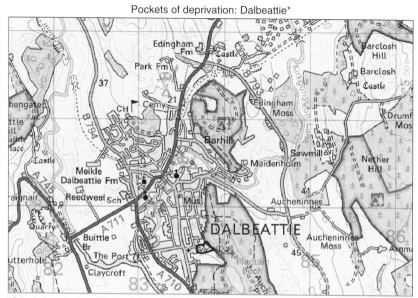

*Criteria for selection: OA Carstairs vs. PPS Carstairs ≥5; Population density ≤100; OA Carstairs ≥7.

(a)

Pockets of deprivation: South Kessock, Inverness*

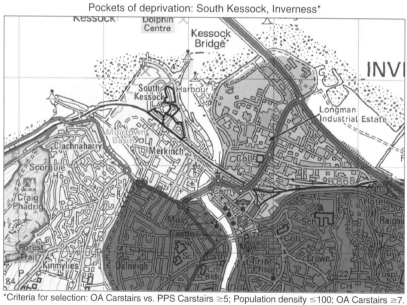

*Criteria for selection: OA Carstairs vs. PPS Carstairs ≥5; Population density ≤100; OA Carstairs ≥7.

(b)

FIGURE 5.4 (See color insert following page 274.)
Examples of pockets of deprivation located in (a) rural settlements, (b) urban areas, and

(*continued*)

Pockets of deprivation: Newbigging, near Blairgowrie*

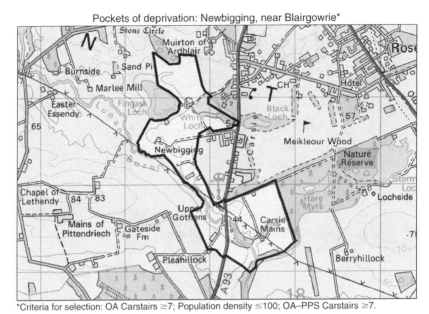

*Criteria for selection: OA Carstairs ≥7; Population density ≤100; OA–PPS Carstairs ≥7.

(c)   © Crown copyright: Ordinance survey. An EDINA digimap/JISC supplied service.

FIGURE 5.4 (continued)
(c) rural areas. (© Crown Copyright/database right 2007. An Ordnance Survey/EDINA supplied service.)

The relationship between the difference calculation and population density for all output areas was statistically significant but weak, with a coefficient of -0.120 ($p < 0.01$). This suggests that in areas where population density is lowest there is more likelihood of deprivation at the output area level to be obscured, if deprivation is analyzed at the pseudo postcode sector level. However, the weakness of the relationship does reflect the existence of pockets of deprivation in more densely populated areas too.

TABLE 5.1

The Five Most Deprived Rural Output Areas Identified using the Output Area Carstairs Score and the Difference between the Output area and Pseudo Postcode Sector Carstairs Scores

Location	OA–PPS	Carstairs	PPS Population Density
Newbigging, near Blairgowrie	14.18	11.60	96.49
Stonefield, near East Kilbride	13.66	8.20	30.90
Stonefield, near East Kilbride	12.97	7.51	30.90
Stonefield, near East Kilbride	12.48	7.02	30.90
Kinross	11.05	7.32	50.27

5.2.5   Carless Carstairs Ranking

Although the Carstairs index of deprivation is widely used and accepted as a suitable measure of material deprivation in Scotland, it has also been criticized for the use of the car ownership variable. Car ownership in rural parts of Scotland is often required as a means for accessing essential resources, rather than an indication of affluence, which is often the case in urban areas where public transport is more readily accessible.

By removing the car ownership variable from the Carstairs index, we aimed to identify rural areas that are deprived regardless of car ownership. The carless Carstairs index ranged from –5.76 to 10.27 with a mean of –0.299 and a standard deviation of 2.75 in pseudo postcode sectors, while the range is from –2.79 to 15.71 with a mean of 0.002 and a standard deviation of 2.21 for the output areas.

We repeated the ranking and sampling procedure outlined in Section 5.2.3 above, and produced a dataset containing 101 deprived pseudo post-code sectors and 380 deprived output areas. As before, we expected that a majority of these deprived output areas should be located in the most deprived pseudo postcode sector sample. This was the case for 293 of the 380 output areas, which were located within 60 of the most deprived pseudo postcode sectors. Most of these pseudo postcode sectors were in Glasgow city, and in the suburbs of Dundee, Edinburgh, and Greenock.

Figure 5.5 shows that surprisingly the majority of sampled output areas outside the deprived pseudo postcode sectors were not in very rural areas, rather they were located in suburban areas. There were seven output areas in pseudo postcode sectors with a population density less than 100 per square kilometer. In this sample, 59 of the output areas were not included in the equivalent Carstairs index-based sample and are mostly located in suburban Glasgow, although they are also situated in the hinterland of Dundee and Edinburgh, as well as Ayrshire, Motherwell, Fife, and Perthshire.

5.2.6   Difference between Output Area and Pseudo Postcode Sector Carless Carstairs Scores

We used the carless Carstairs variable again to calculate the difference between output area and pseudo postcode sector level deprivation. We then extracted a sample of output areas based on similar criteria to those mentioned for the Carstairs difference calculation. However, because there are only a few output areas with a carless Carstairs score greater than 7, we selected output areas with a carless Carstairs score greater than 5.

Seventeen output areas were identified with these criteria, of which nine were also included in the Carstairs difference sample. The other eight were located in Cowie; Stenhousemuir, Falkirk; Broomhouse, Glasgow; and Lochwinnoch and Stonefield, near East Kilbride. Surprisingly, four additional output areas near Stonefield were identified as pockets of deprivation, three of which were very close to the four output areas identified in the previous analysis based on the Carstairs index. Stenhousemuir and

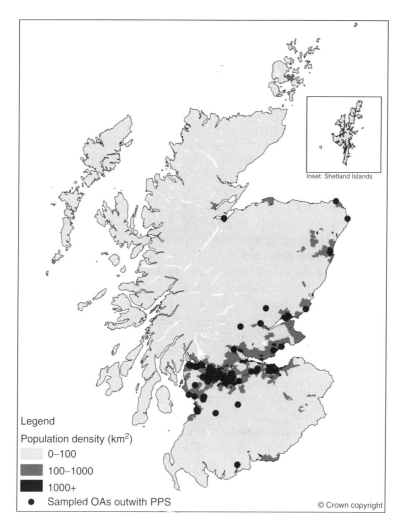

FIGURE 5.5
Deprived output areas located outside the most deprived pseudo postcode sectors, for the carless Carstairs index of deprivation. (From 1991 Census: Digitised Boundary Data (Scotland), 1991 Census: Small Area Statistics (Scotland).)

Broomhouse are both in small suburban output areas, and Cowie and Lochwinnoch are both small output areas in rural settlements, similar to the Dalbeattie example in Figure 5.4a.

5.3 Funding and Homogeneity of Deprivation in Health Boards

By definition pockets of deprivation exist in small administrative areas that are part of a bigger administrative region, such as output areas within pseudo postcode sectors in this work. Therefore, there is potential that resource allocation policies based on the assumption that a region is homogeneous (Beale et al., 2000; Crofts et al., 2000; Field, 2000; Haynes and Gale, 2000; Higgs and White, 2000) will not generate enough resources for the deprived areas that are located within more affluent regions. It is not suggested that resource allocation is based on homogeneity, but funding formulae will respond to overall levels of deprivation without recognizing that pockets of deprivation can occur in less deprived but heterogeneous areas.

To test the hypothesis, we first calculated the standard deviation of the difference calculation for pseudo postcode sectors and then calculated the mean standard deviation for all pseudo postcode sectors located within a given health board as a measure of health board homogeneity. We used the Carstairs index of deprivation because it was commonly used in the early 1990s for resource allocation purposes. The results are presented in Table 5.2 and Figure 5.6.

Table 5.2 lists the population, funding per head of population, and the heterogeneity score we calculated. It was expected that more homogeneous areas would attract higher resource allocations. However, the converse appears to be true whereby the Western Isles, the health board with the most funds per capita, has a much lower heterogeneity score than the Greater Glasgow health board, which has the greatest degree of heterogeneity and whose funding per capita is slightly higher than the national average.

TABLE 5.2

Population, Funding Allocation, Heterogeneity, and Deprivation of Health Boards in Scotland

Health Board	Population	Funding per Head[a]	Heterogeneity[b]	Carstairs Score[c]
Argyll and Clyde	426,900	836	2.47	0.80
Ayrshire and Arran	375,400	809	2.57	0.70
Borders	106,300	830	1.91	−1.90
Dumfries and Galloway	147,300	865	2.01	−1.38
Fife	348,900	757	2.35	−1.31
Forth Valley	275,800	770	2.57	−1.13
Grampian	525,200	749	2.21	−3.97
Greater Glasgow	911,200	868	3.32	4.49
Highland	208,300	820	1.94	−1.75
Lanarkshire	560,800	763	2.89	1.92
Lothian	773,700	758	2.34	−2.61
Orkney	19,550	913	1.64	−1.68
Shetland	22,910	953	1.75	1.14
Tayside	389,800	890	2.29	−0.76
Western Isles	27,940	1,172	2.36	0.67

[a] Allocation of funds for 2000–2001 (*Source*: Scottish Executive Health Department, 1999, *Fair Shares for All* (Edinburgh: The Stationery Office).
[b] Heterogeneity calculated as the mean standard deviation of OA Carstairs deprivation score located within a pseudo postcode sector (PPS).
[c] Carstairs index calculated for health board level.

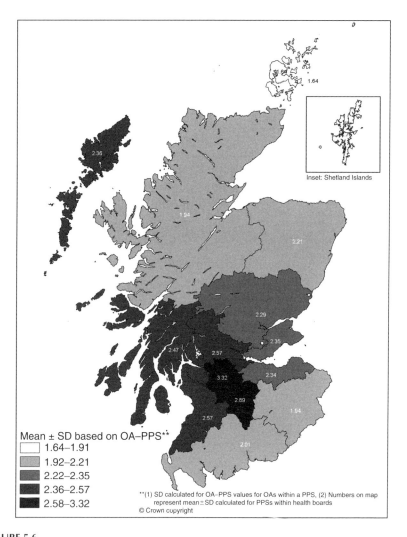

FIGURE 5.6
The homogeneity of health boards in Scotland. Regions with lower values indicate greater homogeneity. (From 1991 Census: Digitised Boundary Data (Scotland), 1991 Census: Small Area Statistics (Scotland).)

The relationship between funding per head and heterogeneity was not statistically significant and very weak (–0.199); however, the relationship remained insignificant, but was strengthened to –0.395 when the outlier (Western Isles) was omitted from the analysis. There is a strong negative relationship between population and funding (–0.515) which is statistically significant at the 0.05 level. This suggests that the cost of healthcare in rural areas is significantly higher than the cost of the same level of healthcare in urban areas, and is a logical assumption. The heterogeneity index, on the other hand is strongly related to population (0.774) at the 0.01 level, indicating that as the level of urbanity increases so too does the level of heterogeneity.

Our findings suggest that resource allocation is related to the heterogeneity of a health board rather than homogeneity. Furthermore, heterogeneity occurs most prominently in the central belt, the most population dense areas of Scotland. The negative relationship between urbanity and homogeneity may be an artifact of the census geography in Scotland, in which homogeneity is considered for the construction of output areas, rather than pseudo postcode sectors or health boards upon which this stage of the analysis was based. Nevertheless, many of the pockets of deprivation that we identified were outside the central population belt and thus in relatively homogeneous areas, which may be overlooked in the resource allocation process. South Kessock in Inverness is an example of pockets of deprivation that may not be considered in resource allocation rounds because the health board in which it is located (Highland) is very homogeneous.

5.4 Conclusions

The focus of this paper was to identify pockets of deprivation in rural parts of Scotland. We used two measures: first by identifying the most deprived output areas that are not situated within the most deprived pseudo postcode sectors; and second by calculating the difference between output area and pseudo postcode sector deprivation.

In addition, we used the Carstairs index of deprivation, and recalculated the Carstairs index excluding the car ownership variable. The car ownership variable in the Carstairs index is often criticized in the literature (Higgs and White, 2000) for the bias that is placed on urban areas where car ownership is an indication of affluence rather than a necessary means of transport which is often the case in rural areas of Scotland. We excluded the car ownership variable from the Carstairs index to create the carless Carstairs index and repeated the two-pocket identification approaches. Population density was used as a surrogate measure of rurality.

The rationale behind the ranking approach was that the most deprived pseudo postcode sectors should contain the most deprived output areas, while the difference between output area and pseudo postcode sector deprivation was used to identify output areas that have significantly different deprivation status than is assigned to the pseudo postcode sector in which they are located.

The carless Carstairs index ranking procedure showed that the majority of output areas outside the pseudo postcode sector sample were also located in suburban areas and only 59 of 380 output areas were not included in the Carstairs output area sample.

The results from the ranking technique suggest that pockets of deprivation do exist for both the Carstairs and carless Carstairs indices of deprivation; however, such pockets tend to exist in suburban areas rather than rural regions.

When the difference between output area and pseudo postcode sector deprivation was examined, we found that although the pattern appeared random, positive differences tended to be located outside the urban areas such as Edinburgh, Glasgow, and Dundee. Nevertheless, there were still cases in these areas where output area deprivation was greater than that reported at the pseudo postcode sector level.

We extracted a sample of the most deprived rural output areas, based on population density less than 100; output Carstairs less than 7 or carless Carstairs score less than 5; and the difference calculation greater than 7 for both analyses. Table 5.1 lists the location and attributes for the five most deprived areas using the Carstairs index of deprivation. Only one of these areas—Newbigging, near Blairgowrie—was a rural area. The other areas located were either small output areas located in suburbs, particularly in South Kessock, Inverness and near Glasgow; or they were output areas situated within rural settlements.

The difference calculation based on the carless Carstairs variable identified 17 pockets of deprivation, nine of which were found in the Carstairs analysis. Of the eight that were not located in the Carstairs analysis, those situated in Cowie and Lochwinnoch were located within rural settlements; Broomhouse and Stenhousemuir were surrounded by relatively urban settlements; and four were found near Stonefield, Hamilton. Surprisingly, in both sets of difference calculation, those output areas that were identified as having the biggest difference between local and pseudo postcode sector deprivation did not have the biggest deprivation values. For example, in the carless Carstairs sample, the difference values ranged from 7.16 to 10.91 and the carless Carstairs scores ranged from 5.02 to 6.53, whereas the maximum carless Carstairs score was 15.71.

The relationship between population density and the difference calculation for the Carstairs and the carless Carstairs indices was statistically significant, although the associations were weak and negative with correlation coefficients of −0.120 and −0.091 for Carstairs and carless Carstairs, respectively. These relationships suggest that as population density increases, the difference between output area deprivation and pseudo postcode sector deprivation decreases, which in turn implies that pockets of deprivation in rural areas do exist.

The implications of our results were also assessed in relation to allocation of resources to health boards in Scotland. We proposed that areas with lower levels of heterogeneity were more likely to receive the most funding, but found that the converse was the case. There was a strong negative correlation between population and funding (−0.515) at the 0.05 level, and the relationship between population and heterogeneity was very strong (0.774) at the 0.01 level. There was a weak negative correlation between funding and heterogeneity, but this was not statistically significant. Nevertheless, many of the pockets of deprivation that we identified were outside the central population belt and may be overlooked in the resource allocation process.

5.5 Future Research

Visual analysis of the results presented in this work (Figure 5.4a–c, in particular) suggests that pockets of deprivation might exist in rural areas, but the tendency is for such pockets to be located in suburban areas.

This research has proved fruitful in the identification of pockets of deprivation, more so in suburban areas than in rural areas. Further research is required to identify other factors that contribute to the notion of pockets of deprivation. In addition, attention will be given to self-reported limiting long-term illness data in an attempt to identify "pockets of poor health" and also to examine the relationship between the pockets of deprivation and pockets of poor health.

While the statistical relationships between population density, the Carstairs and carless Carstairs indices, and the difference between output area and pseudo postcode sector deprivation suggest that rural areas are more likely to contain pockets of deprivation, the locations identified here are often in rural settlements rather than extremely rural areas. This may be partly a matter of scale. Because census data cannot be obtained below the output area level, our entire analysis can only identify pockets of deprivation at the output area (OA) scale, and we can say nothing about the existence of pockets of a smaller size.

While we have acknowledged throughout this work that car ownership in rural areas was considered a necessity rather than an indication of affluence, we have not considered the elimination of other variables in the Carstairs index of deprivation. Further research is required to determine the impact the overcrowding variable has on deprivation in rural areas. There is much debate in the literature with regard to the value of employing a composite index such as the Carstairs index over the use of individual socioeconomic variables. By deconstructing the Carstairs index, with the inclusion of other socioeconomic indicators from the census and other national surveys, such as the Scottish Household Survey, the impacts of individual variables such as overcrowding could be examined further.

Acknowledgments

An ORS Scholarship administered by Universities UK and a Lapsed Bursary at the University of St Andrews fund Daniel Exeter. This work is based on data provided with the support of the ESRC and JISC and uses boundary material which is copyright of the Crown and the Post Office. Census output is Crown copyright and is reproduced with the permission of the Controller of HMSO and the Queen's Printer for Scotland.

References

Beale, N., Baker, N., et al., 2000, Council tax valuation band as marker of deprivation and of general practice workload, *Public Health* 114(4), 260–264.

Carstairs, V. and Morris, R., 1991, *Deprivation and Health in Scotland* (Aberdeen: Aberdeen University Press).

Crofts, D.J., Bowns, I.R., et al., 2000, Hitting the target: the equitable distribution of health visitors across caseloads, *Journal of Public Health Medicine* 22(3), 295–301.

Department of the Environment, Transport and the Regions, 2000, Indices of Deprivation 2000, No. 43 (London, Department of the Environment, Transport and the Regions).

Field, K., 2000, Measuring the need for primary health care: an index of relative disadvantage, *Applied Geography* 20(4), 305–332.

GRO Scotland, 2001, *Urban/Rural Indicator* (Edinburgh: GROS).

Haynes, R. and Gale, S., 2000, Deprivation and poor health in rural areas: inequalities hidden by averages, *Health and Place* 6, 275–285.

Higgs, G. and White, S., 2000, Alternatives to census-based indicators of social disadvantage in rural communities, *Progress in Planning* 53, 1–81.

Kearns, A., Gibb, K., et al., 2000, Area deprivation in Scotland: a new assessment, *Urban Studies* 37(9), 1535–1559.

Mc Laren, G. and Bain, M., 1998, *Deprivation and Health in Scotland: Insights from NHS Data* (Edinburgh: ISD Scotland Publications).

Salmond, C., Crampton, P., et al., 1998, *NZDep96 Index of Deprivation: Instruction Book*, pp. 1–16 (Wellington: Health Services Centre).

Scottish Executive Health Department, 1999, *Fair Shares for All* (Edinburgh: The Stationery Office).

Senior, M., 2002, Deprivation indicators. In *The Census Data System*, edited by P. Rees, D. Martin, and P. Williamson, pp. 123–137 (Chichester: John Wiley & Sons).

Townsend, P., 1987, Deprivation, *Journal of Social Policy* 16, 125–146.

6

Crime Map Analyst: A GIS to Support Local-Area Crime Reduction

Paul Brindley, Max Craglia, Robert P. Haining, and Young-Hoon Kim

CONTENTS

6.1 Introduction

The importance of geographical information systems (GIS) for crime analysis, and strategic and tactical deployment of forces, has been increasingly recognized in both the United States and the United Kingdom. This was forcefully endorsed by former New York mayor, Rudolph Giuliani, during his visit to London in February 2002.

Senior police officers are keen to learn from the New York experience while Mr Giuliani visits London. In the eight years he was mayor of New York crime plunged.

The success was credited to CompStat, the computerised system which keeps track of week-by-week crime figures for each precinct, the basic division of the city's police department.

(The Guardian, 14th February 2002)

CompStat is of course only part of a wider strategy of crime reduction, but it makes the point that the regular analysis of crime for small geographical areas is crucial for the effective deployment of resources, monitoring and evaluating impacts, and sharing intelligence. The increased emphasis by the Home Secretary on increasing detection rates by concentrating police resources into selected hot spot areas goes in the same direction.

The ability to visualize and analyze the data geographically is at the heart of GIS. These types of systems are already widely used in the United Kingdom, but there are significant variations among the forces in extent and purpose of use (Weir and Bangs, 2007). There are therefore opportunities for using GIS more and better, with stronger integration to crime-reduction strategies both in the forces themselves and as part of the wider crime-and-disorder partnerships.

GIS can add value to the data already held by police forces and become a more integrated tool in crime-reduction strategies. There are two essential preconditions to make this happen.

1. *Geo-coded data*
   GIS can only operate effectively if the data to be analyzed have accurate and consistent geographical locations attached to it. Although this may sound a purely technical matter, it is in fact a largely organizational one. It must become a routine to report the location of crime events as accurately as possible, and against a standard gazetteer of locations. Not all forces have adopted such practice and the assigning of coordinates to past crime data provides context to the analysis.

2. *Awareness and training*
   Training staff in the use of GIS, or any new system, is of course time consuming and expensive. The advantage however of adopting off-the-shelf and widely used software is that there are already well-developed courses, training packages, and learning resources, and that there is a support network of millions of users on which to build. This minimizes training costs and makes the most effective use of the investment made. Perhaps, more crucial is ensuring that the necessary awareness exists among senior managers of the value of such investment, and that adequate support is provided.

6.2 Current Crime Pattern Analysis

6.2.1 GIS Crime Systems in the United States

The importance given in the United States to GIS for crime analysis is most clearly demonstrated by the work of the Mapping & Analysis for Public Safety (MAPS) program (formerly the Crime Mapping Research Center) at the National Institute of Justice (http://www.ojp.usdoj.gov/nij/maps). This center was established in 1997 to promote, research, evaluate, develop, and disseminate GIS technology and the spatial analysis of crime. The lessons learned out of the U.S. situation are valuable in the continued development of crime mapping in the United Kingdom.

Personal contact using the CRIMEMAP e-mailing group (crimemap@lists. aspensys.com) was undertaken to disseminate crime mapping and analysis e-mails to all subscribers to discover the main GIS used within U.S. crime mapping. The survey was conducted during 11–31 October 2000, and a total of 93 contacts were collected. Findings support other crime-mapping surveys by several governmental crime-research agencies (Crime Mapping Research Center, 1999; Police Foundation Crime Mapping Laboratory, 2000), whereby over 50% of crime analysis and mapping in the United States was undertaken using just two software applications—ArcView and MapInfo.

6.2.2 Past Crime Mapping and Analysis Research

Since the 1990s, the extensive usage of GIS has enabled police forces to map and analyze crime data efficiently, facilitating crime data analysis (Hirschfield et al., 1995). Computerized mapping technology has broad application areas in various police fields including operational, analytical, and strategic policing (Craglia et al., 2000).

GIS functionality has become widely used in many areas within crime data analysis, such as crime hot spot mapping and cluster detection, repeat victimization, temporal pattern analysis of crime incidents, and police policy making for crime reduction and prevention. For hot spot analysis, Ratcliffe and McCullagh (1999) developed a methodology for detecting various hot spots using a kernel estimate function on the basis of a local spatial autocorrelation statistic (Local Indicators of Spatial Association, LISA) to identify statistical hot spot variation. Crime cluster detection has been carried out within several current crime mapping tools such as STAC and CrimeStat (Bowers and Hirschfield, 1999; Levine, 1999; Craglia et al., 2000). Farrell and Pease (1993) recognized the issue of repeat victimization as a main criminological problem and suggested an implementation strategy for preventing crime repeats. Anderson et al. (1995) also provided strategic guidance for police forces to tackle repeat victimization. Johnson et al. (1997) demonstrated the relationship between repeat victimization and other socioeconomic factors, and explored analytical methods to identify the relationship. However, geo-coding problems have a profound impact upon the reliability of spatial repeat victimization identification (Ratcliffe and McCullagh, 1998a).

The temporal aspect has been identified as a crucial factor to monitor crime incident change. Instead of using general discrete methods (for example, using mid-point of between from-time and to-time interval), Ratcliffe and McCullagh (1998b) introduced a probabilistic rate technique based on aoristic rules to estimate a truer rate of crime incidents. A fuller description of this methodology will be discussed in Section 6.3.3. Practical use of this method was undertaken to explore different temporal patterns of crimes within a number of hotpots (Ratcliffe, 2002). Spatial statistics have been increasingly applied to crime data analysis in order to enhance novel capabilities of GIS-based analysis of crime, such as local spatial statistics for crime pattern analysis (Craglia et al., 2000), urban crime examination (Murray et al., 2001), or detecting temporal changes of crime (Rogerson and Sun, 2001). Anselin et al. (2000) introduced the extensive discussion of spatial analytical techniques and potential of GIS for crime analysis. In addition, for exploring the relationship with socioeconomic area profile and crime incidents, Bowers and Hirschfield (1999) demonstrated an example of GIS applications in crime pattern analysis, and Craglia et al. (2001) reported the strengths of GIS-based spatial analysis with census data for modeling high-intensity urban crime areas. Hirschfield and Bowers (2001) summarized extensive research contributions of GIS and their practical potential in crime data mapping and analysis.

A variety of crime mapping systems, extensions, and software packages for GIS have been developed at practical levels, as summarized in Table 6.1. Some packages were developed for pinpointing crime events and creating thematic choropleth maps, whereas other software systems were developed for locating hot spots and exploring spatial relationship with other socioeconomic data. For example, STAC was a frontier stand-alone hot spot and cluster analysis package developed in the 1980s and still is useful for crime cluster analysis (Craglia et al., 2000). In the 1990s, many mapping packages have been developed as extensions of main GIS commercial software systems using their customization languages such as SCAS, CrimeView and Crime Analysis for ArcView, and Hotspot Detective for MapInfo. To improve their user interface, computer languages and scripts have been integrated such as Visual Basic (SCAS, RCAGIS, and CrimeStat), and MapObject (RCAGIS and Community Policing Beating Book, and MaxResponder). As an alternative, crime-oriented stand-alone mapping software has been developed such as PROphecy and CrimeWatch. However, there has been limited success to tackle crime data analysis for various levels of U.K. police force requirements. Therefore, this chapter demonstrates some of the key functions of GIS crime analysis that can meet various operational, tactic, and strategic police performance in the U.K. police force.

TABLE 6.1

Summary of Main Available Crime Mapping Desktop Software

Name	Source	Primary Functionality
Spatial and Temporal Analysis of Crime (STAC)	Illinois Criminal Justices, 1993 (http://www.icjia.state.il.us/public/index.cfm?metaSection = Data&metaPage = STACfacts, assessed 18th October 2007)	Hot spot analysis package
Spatial Crime Analysis System (SCAS)	CMRC, 1994 (http://www.usdoj.gov/criminal, assessed 18th October 2007).	Query interface; analytical mapping and reporting; installation flexibility; minimum reprogramming (avenue)
Regional Crime Analysis GIS (RCAGIS)	CMRC, 1994 (http://www.usdoj.gov/criminal, assessed 18th October 2007).	Low cost (<$100); analytical mapping and reporting; interface with CrimeStat and MapObject
Community Policing Beating Book	ESRI, 1997 (http://www.esri.com/industries/lawenforce/beatbook.html, accessed on 6th June, 2002)	Simple query; mapping and reporting functions; MapObject application
Crime Analysis Extension	ESRI, 1999 (http://www.esri.com/industries/lawenforce/crime_analysis.html, accessed on 6th June, 2002)	ArcView application using Spatial Analyst extension; Hot spot and cluster analysis
CrimeView	OMEGA, 1999 (http://www.theomegagroup.com/crimeview.htm, accessed on 6th June, 2002)	Query; density mapping and simple analysis; reporting; integrated with ArcView
MaxResponder	ESRI, 1999 (http://www.maxresponder.com/, accessed on 6th June 2002)	Query and mapping function; mobile GIS mapping functions
Hot spot Detective	Ratcliffe, 1999 (http://athene.csu.edu.au/~jratclif/index.html, accessed on 6th June, 2002)	Hot spots; aoristic temporal analysis
Repeat Location Finder	Ratcliffe, 1999 (http://athene.csu.edu.au/~jratclif/index.html, accessed on 6th June, 2002)	Spatial repeat victimization identification
ReCAP-SDE	Virginia Institute for Justice Information, 2000 (http://vijis.sys.virginia.edu/home.htm, accessed on 6th June, 2002)	A stand-alone package; data handling and chart reporting functions
CrimeWatch	Spatial Data Inc., 2000 (http://www.spatial-data.com/pCrimeWatch.htm, accessed on 6th June, 2002)	Database; geo-coding and reporting functions
CrimeStat	Levine, 2000 (http://www.icpsr.umich.edu/NACJD/crimestat.html, accessed on 6th June, 2002)	Well-suited stand-alone software; set of spatial statistical modules for crime analysis; hot spot and clustering functions; compatible of main GIS software packages
PROphecy	ABM, 2000 (http://www.abm-uk.com/uk/index.asp, accessed on 6th June, 2002)	Hot spots; temporal analysis

6.2.3   Background to Crime Map Analyst

This work is based on research undertaken by the authors between 2000 and 2002 for the Home Office and South Yorkshire Police. It illustrates the implementation of some key requirements for a crime data analysis package, which is sensitive to the needs of U.K. police forces and which adds value to the data the police collect themselves. The requirements for the package related to both operational and strategic levels of policing and so the analytical functions are of use in both these policy contexts. They build on existing GIS facilities but also exploit crime analysis tools and data visualization functions, which are not addressed in the current GIS-based crime mapping systems. The results were implemented as an extension to ArcView 3.2, as this was widely used at the time, and have since been ported to ArcGIS 9.2. Of course, these same functions can also be implemented on other packages if required. What is important is the methodology for analysis rather than the software platform on which it is implemented. The next section describes the main functionalities developed in Crime Map Analyst.

6.3   Overview of Crime Map Analyst

Crime Map Analyst (CMA) is an extension for ArcView that enables the user to undertake various functions useful for the analysis of crime data. The functionality is outlined below (Figure 6.1) and is subsequently discussed in greater detail. Particular importance has been attached to

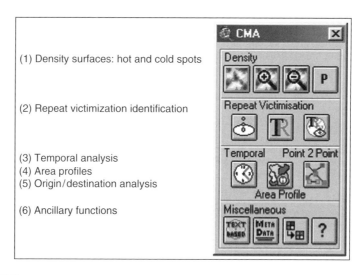

FIGURE 6.1
The main CMA menu.

displaying outputs from the different functional tools against maps that help the user to relate findings to "real" geography.

6.3.1 Density Maps

Crime analysts are increasingly using kernel-smoothing techniques to visualize and interpret crime data (Williamson et al., 2001). The density function takes data that are represented as points on a map and creates a continuous smooth surface of intensity. CMA has the option to construct simple, quadrat densities as well as more sophisticated kernel density maps.

The quadrat density method employed within CMA places a grid over the area of interest. It then simply counts the total number of points that fall within the defined search radius from each grid cell centroid. This method does not take into account the fact that crime incidents are not randomly distributed and association between point locations may exist. Crime events (points) frequently occur at the same locations (repeats) and have a tendency to cluster at specific locations.

In comparison, kernel densities have several practical benefits for creating density maps (Bailey and Gatrell, 1995; Anselin et al., 2000). Significantly, considering spatial repeats and crime clustering, the method takes into account autocorrelation so that points are weighted according to their distance from the grid cell centroid. Also, they do not have to adhere to socially constructed boundaries (for example, police beats, enumeration districts, or wards) like thematic maps that are using administrative boundaries. They are free from such constraints of shape unlike administrative boundaries or hot spots derived using circles or ellipses. Also, altering the number of grid cells and their dimensions can alter the spatial scale at which the data is being investigated to fit the needs for which analysis is required.

A further benefit is that the levels of crime intensity are easily visible, unlike with pin mapping techniques where crimes at the same location obscure the intensity of crimes at the same location, frequently causing data overload (Block, 1998). However, kernel densities are dependant on appropriate parameters used. In particular, the search radius is critical because it determines the level of smoothing applied (Anselin et al., 2000).

All the parameters within CMA required before creating a kernel density are determined by the extent of the area of interest. This makes the process very simple for a user with little or no understanding of the method to produce meaningful results. The parameters can be accessed within CMA so that more experienced users can adjust them according to their exact need.

The user simply defines the area of interest by drawing a grid over the desired location. The grid cell sizes are set so that the smallest extent (either horizontal rows or vertical columns) is set at a minimum of 400 grid cells and the other, larger extent is then scaled in the same proportions. The search radius is predefined as a percentage of the perimeter of the area defined (default 0.4% of the perimeter). The default percentage used is by no means the perfect solution, and it was considered outside the scope of this project, especially as more experienced users can easily adjust the search

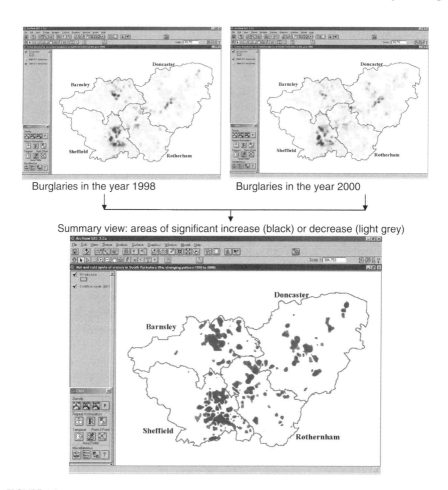

FIGURE 6.2
Construction of hot and cold spots for domestic burglary in South Yorkshire between 1998 and 2000. (From 1991 Census: Digitised Boundary Data (England and Wales).)

radius to improve results for their specific data. However, the default percentage was found to provide good result given a variety of spatial scales.

By comparing density maps for differing time periods, it is possible to identify hot and cold spot crime locations as illustrated (Figure 6.2). This function is also of value in providing a synoptic view of recent changes and helping to evaluate the impact of different policing strategies.

A novel feature of the system is that it enables the analyst to look in greater detail at particular locations on the map. The user identifies an area of interest with the mouse, and the Density-Zoom-in function creates a new grid with as much detail as required (Figure 6.3). This dynamic zooming capability is particularly useful in exploring the areas of high density within larger scaled hot spots.

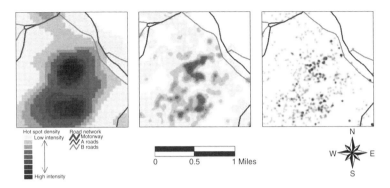

FIGURE 6.3
Dynamic zooming capability. (From 1991 Census: Digitised Boundary Data (England and Wales).)

It is also possible to create densities based on a population field. This can be used to weigh some points more heavily than others, or to allow one point to represent several observations. For example, one address might represent a condominium with six units, or some crimes might be weighted more severely than others in determining overall crime levels.

A specific function in the menu (contained within the ancillary functions) has been created to facilitate the display of density hot spots over raster images, without obscuring the background. This helps the analyst to recognize areas and focus on patterns of events in terms of specific areas or neighborhoods.

6.3.2 Repeat Victimization Identification

Repeat victimization is where the same offense occurs a number of times against the same victim (be this a person or entity, such as a house or vehicle). According to Farrell and Pease (1993), 10% of the victims of crime account for up to 40% of crimes in a given year. The rationale is to reduce crime by targeting prevention at repeat victimization locations. However, defining repeat cases is notoriously difficult (Ratcliffe and McCullagh, 2001). CMA uses two different functions to identify repeats: spatially or textually defined repeats.

When performing a search for *spatial repeats*, the program examines a set of points to find those with the same geo-coordinates. This detects repeats at the same *place*. There is also an option (radius) to search for points that are in a close proximity to others. This may assist users if the geo-referencing of offenses is not entirely accurate. Another use of the radius tool is to search for "clusters" where numerous offenses are committed over a limited geographical space (e.g., within 50 or 100 m from the identified location).

To investigate repeats for the same *person* (such as a vehicle owner) or entity (such as a vehicle registration) a textual repeat finder can be employed, which searches for identical text strings within a field. This is of course a standard facility on any database package, but the advantage of this method within a GIS package is that the outcome is immediately localized on a map. There are opportunities to extend this function to explore spatial patterns of offenses based, for example, on the description of the offender. This would be of particular value in the current priority to reduce street crime.

A third function reveals if the events are within a certain time period of each other (e.g., within 6 months), adding time as a variable in the identification of repeat victimization. A comparison of the results using the different identification procedures is shown in Figures 6.4 and 6.5. The combination of spatial and textual repeat functions can also assist in assessing the quality of the data. For example, it may highlight instances in which different addresses may have been given the same location or identical addresses appear "displaced."

6.3.3   Temporal Analysis

Police records attach a start time and an end time to each crime event. This is to characterize the uncertainty that exists over exactly when the offense occurred. Most current analysis tends to employ some measure of the middle point between the start and end times. A more accurate method—termed aoristic search—considers that the offense might have occurred at any instance between the start and end times (see Ratcliffe and McCullagh, 1998b for further details).

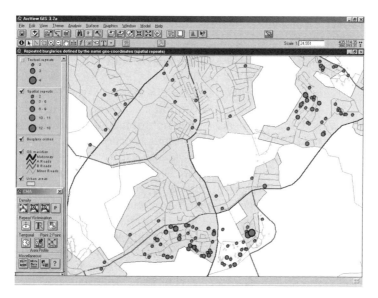

FIGURE 6.4
Spatially defined domestic burglary repeats. (© Crown Copyright/database right 2007. An Ordnance Survey/EDINA supplied service.)

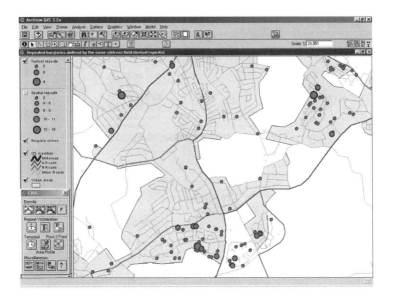

FIGURE 6.5
Textual defined domestic burglary repeats. (© Crown Copyright/database right 2007. An Ordnance Survey/EDINA supplied service.)

The greater the uncertainty within the search domain, the more aoristic analysis becomes desirable (Ratcliffe, 1999). Aoristic analysis techniques can smooth irregularities in time series caused by data uncertainty, as shown by Ratcliffe (2000). CMA uses probabilistic aoristic temporal analysis where each case is weighted by the probability that the crime might have occurred at that particular instance. For example, a crime that may have occurred equally on either Monday or Tuesday (e.g., between 8 PM on Monday and 4 AM Tuesday) would have a probability of 0.5 assigned to both Monday and Tuesday. However, if the crime occurred between 10 PM on Monday and 4 AM on Tuesday, in the absence of any further relevant information we might assume that it was twice as likely to have occurred on Tuesday, and hence, a probability of 0.66 is allocated to Tuesday and 0.33 to Monday.

The temporal units (minutes, hours, and days) used for analysis are dependant on the temporal level desired. When undertaking hourly probabilistic aoristic analysis, minutes are used to determine in which hours the crime event might have occurred. For example, a crime that occurred between 11:30 AM and 12:15 PM would be defined as being twice as likely to have occurred within the hour of 11 AM to 12 PM compared to the hour between 12 PM and 1 PM. Thus, a likelihood of 0.66 would be assigned to 11 AM while a figure of 0.33 would be attributed to 12 PM. Day of the week analysis is determined using hours as the temporal units, whereas monthly calculations are based on the number of days within the two search times.

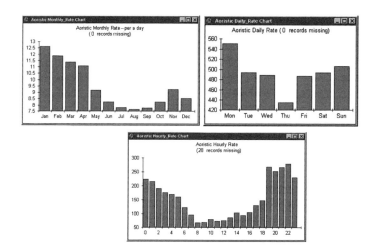

FIGURE 6.6
Examples of CMA temporal analysis for car crime in Sheffield, 1998–2000.

This function allows the user to calculate and graph the number of crimes that occurred within each month, day of the week, or hour (Figure 6.6), in a way that better reflects the uncertainty associated with the data than the "middle-point" approach. Monthly rates can be calculated as either independent of the number of days within each month, or as the number of crimes per day for each month. This overcomes the inaccuracy of some monthly temporal analysis where certain months (e.g., February) may appear to have lower crime levels simply because there is less opportunity due to the fewer number of days within the month. The function can convert any string or numeric values to the date format used within ArcView to facilitate data integration.

6.3.4   Area Profiles

There is growing importance attached to area-based performance indicators in almost all government policy areas, including crime reduction. However, performance indicators need to be evaluated against the local characteristics because these can vary quite substantially. With this in mind, we have developed this particular function so that a summary profile of the charac-teristics of an area can be called up at any time. Whether the user is interested in a police beat, ward, local neighborhood, or other areal units, this function sums up the key features in the database and can also integrate crime data with underlying social and economic characteristics as indicated in the example in Figure 6.7. Rates can also be compared for different underlying geographies thus providing further insights into local hetero-geneity.

The use of rates (e.g., burglaries/number of households) is of parti-cular value for strategic analysis as it highlights areas where there is a

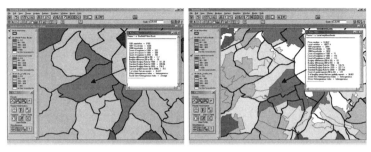

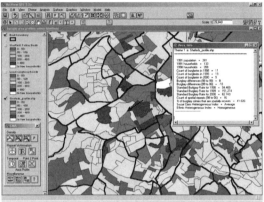

FIGURE 6.7 (See color insert following page 274.)
Comparison of standard burglary rates (SBR) for 2000 at varying spatial frameworks (police beats, neighborhood areas, and enumeration districts). (From 1991 Census: Digitised Boundary Data (England and Wales).)

disproportionate incidence of crime given an equal amount of opportunity. When rates are standardized for socioeconomic characteristics such as deprivation, or geo-demographic profile (such as MOSAIC, or ACORN), it is also possible to highlight areas where, all other considerations being equal, crime rates may be lower than expected owing to factors such as strong social cohesion, or alternatively be higher than expected owing to the breakdown of social relationships in the neighborhood.

6.3.5 Origin/Destination Analysis

This function may be of particular use for the analysis of travel to crime or vehicle thefts (facilitating suitable positioning of Automatic Number Plate Recognition systems, ANPR). It identifies the most likely direction of travel by dropping points along a hypothetical line from an origin (e.g., where a vehicle was stolen) to a destination (where the vehicle was abandoned) thereby connecting the two sites. CMA then measures the distance between the pairs of points and constructs a surface density of the lines as illustrated in Figure 6.8.

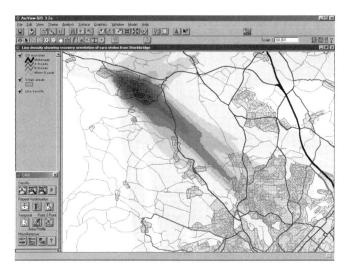

FIGURE 6.8

Density surface of car theft recovery orientation constructed within CMA for cars stolen from the town of Stocksbridge in the northwest of the map. (© Crown Copyright/database right 2007. An Ordnance Survey/EDINA supplied service.)

The resulting map does not depict where people have traveled, it simply represents the orientation that they took. For example, in Figure 6.8 we are not certain which road the stolen cars were predominately driven along but it demonstrates that the majority of thieves who stole cars from the town of Stocksbridge traveled to Sheffield, in a southeastern direction. This in turn may help in deciding the more probable route taken and where to locate ANPR equipment.

6.3.6 Ancillary Tools

In addition to the functions described in the previous sections, CMA offers a number of ancillary tools like a metadata editor to document the data files created, a Help Menu with discussion of the methods deployed and practical advice, and a Text Search Tool, which combines geographical searches to the traditional SQL. This function allows the user to perform searches to look for similar events within a user-defined radius from a selected crime event by searching for identical text strings within selected fields. One of the potential uses of such an operation is to search for all crimes within a given area, where the modus operandi are identical. In the example shown in Figure 6.9, there were 29 Vauxhall Astras stolen within a 2-km radius of the selected crime event; of these seven were saloons, and out of the seven, three were blue in color.

Extending this function to search in a rectangle, ellipse, or even within a user-defined buffered graphical line could identify similar crimes along a stretch of road, and given the current emphasis on tackling street crime, this would seem to be of particular value.

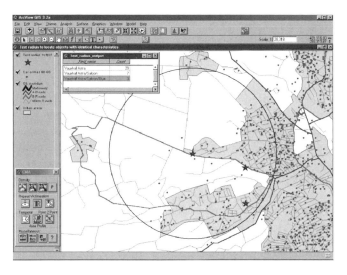

FIGURE 6.9
Example of a CMA constructed text radius search, illustrating similar vehicle types stolen within a 2-km radius of the existing case. (© Crown Copyright/database right 2007. An Ordnance Survey/EDINA supplied service.)

6.4 Conclusions

There is an increasing recognition of the need to improve detection rates by targeting intervention in selected problem areas. This requires not only a more concentrated deployment of resources but also a more integrated use of computer programs that can analyze crime patterns over time for small geographical areas. GIS provide the methodological framework to do this.

GIS are already widely available to police forces in the United Kingdom. However, anecdotal evidence suggests that there are significant variations among the forces on the extent to which GIS are used for more than command-and-control, and the visualization of 999 calls.

The U.S. experience indicates that there are significant benefits in integrating GIS throughout the crime detection and analysis chain, with regular reporting of key statistics at small geographical area, such as in New York's COMPSTAT.

In the United Kingdom, there are particular benefits to be accrued by reporting and analyzing crime spatially, given the excellent coverage of digital maps throughout the country, and the extent to which interagency partnerships at the local level have become one of the main vehicles to deliver government policy, from crime reduction to youth policy. "Joined-up" government requires a common understanding of the problem among the agencies that form the partnerships. GIS offers enormous opportunities for integrating the data coming from different agencies that pertain to a common location, and supporting joined-up auditing of the problem, spatial targeting of resources, and monitoring and evaluation of performance. With this in mind, the Home Office considered CMA as one of the possible tools to support the work of its regional research teams, police forces, and crime-and-reduction partnerships as part of its Partnership Business Model initiative. CMA was evaluated by different police forces and partnership analysts and found to be particularly interesting in respect to the ease with which hot spot maps could be created but above anything else for its aoristic time analysis (see Section 6.3.3) which is currently not available in other software packages available in the market. Similarly, South Yorkshire police who had been heavily involved in the development of CMA through the feedback provided by some of its data analysts, has decided to adopt it as part of its suite of analytical tools, and has developed it further inhouse, updating also the GIS platform to ArcGIS.

As indicated in this overview, CMA essentially consists of a set of exploratory, map-based tools. Its primary strengths are first that it is sensitive to the nature of police data and the uncertainties, which are a feature of such data. Second, it is sensitive to the U.K. context such as the availability and formats of other, nonpolice, data sets to which police data need to be linked, if their full value for policing is to be realized. In terms of the tools that are included, while they are not exploratory in the sense of employing resistant statistics (in the statistical sense of that word meaning statistics that are not unduly influenced by extreme data values), they are in the sense of being highly visual, staying close to the original data and making few assumptions in the processing of the data. These features are important from the perspective of interpreting results particularly by the police themselves as indicated by the positive feedback received from South Yorkshire police and other forces which have evaluated the system.

CMA is a software system that is probably most useful at the present time at the tactical, rather than the strategic, level of policing. If it is to enhance its usefulness at the strategic level, there are various options for future development. At this level, the need is to get behind the actual data sets in order to identify the underlying risks (of crime, or being a victim of crime) and the reasons behind those risks as they vary spatially. This suggests the need to be able to calculate rates and ratios while controlling for different population attributes. Then it becomes necessary to add resistant and robust mapping tools that can support the data analysis (Haining, 2003, pp. 226–237). This can extend to robust tools for detecting clusters and hot spots, possibly including simple significance testing to assist the user to distinguish between real and apparent concentrations. Thereafter, the user will want to be able to drill down into identified areas, while in other circumstances may want to explore for possible relationships through, initially, simple graphical plots (e.g., scatterplots) and later through numerical measures (e.g., correlations).

As with any package there is always scope for improvement and further development, and CMA is no exception. First, there are additional functionalities that could be incorporated. These include (but are not limited to) buffering based on line features (such as roads) rather than just simple concentric rings, and also spatial visualization of temporal analysis to explore the associations between spatial and temporal patterns (e.g., through the use of animations). CMA's usefulness for tactical policing may be further enhanced by using interactive and dynamic visualization techniques (e.g., brushing and animation). These tools can be applied to maps (e.g., geographical brushing) as well as to graphs (Haining, 2003, pp. 188–225). Second, the incorporation of interactive visualization between charts or graphs and maps can enhance interpretation of the data, especially at the operational or tactical levels. An example would be by highlighting the Monday crime column in the aoristic histogram, all crimes that may have occurred on a Monday are also automatically highlighted, spatially, on the map. This could then also be applied to select extreme values on both boxplots and scatterplots. Finally, although CMA was developed within ArcView 3.2, there is a need to be more universally applicable. A stand-alone system or web-based application would help enhance user practicability at operational and tactical levels. For example, online access of CMA could play a role in providing new insights into and an understanding of crime mapping and analysis.

Acknowledgments

The authors wish to appreciate South Yorkshire police for the provision of data used within the project. This work is based on data provided with the support of the ESRC and JISC and uses boundary material which is copyright of the Crown and the ED-Line Consortium.

References

Anderson, D., Chenery, S., and Pease, K., 1995, Biting back: tackling repeat burglary and car crime. *Police Research Group: Crime Detection and Prevention Series*, Paper 58 (London: Home Office).

Anselin, L., Cohen, J., Cook, D., Gorr, W., and Tita, G., 2000, Spatial analyses of crime, In *Criminal Justice 2000: Volume 4. Measurement and Analysis of Crime and Justice*, edited by D. Dufee (Washington, DC: National Institute of Justice).

Bailey, T.C. and Gatrell, A.C., 1995, *Interactive Spatial Data Analysis* (London: Harlow).

Block, C.R., 1998, The geoarchive: information foundation for community policing, In *Crime Mapping and Crime Prevention*, edited by D. Weisburd and T. McEwen (New York: Criminal Justice Press).

Bowers, K. and Hirschfield, A., 1999, Exploring links between crime and disadvantage in north-west England: an analysis using geographic information systems. *International Journal of Geographical Information Science* 13(2), 159–184.

Craglia, M., Haining, R., and Wiles, P., 2000, A comparative evaluation of approaches to urban crime pattern analysis. *Urban Studies* 37(4), 711–729.

Craglia, M., Haining, R., and Signoretta, P., 2001, Modelling high-intensity crime areas in English cities. *Urban Studies* 38(11), 1921–1941.

CMRC, 1994 (http://www.usdoj.gov/criminal, assessed 18th October 2007).

Crime Mapping Research Center, 1999, *The Use of Computerized Crime Mapping by Law Enforcement* (Washington, DC: National Institute of Justice).

Farrell, G. and Pease, K., 1993, Once bitten, twice bitten: repeat victimisation and its implications for crime prevention. *Police Research Group: Crime Prevention Unit Series,* Paper 46 (London: Home Office).

Haining, R.P., 2003. *Spatial Data Analysis: Theory and Practice* (Cambridge: Cambridge University Press).

Hirschfield, A. and Bowers, K. (editors), 2001, *Mapping and Analysing Crime Data: Lessons from Research and Practice* (London: Taylor & Francis).

Hirschfield, A., Brown, P., and Todd, P., 1995, GIS and the analysis of spatially-referenced crime data: experiences in Merseyside, UK. *International Journal of Geographical Information Systems* 9(2), 191–210.

Johnson, S.D., Bowers, K., and Hirschfield, A., 1997, New insight into the spatial and temporal distribution of repeat victimisation. *British Journal of Criminology* 37(2), 224–241.

Illinois Criminal Justices, 1993 (http://www.icjia.state.il.us/public/index.cfm?metaSection=Data&metaPage=STACfacts, assessed 18th October 2007).

Levine, N., 1999, *CrimeStat: a Spatial Statistics Program for the Analysis of Crime Incident Locations* (Washington, DC: National Institute of Justice).

Murray, A.T., McGuffog, I., Western, J.S., and Mullins, P., 2001, Exploratory spatial data analysis techniques for examining urban crime. *British Journal of Criminology* 41, 309–329.

Police Foundation Crime Mapping Laboratory, 2000, *Users' Guide to Mapping Software for Police Agencies* (Washington, DC: U.S. Department of Justice).

Ratcliffe, J.H., 1999, The genius loci of crime: revealing associations in time and space, Unpublished Ph.D. Thesis, University of Nottingham.

Ratcliffe, J.H., 2000, Aoristic analysis: the spatial interpretation of unspecific temporal events. *International Journal of Geographical Information Science* 14(7), 669–679.

Ratcliffe, J.H., 2002, Aoristic signatures and the spatio-temporal analysis of high volume crime patterns. *Journal of Quantitative Criminology* 18(1), 23–43.

Ratcliffe, J.H. and McCullagh, M.J., 1998a, Identifying repeat victimisation with GIS. *British Journal of Criminology* 38(4), 651–662.

Ratcliffe, J.H. and McCullagh, M.J., 1998b, Aoristic crime analysis. *International Journal of Geographical Information Science* 12(7), 751–764.

Ratcliffe, J.H. and McCullagh, M.J., 1999, Hotbeds of crime and the search for spatial accuracy. *Journal of Geographical Systems* 1, 385–398.

Ratcliffe, J.H. and McCullagh, M.J., 2001, Crime, repeat victimisation and GIS, In *Mapping and Analysing Crime Data: Lessons from Research and Practice*, edited by A. Hirschfield and K. Bowers (London: Taylor & Francis).

Rogerson, P. and Sun, Y., 2001, Spatial monitoring of geographic patterns: an application to crime analysis. *Computers, Environment and Urban Systems* 25, 539–556.

The Guardian, 2002, Livingstone's London "needs New York policing levels". *The Guardian*, 14th February 2002.

Weir, R. and Bangs, M., 2007, The use if geographic information systems by crime analysts in England and Wales. *Home Office Online Report 03/07* (London: The Stationary Office).

Williamson, D., McLafferty, S., McGuire, P., Ross, T., Mollenkopf, J., Goldsmith, V., and Quinn, S., 2001, Tools in the spatial analysis of crime, In *Mapping and Analysing Crime Data: Lessons from Research and Practice*, edited by A. Hirschfield and K. Bowers (London: Taylor & Francis).

7

Using GIS to Identify Social Vulnerability in Areas of the United Kingdom That Are at Risk from Flooding

Tom Kieron Whittington

CONTENTS

7.1 Introduction

Around 5 million people in 2 million properties live in flood-risk areas in England and Wales (Environment Agency, 2000). Property worth over

£200 billion and agricultural land worth approximately £7 billion are potentially at risk of flooding (HR Wallingford, 2000). The floods of Easter 1998
and autumn 2000 gave the United Kingdom an important reminder of a
hazard that, though ever present, has been neglected by society in recent
times. Many organizations are encouraged to deal with the problem, which
is predicted to increase in frequency in the future due to climate change and
continued urbanization of the floodplain (Price and McInally, 2001). There is
a rise in the philosophical approach of "living with the hazard" (Smith and
Ward, 1998) that focuses on flood warning and emergency planning, than
flood prevention. Initiatives to help communities to help themselves are
therefore high on the agenda but require a clear understanding of the social
variability and different needs of communities at risk.

It is the high profile of the field of research that motivates this project into
establishing how geographical information systems (GIS) may be used to
improve flood warning, and emergency planning and response in the
United Kingdom. To determine how the technology could be best put to
use with immediate effect, a requirements study has been accomplished
from a literature review and interviews with the main organizations
involved in flood warning, planning, and research in the United Kingdom.
The conclusion of the requirement study identifies that the spatial distribution of vulnerable groups living within the floodplain is a prime target for
research, and this group would benefit greatly from GIS investigation.

This research attempts to bring together social-vulnerability studies and
flood-probability data with GIS technology to produce a high-resolution
index of flood vulnerability (IFV). A number of applications demonstrate
how the index and some of the data layers used in its creation may be used to
improve the efficiency and quality of flood managers' decision-making.
During a flood emergency, planners can quickly identify different groups
of people with different social needs and can thus disseminate resources
appropriately. Alternatively, flood-warning education can be adapted for
different communities identified by their postcode. The final index produced
is a prototype tool, which requires refinement, but demonstrates how existing studies could be improved with the inclusion of GIS technology.

7.1.1   Background

The next century may see apparent increases in CO_2 due to human activities
resulting in climate change and consequently flooding from increased
intensity and frequency of rainfall and sea-level rise (Price and McInally,
2001). Most flood-prevention schemes can be expected to fail if a high-level
flood scenario occurs, and there are many locations where an engineering
solution is impractical or could lead to considerable damage to the environment (Borrows, 1999; Environment Agency, 2001). The expansion of
urbanized areas can create the risk of more-frequent flood situations
where increased precipitation results in greater runoff (Environment
Agency, 2001; Price and McInally, 2001). Around 5 million people in

2 million properties are vulnerable to flood risk in England and Wales (Environment Agency, 2000); the flood levels during the October 2000 floods were the highest on record in many locations, and 10,000 properties were affected (Environment Agency, 2001).

The Better Regulation Task Force (2000) recommends that policy makers consider vulnerable people at all stages of their work, with greatest consideration to include vulnerability impact assessments. Knowledge of where social differences lie within communities potentially at risk from flooding and the general nature of their circumstances is needed to better target public awareness information and to respond appropriately to emergency situations (Morrow, 1999; Tunstall, 1999; Blyth et al., 2001; Environment Agency, 2001). One of the main reasons for targeting vulnerable groups is the desire to concentrate on the worst affected areas and population (Jaspars and Shoham 1999); emergency planners need to know who they are and where they are concentrated (Morrow, 1999). Performance of flood forecasting and warning systems appears to be poor in the United Kingdom (Haggett, 1998; Horner, 2000; Penning-Rowsell et al., 2000; Environment Agency, 2001) but will continue to improve, provided that the right information can be delivered in advance to the right people (FHRC, 2001).

There is a need to develop accurate flood-hazard maps and flood-rescue action plans for hazard-prone areas (Rantakokko, 1999). The Environment Agency has created the indicative floodplain map using historical and rainfall catchment models (e.g., ISIS and MIKE11). The mapped floodplain boundaries are disaggregated to unit postcodes to assist public identification of risk, but there is a mismatch between postcode units and identified flood-risk areas (Plougher, 2000). Boyle et al. (1998) discriminate the floodplain into units defined by the quantification and spatial variability of flood hazard. Flood-probability contours are created with hydrological modeling of flood flow rates associated with different return-periods (the flood frequency in years). The use of GIS in this type of hazard exposure provides an efficient and accurate assessment for areas prone to flooding (Boyle et al., 1998), but it does not consider the socioeconomic variability that may also be associated with that location.

7.1.2   Requirement Study

It is the intention of this research to enhance the flood-warning and emergency-planning industry with GIS technology. There are a number of organizations in the United Kingdom with different responsibilities within the flood industry (Table 7.1); consultants and research institutes are assigned projects to develop the roles of these organizations. As the industry does not have a single function, it was necessary to identify one aspect of technical research that could reasonably be undertaken with limited time and data, and yet provide a useful service. A GIS requirement study has been administered to evaluate the research needs of flood warning and emergency planning, but these particular needs are difficult to define

TABLE 7.1

Flood Hazard Responsibilities in the United Kingdom

Organization	Responsibilities
Legislative bodies	Ministry of Agriculture, Fisheries and Food (MAFF), now Department of Environment, Food and Rural Affairs (DEFRA) have policy responsibility for flood and coastal defense in England and administers the legislation that enables work to be carried out. The Flood and Coastal Defence Programme is aimed at reducing risk to people and the developed and natural land by financially supporting, advising, and guiding flood and coastal defense operating authorities, and funding research programs. Following the 2000 floods, better definition of flood or erosion risk areas was identified as being of particular necessity to flood planning (HR Wallingford, 2000)
Meteorological office	Continuous monitoring of weather conditions and rainfall patterns from remote sensing and ground-based measurements falls under the authority of the meteorological office (MetOffice). Computer models simulate river discharge based on rain-gauge readings over time-periods in various parts of different catchments. If a flood situation arises, the MetOffice alerts the EA and local authorities to the possible threat
Environment Agency (and the Scottish Environmental Protection Agency)	The Environment Agency (EA) has the lead role in disseminating flood-warning messages with the help of local authorities; flood defense accounts for about 50% of the Agency's annual budget (EA, 1996). There are a number of departments within EA to deal with various aspects of flood warning. The National Flood-warning Centre continuously monitors changes in the conditions of rivers, such as the effect of development on channel flow; and it is the interface between the agency and the public. The Emergency Management Team develops contingency planning policy on which it advises local authorities and coordinates the information flow during an emergency. The EMT liase with weather service providers and issue warnings and press releases during an emergency and are responsible for education campaigns prior to any particular event. EA regional offices deal with more localized issues with respect to flood-alleviation schemes and flood warning. There is some variation between EA regions in the scope and sophistication of the facilities available to support operational decision-making (Haggett, 1998)
Police and local authorities	Local planning and emergency considerations: It is the role of the police to organize localized planning and response on behalf of LA and EA guidelines. Once alerted to the onset of a flood event a command-and-control center is set up by the police, local warnings are issued, and emergency response teams dispatched. The police have roles in the evacuation of people at-risk and traffic management (Smith and Ward, 1998). Without detailed information on the social characteristics of different threatened communities, emergency planners and services can have great difficulty in reaching specific communities in need (Haggett, 1998). The police have endeavored to improve reliability and speed of flood-warning dissemination and focus on emergency response through the use of technology (Horner, 2000)

TABLE 7.1 (continued)

Flood Hazard Responsibilities in the United Kingdom

Organization	Responsibilities
Insurance companies	In different European countries, there are varying methods for insuring against flood damage. Countries with a high risk of flooding have responded in different ways. Austria and Belgium have state-funded schemes; Italy has insurance available that is rarely bought; and Netherlands has no insurance cover available other than for some industrial risks. The United Kingdom, with medium risk, is the only country in EU with 100% private insurance solutions (Ebel, 1999). Flood aid and insurance do nothing to mitigate and reduce the risk of future disasters, and such measures may be counterproductive if they continue to encourage settlement into high-risk areas (Smith and Ward, 1998). The insurance industry has recently discovered the value of flood-risk maps that combine flood depth with information on building types within a given area, usually defined by its postcode (Rodda, 2001)

owing to the dearth of appropriate literature. To overcome this problem, a series of interviews has been conducted with many of the key organizations from the flood industry in the United Kingdom to supplement findings from relevant literature. The object of the interviews are twofold: to collate an audit of current GIS to consider how the technology is used in flood warning and emergency planning, and identify areas where GIS are not meeting their potential; and to consider the opinions of some of the leading practitioners in these areas, which would significantly benefit the flood-warning and emergency-planning process.

The audit shows that most systems deal with flood warning and prevention, or emergency planning, but very few systems are used for emergency response that requires precise and up-to-date information for efficient decision-making on how to respond to changes in circumstances (Table 7.2). No systems-in-use consider all aspects of the flood industry, and nearly all the use of GIS is only involved with research, possibly indicating the deficiency of actual running systems. Overall the systems are sparse in utilizing socioeconomic data and assessing floods according to population and commodity risk. Advances of GIS may involve more social considerations of flooding as the necessity for monitoring flood-risk comes from the effect it may have on society. It may also be fruitful to consider a system that not only provides flood warning but also assists in the emergency planning and response of a flood in real time, another use of GIS which is not being optimally exploited. Real-time GIS may be useful in modifying emergency plans appropriately during a flood event as new data becomes available.

Interviews included advances for research in the flood industry that organizations felt were important, and these were found to generally concur with ideas expressed in the literature. The so-called "living with hazards" philosophy (Alexander, 1997; Smith and Ward, 1998) has created the need for more effective initiatives to be researched within the roles of warning,

TABLE 7.2

Audit of GIS in the Flood Industry (England and Wales)

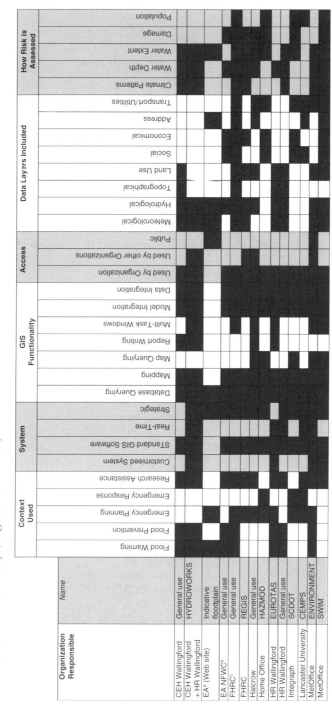

The author would like to apologize for any systems that are omitted. The matrix has been constructed using knowledge gained from the requirements study and may not be definitive. The list only includes systems which were in use at the time the requirements study was undertaken, and not ones which were then in development. (e.g., FLOODWORKS for DEFRA).

Notes: [a]Environment Agency, [b]National Flood-warning Centre, and [c]Flood Hazard Research Centre, University of Middlesex planning, and response. Emergency planners, police, and local authorities would benefit from GIS, but the technology does not currently play an important role. Traditionally, GIS technology has been used in supporting surface-water modeling and flood-hazard exposure analysis by providing the ability to integrate modeling results with other layers of information to enhance the decision-making process (Boyle et al., 1998).

However, GIS provides the means of integrating different phenomenon such as social and geographical data, in order to increase the overall understanding of the relationship between society and disaster (Dash, 1997). Technological solutions could be sought to improve the dissemination process with decision-support tools (Haggett, 1998) but for many flood planners the advantages of GIS are not considered or fully understood. For instance, the City of Edinburgh Council stores all information for responding to a flood within one paper file binder. During a hazard situation, members of a committee sit around a table and discuss appropriate courses of action. Benefits of GIS and automated information are seen as negligible because flooding *seldom occurs twice*. The Thames police also resolve response management without the use of GIS, where and when it is necessary; a system that the users consider sufficient (Whittington, 2001).

Conclusions of the requirements study identify communicating the flood message (education prior to and alerting during an emergency) and improving disaster response as two important advances in flood research. Both of these initiatives are common to the need for discriminating between different groups of people living within the floodplain.

There are two ways that communities could be classified within the floodplain. First, since disaster vulnerability is partially socially constructed (Morrow, 1999), identifying different levels of social vulnerability could improve efficiency of warning and emergency response. For example, awareness that 75% of a community are non-English speakers could result in more appropriately tailored education and warning, whereas the knowledge that 50% of a community are aged over 75 years could help allocate the dispatch of sufficient help.

Secondly, flood return periods (or scenarios) are effectively an estimated probability of a flood occurring at any year. The flood risk map for England and Wales (Morris and Flavin, 1996) and the indicative floodplain map (Environment Agency, 2000) consider the maximum predicted flood but do not distinguish degrees of risk. Flood risk can vary within the area of a community and a detailed understanding of flood-risk variability could be extremely valuable to flood planners. Flood-vulnerability mapping could help prioritizing emergency dissemination where accurate flood data identifies where and when different flood scenarios are going to occur.

7.1.3 Aims

The conclusions of the requirement study are the basis for the direction taken by this research. Social vulnerability and flood vulnerability are calculated from socioeconomic and flood-depth data. The broad approach is to merge the physical assessment of flood risk with the social assessment of vulnerability to produce an IFV to the unit postcode level. The index is used to create a prototype decision-support tool that utilizes GIS functionality to improve the efficiency of flood-warning and emergency planning, as well as having scope for emergency response. From this research, it is hoped to convince emergency planners of few GIS benefits. The vulnerability-index is not required in this research to be highly accurate; it is the concepts that are under scrutiny. It is thought that the type decision-support tool is useable, given that sufficiently accurate data is available.

The main research questions considered are

- How can GIS be used to assist the identification of vulnerable groups within communities in areas prone to flooding?
- What data provides the best indicators of social vulnerability?
- What data would best serve the definition of flood-risk spatial variability?
- Does the integration of the two datasets improve the vulnerability index?

7.2   Methods

7.2.1   Index of Social Vulnerability

The model is tested on a 30-km stretch of the River Thames between Eton and Walton-on-Thames. Socioeconomic data is used to map an index of social vulnerability (ISV) at the level of enumeration districts (EDs). Social vulnerability to a disaster may not only be affected by poverty but is the overall ability to respond to a hazard situation (Tunstall, 1999). Certain physical and social attributes (age, race, and gender) and living arrangements (single parent households), where the relationship with social class is not so well-defined, are likely to have as much, if not more effect on vulnerability as poverty. Far from being mutually exclusive, these factors tend to occur in combinations that intensify risk exponentially (Morrow, 1999). Data was obtained to reflect a wide range of attributes that may increase (or decrease) social vulnerability of a small community. From discussion with experts, criteria expressed in the literature (e.g., Morrow, 1999; Tapsell, 1999, 2000; Tunstall, 1999; Dralsek, 2000; Environment Agency, 2000, 2001) and information that can reasonably be thought to determine social vulnerability, a restricted set of 10 domains have been chosen. The 1991 U.K. Census provides nine domains and the 10th one from the index of multiple deprivation; the calculations of a domain are constrained from the available source data (Table 7.3).

TABLE 7.3
Domains Used for Calculating the Index of Social Vulnerability

Domain	Reason for Inclusion	Values Used	Reason for Values Used	Weighting Applied	How Index Is Calculated	Source
Long-term illness [ED]	People with illness may be restricted in their ability to respond effectively without assistance (Tunstall, 1999)	% of long-term illnesses to total ED population	Assumes all ill persons are equally dependent on others and thus equally as vulnerable	None, all people with long-term illness are equally vulnerable	% is standardized	Census (1991)
OAPs [ED]	OAPs may be restricted in their ability to respond effectively (Tunstall, 1999)	% of OAPs that are: (a) ill, (b) >75 years, and (c) live alone	All OAPs are potentially vulnerable but being >75 years, ill, or living alone will increase the vulnerability	(a)–(c) are given equal weighting (as cannot distinguish further variability) and multiplied together	%OAPs × % >75 × %live alone × %ill Results standardized	Census (1991)
Single parents [ED]	Single heads of households are disproportionately disadvantaged following flooding because they are responsible for dependants who may not be able to act appropriately themselves (Morrow, 1999; Tapsell, 1999; Dralsek, 2000)	Number of parents with children aged: (a) 0–4 (b) 0–4 and 5–15 (c) 5–15	Different aged children may have different abilities and having more than one child may affect how easily parents can cope in a hazard situation. Younger children are considered the most vulnerable, especially those that cannot walk	When added, (a)–(c) makes up number of single parents. The weightings are applied by multiplying (a)–(c) by individual significance and totalling value; e.g., youngest children are given 3× vulnerability of older ones	3[%0–4] + 2[%0–4 and 5–15] + 1[%5–15] Results standardized	Census (1991)

(continued)

TABLE 7.3 (continued)

Domains Used for Calculating the Index of Social Vulnerability

Domain	Reason for Inclusion	Values Used	Reason for Values Used	Weighting Applied	How Index Is Calculated	Source
Non-U.K. born [ED]	People born outside the United Kingdom may not speak English as their first language or have cultural differences that affect understanding of flood warnings or methods of response (Morrow, 1999)	% of non-U.K. born to total ED population	Assumes all people born outside United Kingdom are equally vulnerable and just considers the proportion of the total population. (Data does not identify particular English-speaking nations)	None, all people born outside the United Kingdom are potentially equally vulnerable	% is standardized	Census (1991)
Business properties [ED]	Less households may be threatened in areas where there are more business properties	% of business properties to total ED properties	Has a negative impact on social vulnerability due to smaller number of residential properties	None, business properties have an equal effect on reducing social vulnerability	% is standardized and multiplies by –1 to consider negative impact	Census (1991)
Children [ED]	As dependants are vulnerable themselves and increase the vulnerability of others (Morrow, 1999), they may not understand how to respond	% of children to total ED population	Assumes all persons are all dependent on others and therefore equally vulnerable. Further breakdown of ages is considered for single parents	None, all children are equally vulnerable (conflicts with method for calculating single parent vulnerability)	% is standardized	Census (1991)

Migrants [ED]	Did not live in ED one year ago and may therefore be unaware of the hazard and methods of response	% of migrants to total ED population	Assumes all migrants have equal ignorance of hazard warning and response issues	None, all migrants have equal ignorance of hazard warning and response issues	% is standardized — Census (1991)
No central heating [ED]	May inhibit flood recovery as houses become damp and is also a further indication of poverty	% of "no central heating" to total ED properties	Assumes properties without central heating are equally vulnerable. No data to indicate the contrary	None, all properties without central heating are equally vulnerable	% is standardized — Census (1991)
No car [ED]	May inhibit flood evacuation irrespective of social class but is also a further indication of poverty.	% of "no car" properties to total ED properties	Assumes properties without cars are equally vulnerable. No data to indicate the contrary	None, all properties without cars are equally vulnerable	% is standardized — Census (1991)
Deprivation index [ward]	Deprivation has strong effect on vulnerability due to lack of education, health, and financial situation. This index is produced from a detailed research (DETR, 2001)	Income, employment, health and disability, education and training, access to service, and housing data are all used as separate domains to create index at ward level	None, the deprivation index is already calculated	Values calculated from wards to EDs and standardized	DETR (2001)[a]

All domain values are standardized to a range of 0 to 100.

[a] A full account of this methodology is provided by Whittington (2001).

TABLE 7.4

Weight Assigned to Each
Vulnerability Domain

Domain	% Weight
Deprivation	18
Long-term illnesses	18
OAPs	18
Single parents	10
Non-U.K. born	10
Businesses	10
Children	10
Migrants	2
No central heating	2
No car	2

Calculated domains are combined to form an overall index with consideration of how much of an impact each domain has on socioeconomic vulnerability. Simply summing the different domains together does not allow for the fact that some domains may have more of an impact on vulnerability than the others. The index of multiple deprivation (DETR, 2001) combines different domains by attributing each domain a percentage weight according to importance. To be consistent with the DETR study, the domains are subjectively ranked, attributed a percentage weight (Table 7.4), and the value for each ED is adjusted accordingly. Finally, the domain values are summed to create the index and adjusted to values between 0 and 100 to create the ISV, which is calculated from the equation:

$$\text{Social vulnerability} = \sum_{i=1}^{10} W_i \times D_i$$

The ISV and individual domain values are assigned to ED boundary data using the unique ED-id. This enables mapping the data and overlay analysis with flood data.

7.2.2 Index of Flood Probability

The desired index of flood probability (IFP) is a set of contours that define the extent of different flood return-periods. The functionality of ArcInfo GIS requires a continuous data surface to be able to calculate the contours accurately. An interpolation method is thus required, which models the entire extent of a return-period based on actual flood-depth data recorded from historic events. Flood-depth data and a digital elevation model (DEM; with resolution 50 m horizontal and 1 m vertical) are used as model input data to map the predicted extent of flooding caused by different flood return-periods and combining these maps produces an IFP. The flooding process is simplified by an assumption that states: flood depth is reduced as the elevation above the river increases. The method requires the calculation of the height above the river (HAR) for the whole of the floodplain.

River height is calculated by assigning elevation values to each point along a digitized river-line using a DEM. Converted to a raster grid, the river appears as a series of cells with different elevations. A cost-allocation command allocates cells in a new grid the elevation value of the nearest river cell. HAR is calculated by subtracting the height of river grid from the DEM (Figure 7.1).

Actual flood data used (*Source*: FHRC) consists of actual water-depths for five flood return-periods (10, 25, 50, 100, and 200 years) recorded at the location of 10,000 properties in the study area. A method was required to interpolate flood depth by incorporating the sample data with the HAR grid. Statistical interpolation techniques were considered (e.g., kriging and inverse distance weighting) but thought inappropriate due to the sporadic and clustered spatial distribution of the sample data. Regression analysis attempted to correlate flood-depth and HAR as a model of flood extent but produced inconclusive results owing to an inadequate resolution of DEM (Whittington, 2001).

An alternative approach that produces a satisfactory result uses the maximum depth of flood for each scenario identified from the flood data and assumes that, given an isotropic surface, the flood depth would be the same everywhere. By subtracting the maximum flood depth from HAR, any positive values in the resulting grid indicate a flood. The positive values are converted to 1 (for a flood) and negative to 0 (no flood), repeating the process for all scenarios. The five scenario grids are literally added

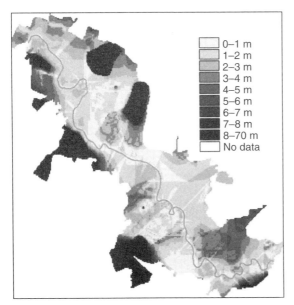

	0–1 m
	1–2 m
	2–3 m
	3–4 m
	4–5 m
	5–6 m
	6–7 m
	7–8 m
	8–70 m
	No data

FIGURE 7.1 (See color insert following page 274.)
Modeled height above the river (HAR). (© Crown Copyright/ database right 2007. An Ordnance Survey/EDINA supplied service.)

TABLE 7.5

Cell Values and Probability of Each Scenario

Flood Return Period (in Years)	Cell Value	Probability
10	5	0.1
25	4	0.04
50	3	0.02
100	2	0.01
200	1	0.005
No flooding	0	0

together producing a grid with values from 0 to 5 depending on the scenario (Table 7.5). Converted to a vector coverage creates a series of flood scenarios. The IFP is created by replacement of the cell value with the appropriate probability value (Table 7.5).

7.2.3   Combined Index of Flood Vulnerability

Overlay analysis makes it possible to multiply the ISV of any location with the IFP of the same location to create a new coverage. The values are calculated and scaled by 100 with the equation:

Flood vulnerability = 100 ([Social vulnerability] × [Flood probability])

As ISV is calculated at ED level, the new coverage shows spatial variability of combined index of flood vulnerability (CIFV) within an ED due to the influence of flood probability that is not constrained by EDs. To increase this resolution to unit postcode, flood vulnerability is intersected with postcode data to produce a cover in which each postcode is divided into areas of different vulnerability (Figure 7.2a). Each postcode requires a homogeneous vulnerability value; the values are dissolved (Arc/Info command) to eliminate the variance within one postcode (Figure 7.2b).

7.3   Results

7.3.1   Index of Social Vulnerability

Once standardized the results of the domains and ISV cover the same range of values. Comparison of values for individual EDs can thus be made between domains or ISV. Settlements are recognizable from the maps because EDs are smaller in urban locations. One may speculate that vulnerable people tend to live in areas of dense populations, as all EDs have approximately the same number of residents but at different locational

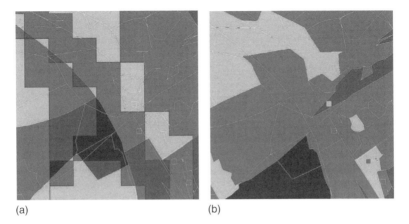

(a)   (b)

FIGURE 7.2 (See color insert following page 274.)
Associated vulnerability index with postcodes (a) before and (b) after dissolving (yellow lines indicate postcode boundaries). (© Crown Copyright/database right 2007. An Ordnance Survey/EDINA supplied service.)

densities. The actual range of values produced in the ISV does not merit detailed analysis in this research, because the importance here is more on how the data can be put to valuable use in the flood industry.

There is correlation between some of the domains (e.g., deprivation, long-term illness, and single parents) where the top and bottom of the maps show similar patterns of high values (Figure 7.3). This may be because these domains are all plausible measures of poverty. However, high values in one domain do not necessitate high values in another [old age pensioners (OAPs); Figure 7.3]. This justifies the range of domains used in determining social vulnerability. Each domain offers a different perspective on vulnerability and thus has the effect of diluting the ISV for each ED, since only areas that contain vulnerability over a number of domains will be accentuated in the ISV. If flood planners require more detail on any of the domains than is available in the ISV, they can view the domain maps independently. This is illustrated by considering how analyzing the spatial distribution of a single domain, such as non-U.K. born, could identify communities with different flood education needs.

The weights used in combining the ISVs to produce the ISV (Figure 7.4) ensure that the domains thought to be most influential on vulnerability have the greatest effect. However, the areas with the highest vulnerability seem to be those that are also the most socially deprived (compare elements of Figure 7.3 with Figure 7.4).

The location of the river currently has no influence on the range of values. This is to be expected and brings to attention that ISV is not calculated in the context of the flood hazard and could possibly be used in studies for different hazards.

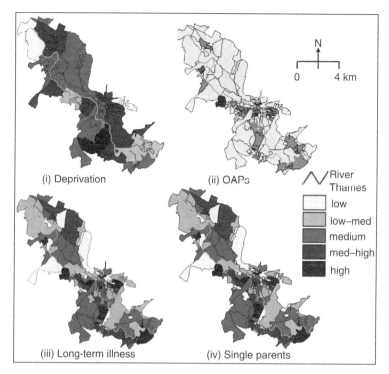

FIGURE 7.3 (See color insert following page 274.)
Output index for four of the separate domains that are combined to form an overall index of social vulnerability. (© Crown Copyright/database right 2007. An Ordnance Survey/EDINA supplied service.)

7.3.2 Index of Flood Probability

The IFP clearly shows areas defined by each of the different probabilities calculated (Figure 7.5). It is assumed that, for a lower probability flood, the extents of the probabilities greater will also be flooded. The extents of each scenario modeled seem to create a reasonable result for such a low-lying floodplain. In each case, as the flood probability is halved, the extent of the flood is approximately doubled. The IFP also produces the kind of expected result, since the probability of a flood is generally reduced away from the river as the elevation increases.

However, this index differs from a DEM in that relative to the surrounding land the elevation of the river is considered to be constant from the top of the catchment to the bottom. The image is particularly useful at showing in some areas the flood probability extends over large areas. During a flood, emergency planners could use the IFP to determine the best routes for evacuation, considering the shortest routes to land at elevations above the flood, rather than evacuating through the low-lying areas. The IFP also shows localized areas safe from lower level flood scenarios, which are surrounded by areas that would be flooded. These areas may be useful for

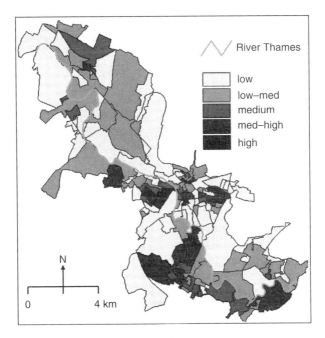

FIGURE 7.4 (See color insert following page 274.)
Index of social vulnerability. (© Crown Copyright/database right 2007. An Ordnance Survey/EDINA supplied service.)

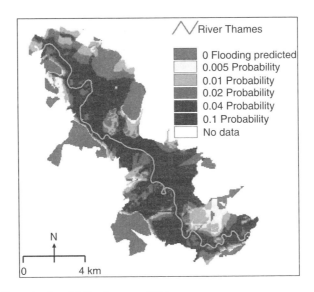

FIGURE 7.5 (See color insert following page 274.)
Index of flood probability. For each probability, any darker shading of blue is included in the full extent of the prediction. (© Crown Copyright/database right 2007. An Ordnance Survey/EDINA supplied service.)

communities' rest centers if an entirely risk-free location is difficult to get to. Importantly, it is evident that the modeled IFP is a useful product in itself as well as being incorporated into the CIFV.

7.3.3 Combined Index of Flood Vulnerability

The CIFV shows the actual spatial variability calculated when flood probability is amalgamated with social vulnerability at the unit postcode resolution (Figure 7.6). Combining the indices provides detail of how vulnerability is spatially distributed in the floodplain. Some of this detail may be lost when analyzed in unit postcodes because an individual postcode may contain variability in the IFP that would therefore affect the overall vulnerability. When calculated to the postcode, the method selects the first vulnerability it comes across in the database, which may not be the greatest vulnerability. Any loss of detail is a disadvantage of the model, but the rationale for analyzing vulnerability at the postcode level is based on them being the smallest demographic unit currently available. Furthermore, unit postcodes are linked to mailing addresses that are used for sending appropriate literature from flood planners. If the improved detail better suits a particular problem, a flood manager could refer to a more detailed map (e.g., CIFV before conversion to postcode resolution).

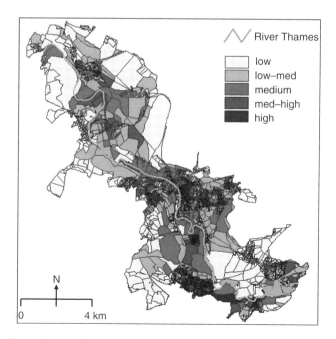

FIGURE 7.6 (See color insert following page 274.)
Combined index of flood vulnerability at the unit postcode resolution. A map viewer or GIS can be used to enlarge areas that are not clear at this scale. (© Crown Copyright/database right 2007. An Ordnance Survey/EDINA supplied service.)

Regions of high vulnerability can be visualized with relative ease (Figure 7.6). Some of the highest values of CIFP are not immediately adjacent to the River Thames, because the ISV used in part to create it is not itself related to the hazard. However, the most prominent areas of medium to high values of vulnerability are in areas predicted for a 10-year flood (cf. Figure 7.5). These observations show that both the ISV and IFP seem to have had an influence on the final CIFV. The model appears to produce the kind of data product intended because there is a clear range in vulnerability to assist flood managers in discriminating the social diversity of a floodplain to suit their purpose (albeit flood warning, emergency planning, or response).

7.4 Discussion

7.4.1 Evaluation of Results

The results are not analyzed in terms of how accurately they represent what is occurring in the study area. It is recognized that an extensive study would be necessary to determine which methods for calculating social vulnerability and flood probability are suitable. What are evaluated here are the major limitations of the methods used and how valuable the incorporation of ISV and IFP data are to the flood industry for which the research is aimed. More emphasis is placed on determining how to create a prototype CIFV, than attempting to create a useable index. However, it is useful to consider where the methodology might be improved.

Without comparing the ISV calculated with the social variability of an actual ED, it is difficult to evaluate the reliability of the method. However, the main problems with the index are identified as follows:

- Inconsistency in depth of domains considered (e.g., OAPs are considered in greater detail than children).
- Deficiency of information on how vulnerability varies within and between domains. For example, single parents are considered in this study to be no more or less vulnerable if they are male or female, while other studies have suggested females suffer higher levels of vulnerability (Tapsell, 1999, 2000; Dralsek, 2000).
- Weightings are assigned subjectively, and may therefore be inaccurate.
- Census data used here is 10 years old and there can be many changes in social demographics over that period of time.

The methodology used to interpolate IFP has been evaluated with the actual data recorded in the field. The flooded properties for different scenarios do not sufficiently match the areas predicted by interpolation. The main problems with the estimation of IFP are identified as follows:

- The relationship between elevation and flood depth may not be straightforward.

- The relationship presumes that flooding will occur when a location is within a threshold HAR and does not consider the effect an undulating surface may have on the spread of floodwater; enabling reduced flooding at lower elevations.

- Channeling may focus floodwater into other areas and cause deeper flooding at higher elevations.

- There is discrepancy in data accuracy when calculating HAR from a DEM resolution of 50×50 m and vertical accuracy of 1 m, yet calculating depth of flooding from depths recorded from field study to an accuracy of 1 cm.

- Calculating HAR assumes that flooding will only occur from the point on the river that is nearest in Euclidean space. This leads to what appear to be anthropogenic lines appearing on the predicted flood probabilities but are actually spurious artifacts produced by the data processing.

The CIFV is useful for flood warning and emergency planning concepts because its consideration of the potential frequency of flooding for different social groups could assist with contingency plans. However, the index is limited in its real-time application for emergency response because it does not clearly show what areas are vulnerable for a particular scenario. In fact, overlaying the social-vulnerability and flood-probability datasets in a GIS can easily identify areas affected by a particular flood scenario (Figure 7.7). The potential of the indexes may be realized with the demonstration of possible applications.

FIGURE 7.7 (See color insert following page 274.)
Identifying vulnerable areas during a 25-year flood. To the left is a close-up of the coarser resolution map to the right. The green masks areas of the ISV that are safe and the Landline data helps see which particular properties are most vulnerable. Vulnerability is increased as the red shading becomes darker. The spatial units are not relevant for this demonstration. (© Crown Copyright/ database right 2007. An Ordnance Survey/EDINA supplied service.)

7.4.2   Interface Potential

The data may be most useful when viewed in conjunction with a GIS. The maps become dynamic because different data associated with the maps can be queried and displayed; particular properties, postcodes, or EDs can be selected as required. Examples of applications show that the indices can be modified in an interface to provide the information that is required by a particular user; there are many possibilities of how the data could be presented. It is important that the interface is easy for its users to understand so that with little training the data may be placed into the hands of flood-planners or the emergency services.

One element identified from the requirement study showed the need for improvements in flood warning for vulnerable groups. The Environment Agency has current initiatives in varying types of warning to suit the needs of specific groups (Environment Agency, 2001). The individual domains or the social-vulnerability index can be chosen as required to help identify certain groups and any attributes that go with them, such as postcode which may be used in advance of a flood-situation for sending educational brochures (Figure 7.8).

In the event of a flood and with data indicating the return-period in occurrence, emergency response teams can quickly identify the vulnerable properties and postcodes that require immediate assistance, discriminating them from vulnerable areas that are not at risk in the present situation

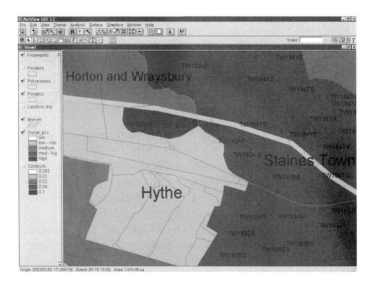

FIGURE 7.8 (See color insert following page 274.)
ArcView interface for flood-warning education (non-U.K. born displayed). The Environment Agency, local authorities, or police can easily identify enumeration districts or unit postcodes with particular flood-warning needs. The example given might result in flood-warning that does not rely on a full comprehension of the English language. The spatial units are not relevant for this demonstration. (© Crown Copyright/database right 2007. An Ordnance Survey/EDINA supplied service.)

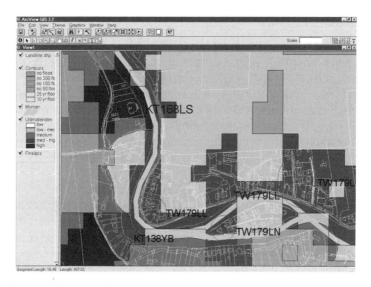

FIGURE 7.9 (See color insert following page 274.)
ArcView interface for emergency planning and response for a flood return-period identified as 25 years (green masking indicates regions of no flooding). Emergency services can easily identify the most vulnerable properties. The spatial units are not relevant for this demonstration. (© Crown Copyright/database right 2007. An Ordnance Survey/EDINA supplied service.)

(Figure 7.9). This interface could easily include police, fire station, and hospital catchments to assist dissemination tactics. Landline and postcode data place the map of flood vulnerability in context and planners can update the system if new information is available. For example, if during a flood it becomes apparent that the situation is actually a 50- and not a 25-year flood, plans can be modified accordingly.

7.4.3   Modifiable Areal Unit Problem (MAUP)

The main limitation to this research, other than the reliability of the methodology used is with modifiable areal unit problem (MAUP). This is described by Heywood (1998) as a problem arising from the imposition of artificial units of spatial reporting on continuous geographical phenomena, resulting in the generation of artificial spatial patterns. MAUP is prominent in the application of socioeconomic data because it suggests that values recorded are unique to the area in which they are defined (in this case, EDs and postcodes), whereas in reality changes in socioeconomic patterns are continuous. The problem is increased in this research when amalgamating the flood-vulnerability index into unit postcodes inferring that the whole postcode has the same flood-vulnerability characteristics when in fact the overall value may be less than the greatest vulnerability recorded for that postcode. If planners are to accurately identify vulnerability of postcodes they need to use data sources to the same detail, data that is not currently available.

7.5 Conclusions

7.5.1 Answering the Research Questions Posed

(1) Can GIS be used to assist the identification of vulnerable groups within communities in areas prone to flooding?

It can be argued that the approach used in this research demonstrates a valid way of identifying spatial differences in vulnerability based on the combination of different data. The manipulation of this data through database and visual commands with a GIS produces simple but effective results. The advantage of GIS also arises from the ability to query the data through dynamic mapping and a graphical user interface, creating customizable products satisfying different user needs.

(2) What data provides the best indicators of social vulnerability?

Census data has disadvantages. One problem is with the MAUP where assumptions are made about the even distribution of certain variables within an ED. There is also the issue of having a specific set of criteria for identifying vulnerability accurately and keeping the data up-to-date, for which more in-depth research is required. The rule-based approach is justifiable because identifying the actual location of all vulnerable people would be an inefficient task; identifying social patterns and automating the analysis therefore seems to be more efficient. However, unless the results are accurate, highly vulnerable people may be neglected. The success of the index developed here can only be truly evaluated if its results are compared to actual findings from fieldwork in different postcodes. The data used in this study is adequate for demonstrating the potential of combining social and flood data.

(3) What data would best serve the definition of flood-risk spatial variability?

While a model of flood probability is useful in the creation of a CIFV, in identifying the variability of social vulnerability within the floodplain, the method used here is flawed by the accuracy of the DEM used. More accurate data, such as DEMs produced from LIDAR data, could potentially improve the results using the same methodology used here. Alternatively, a better method could be established (see, e.g., Boyle et al., 1998).

(4) Can the two proceeding types of index be integrated to improve the vulnerability analysis? How is this achieved?

The main advantage of combining the indices is the potential for immediately identifying vulnerable groups in the particular scenario that is occurring, assuming it is possible to estimate which scenario is occurring in real-time. In flood-warning education, communities can be told how often they should expect to be flooded, helping them come to terms with the hazard and prepare for it. The index is improved because social vulnerability does not consider the intensity of the hazard threat; including flood data places the vulnerability in context. The methodology could potentially be applied to any hazard where different hazard probabilities can be calculated.

7.5.2   Developing the Model for Use within the Flood Industry

The inclusion of other data sources would improve the accessibility of this research. It would be useful to know the locations of people with disabilities but this could be difficult to achieve due to issues of confidentiality. Likewise, the locations of hospitals, fire stations, schools, power and water-works would help planners quantify the risk to services and utilities.

It has been shown how the use of a few simple datasets can produce results that improve flood-warning, emergency planning, and response. The Environment Agency has identified vulnerability studies as an important development for the flood industry (Environment Agency, 2001) but conversations with planners of emergency dissemination have showed a surprisingly negligent attitude to the need for computer-based systems. It is for these reasons that this research has attempted to use methods that apply to the various organizations involved in the hope that, while of immediate use to the Environment Agency in flood warning, it can also demonstrate to others that the system is worthwhile.

This research produces a prototype that would be inappropriate to use without further development. However, if it can be used to illustrate to the flood industry what could be done if more accurate vulnerability indexes and flood-probability contours were produced, and further vulnerability research regards the findings as a useful contribution to further work, the present research will have been successful.

References

Alexander, D., 1997, The study of disasters, 1977–1997: some reflections on a changing field of knowledge. *Disasters* 21(4), 284–304.

Better Regulation Task Force, 2000, *Protecting Vulnerable People* (London: Better Regulation Commission) (http://www.brc.gov.uk/publications/vulnerablesentry.asp).

Blyth, K., Baltas, E., Benedini, M., and Givone, P., 2001, *Risk of Inundation—Planning and Response Interactive User System*, Final Report, EC Project EN4302, CEH (Wallingford: CEH—Institute of Hydrology) (http://www.nwl.ac.uk/ih/www/research/images/EN4302_FINAL_REPORT.pdf).

Borrows, P., 1999, Issues for flood-warning in extreme events, In *Public Needs and Perceptions*, edited by K. Blyth, E. Baltas, M. Benedini, and P. Givone, *Report of RIPARIUS Expert Meeting 1*, January 1999, pp. 8–14 (Wallingford: CEH—Institute

of Hydrology) (http://www.nwl.ac.uk/ih/www/research/images/RipariusEM1.
pdf).

Boyle, S.J., Tsanis, I.K., and Kanaroglou, P.S., 1998, Developing geographical
information systems for land use impact assessment in flooding conditions.
Journal of Water resources Planning and Management 124(2), 89–98.

Dash, N., 1997, The use of geographical information systems in disaster research.
International Journal of Mass Emergencies and Disasters 15(1), 135–146.

DETR, 2001, *Indices of Deprivation 2000, Regeneration Research Summary* (London:
Department of Environment, Transport and the Regions).

Dralsek, T.E., 2000, The social factors that constrain human responses to flood-warnings, In *Floods Hazards and Disasters*, edited by D.J. Parker (London: Routledge).

Environment Agency, 2000, Environment Agency (http://environment-agency.gov.uk).

Environment Agency, 2001, *Lessons Learned: Autumn 2000 Floods* (London: Environment Agency Publication).

Ebel, U., 1999, River flooding and insurance in the European Union, In *Flood
Protection in Finland*, Rantakokko, 1999. In *Flood Mitigation through Improved
Communication and Data/Information Availability. Report of RIPARIUS Expert
Meeting 2*, edited by K. Blyth, E. Baltas, M. Benedini, and P. Givone, pp. 19–22
(Wallingford: CEH—Institute of Hydrology) (http://www.nwl.ac.uk/ih/www/
research/images/RipariusEM2.pdf).

FHRC, 2001, Flood Hazard Research Centre (http://www.fhrc.mdx.ac.uk).

Haggett, C., 1998, An integrated approach to flood forecasting and warning in
England and Wales. *Journal for Chartered Institute of Water and Environmental
Management* 12(6), 425–432.

Heywood, D.I., Cornelius, S., and Carver, S., 1998, *Introduction to Geographical Information Systems* (Harlow: Addison Wesley Longman).

Horner, M.W., 2000, Easter 1998 floods. *Journal for Chartered Institute of Water and
Environmental Management* 14, 415–418.

Jaspars, S. and Shoham, J., 1999, Targeting the vulnerable: a review of the necessity
and feasibility of targeting vulnerable households. *Disasters* 23(4), 359–372.

HR Wallingford, 2000, *National Appraisal of Assets at Risk for Flooding and
Coastal Erosion—Technical Report*. HR Wallingford, SR 573 (Wallingford:
HR Wallingford).

Morris, D.G. and Flavin, R.W., 1996, *Flood Risk Map for England and Wales*
(Wallingford: Institute of Hydrology).

Morrow, B.H., 1999. Identifying and mapping community vulnerability. *Disasters*
23(1), 1–18.

Penning-Rowsell, E.C., Tunstall, S.M., Tapsell, S.M., and Parker, D.J., 2000, The
benefits of flood-warnings: real but elusive and politically significant. *Journal
of the Chartered Institute of Water Engineering and Management* 14, 7–14.

Plougher, A., 2000, The future of telematics and telephony in flood-warning, In
Communicating the Flood Message. Report Of RIPARIUS Expert Meeting 4, edited
by K. Blyth, E. Baltas, M. Benedini, and P. Givone, pp. 20–38 (Wallingford:
CEH—Institute of Hydrology) (http://www.nwl.ac.uk/ih/www/research/
images/RipariusEM4.pdf).

Price, D.J. and McInally, G., 2001, *Climate Change: Review of Levels Protection Offered
by Flood Prevention Schemes* (Edinburgh: Babtie Group, Scottish Executive Central
Research Unit).

Rantakokko, K., 1999, Flood protection in Finland, In *Flood Mitigation through
Improved Communication and Data/Information Availability. Report of RIPARIUS*

Expert Meeting 2, edited by K. Blyth, E. Baltas, M. Benedini, and P. Givone, pp. 9–18 (Wallingford: CEH—Institute of Hydrology) (http://www.nwl.ac.uk/ih/www/research/images/RipariusEM2.pdf).

Rodda, H.J.E., 2001, Insuring against disaster. *GeoEurope* (January), 48–49.

Smith, K. and Ward, R., 1998, Mitigating and managing flood losses. In *Floods: Physical Processes and Human Impacts* (Chichester: John Wiley & Sons).

Tapsell, S.M., 1999, The health effects of the 1998 Easter flooding in Banbury and Kidlington, England, In *Flood Hazards and Disasters*, edited by D.J. Parker (London: Routledge).

Tapsell, S.M., 2000, *Follow-up Study of the Health Effects of the 1998 Easter Flooding in Banbury and Kidlington*, Final Report to the Environment Agency. (Enfield: Flood Hazard Research Centre).

Tunstall, S., 1999, Audit of best practice in flood-warning dissemination, In *Flood Mitigation through Improved Communication and Data/Information Availability. Report of RIPARIUS Expert Meeting 2*, edited by K. Blyth, E. Baltas, M. Benedini, and P. Givone, pp. 33–66 (Wallingford: CEH—Institute of Hydrology) (http://www.nwl.ac.uk/ih/www/research/images/RipariusEM2.pdf).

Whittington, T.K., 2001, Using GIS to identify social vulnerability in areas in the UK that are at risk from flooding: Unpublished MSc Dissertion, University of Edinburgh, Edinburgh.

8

Pattern Identification in Public Health Data Sets: The Potential Offered by Graph Theory

Peter A. Bath, Cheryl Craigs, Ravi Maheswaran, John Raymond, and Peter Willett

CONTENTS

8.1 Introduction

Pattern identification is an important issue in public health, and current methods are not designed to deal with identifying complex geographical patterns of illness and disease. Graph theory has been used successfully within the field of chemoinformatics to identify complex user-defined patterns, or substructures, within molecules in databases of two-dimensional (2D) and three-dimensional (3D) chemical structures. In this paper we describe a study in which one graph theoretical method, the maximum common substructure (MCS) algorithm, which has been successful in identifying such patterns, has been adapted for use in identifying geographical patterns in public health data. We describe how the RASCAL (RApid Similarity CALculator) program (Raymond and Willett, 2002; Raymond et al., 2002a,b), which uses the MCS method, was utilized for identifying user-specified geographical patterns of socioeconomic deprivation and long-term limiting illness. The paper illustrates the use of this method, presents the results from searches in a large database of public health data, and then discusses the potential of graph theory for use in searching for geographical-based information.

8.1.1 Background

The need to identify patterns of illness and disease is not uncommon in public health, for example the identification of disease clusters and tendencies toward clustering, such as outbreaks of communicable disease (e.g., tuberculosis), and higher than expected prevalence/incidence of diseases (e.g., childhood leukemia). The basic building blocks or units for such patterns may be individuals or geographical units, but the key factor is the association between units in terms of time, space, or other complex links. However, searching for patterns of disease using geographical-based data can help not only to identify disease clusters in a geographical area but also can be helpful in seeking to identify potential causes of such outbreaks, which may be geographical features themselves or be characteristics of a geographical area.

Cluster detection, particularly the identification of geographical disease clusters, has been the subject of intensive research within public health and geographical information sciences (Openshaw et al., 1988; Knox, 1989; Besag and Newell, 1991; Alexander and Cuzick, 1992; Kulldorff, 1999). Within the domain of public health and spatial epidemiology, Besag and Newell (1991) classified tests for disease clustering into two groups. The first comprises general or nonspecific tests that examine the tendency for diseases to cluster. The second group comprises specific tests that assess clustering around predefined points, e.g., nuclear installations, or assess the locational structure of clusters. Among the better-known cluster detection methods are Openshaw's Geographical Analysis Machine (Openshaw et al., 1988), Kulldorff's spatial scan statistic (Kulldorff, 1999), Knox's test (Knox, 1989), and Besag and Newell's method (Besag and Newell, 1991). Issues related to clustering and cluster detection are discussed in detail in recent comprehensive publications in the subject area (Lawson et al., 1999; Elliott et al., 2000). The methods described, however, are all concerned with statistical probability and estimation of effect size. They were not designed to handle complex pattern searching queries, and there are currently no satisfactory methods available for this purpose.

In the domain of geographical information science, the ability of current software systems to recognize the relationship between neighboring areas is determined by whether the software has the property of topology, and in particular the branch of topology called pointset topology. Pointset topology is concerned with the concepts of sets of points, their neighborhood, and nearness (Worboys, 1995). It is this concept that allows for the analysis of contiguous areas. Many current GIS, such as *ArcView 3.2* (2002), do not have this property and so cannot deal with contiguous problems such as identifying complex geographical patterns involving neighboring areas. More sophisticated software such as ArcInfo7, however, has topological properties and in theory can identify complex patterns of adjacent neighbors (*ArcInfo 8.2*, 2002). However, three major difficulties are associated with this type of searching. The first problem is that any complex geographical pattern search must be programmed into the software separately, which is time-consuming and requires a high level of programming expertise. The other two problems are that the resulting programs are computationally very intensive and generate very large result files.

In this paper, we describe early work in developing and using techniques that are successfully used in computational chemistry for identifying geographical patterns in public health data.

8.1.2 Computational Chemistry and Graph Theory

In the field of computational chemistry, sophisticated techniques have been developed for the efficient storage and retrieval of various types of chemical information. Highly specified, sophisticated, and flexible searches can be carried out within large databases of molecular structures using techniques derived from graph theory, a branch of mathematics. Graph-theoretical methods of storing 2D and 3D chemical structures have been developed within the Chemoinformatics Research Group in the Department of Information Studies at the University of Sheffield (Willett, 1995, 1999).

Graph theory is used to describe a set of objects, or nodes, and the relationships, or edges, between the nodes. In computational chemistry, nodes are used to represent the atoms in chemical structures. The edges represent the bonds in 2D chemical structure representations and interatomic distances in 3D chemical structure representations of the molecule. The resulting graph is called a connection table and contains a list of all the (non-hydrogen) atoms within the structure and their relationships to each other, in terms of bonds (2D) or distances (3D) (Willett, 1995, 1999). Thus, information about molecules can be stored on databases and retrieved using algorithms developed to identify identical structures (called isomorphism).

There are three types of isomorphism used to compare pairs of graphs:

- *Graph isomorphism*, used to check whether two graphs are identical
- *Subgraph isomorphism*, used to check whether one graph is completely contained within another graph
- *Maximum common subgraph isomorphism*, used to identify the largest subgraph common to a pair of graphs

Algorithms using these types of isomorphism have been developed and used successfully within chemistry to represent and search large files of 2D and 3D structures. The principle of representing information in terms of nodes and edges is not, however, exclusive to computational chemistry and has been used in other areas. If one considers the map of the London Underground as an example of a geographical map, it can be regarded as a graph, with the nodes of the graph representing the stations, and edges representing connecting stations; for example, Russell Square and Covent Garden are on the same underground line, the Piccadilly line. Most other geographical maps or spatially distributed data could be represented in this way.

The aim of the study was to assess the ability of the graph-theoretical methods, used in computational chemistry, to identify a series of increasingly complex patterns of geographical areas that are of interest in public health. We were particularly interested in identifying areas of deprivation and areas of deprivation that have poor health. We briefly describe the MCS algorithm and the structure of the data files that were developed for searching the geographical data. After presenting the results of the searches, we discuss the utility of the method for identifying geographical patterns for public health.

8.2   Methods

8.2.1   Program

The RASCAL program, which is an example of a maximum common subgraph isomorphism method, has been used previously within chemoinfomatics, was modified to enable the program to be used with geographically based public health data, so that the nodes were geographical area and the edges were the association between these areas. Just as the chemical structures can have information associated with them, such as atomic type, geographical areas can also have information associated with them, such as deprivation, census variables, and mortality and morbidity information. The modified program had previously been validated using a test data set (Bath et al., 2002a).

The modified RASCAL program can identify all geographical patterns within the area of interest that match a predefined geographical pattern, in terms of variable criteria and area adjacency. The program requires two distinct pieces of information about each geographical area: variable information that will be used in the selection criteria and information about which areas are neighboring.

8.2.2   Data

8.2.2.1   Geographical Area

The geographical area used in the study was the area previously covered by the Trent Region Health Authority, which includes South Yorkshire, Derbyshire, Leicestershire, Nottinghamshire, Lincolnshire, and South Humberside

FIGURE 8.1
Map of Trent region showing the enumeration districts for the 1991 census. (From 1991 Census: Digitised Boundary Data (England and Wales).)

(Figure 8.1). The areas of interest were the 10,665 enumeration districts (EDs) that make up Trent region. EDs are the lowest level of census geography in England and Wales representing on average 200 households in 1991.

Information on two census-derived variables was used in the study: deprivation and standardized long-term limiting illness ratio for people aged under 75 years (SLTLI<75).

8.2.2.2   Deprivation

The Townsend Material Deprivation Index (Townsend et al., 1988) was calculated for each ED within the Trent region and this index was used to assign each ED with a deprivation quintile variable. The Townsend Material Deprivation Index is a composite score made up of the summation of four standardized variables taken from the 1991 Census small area statistics (SAS). The census variables are: unemployment, overcrowding, lack of owner occupied accommodation, and lack of car ownership. This index was chosen because previous studies have suggested that it is a reasonable measure for explaining material disadvantage (Morris and Carstairs, 1991). A high positive score indicates relatively high levels of deprivation within an area whereas a high negative score indicates relatively high levels of affluence within an area.

The Townsend Material Deprivation Index was calculated for each ED within Trent, standardized to Trent. In total, 195 EDs could not be allocated

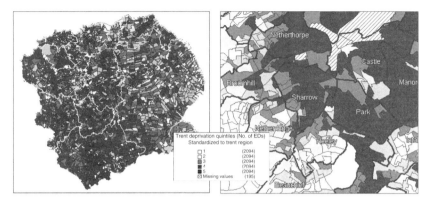

FIGURE 8.2
Maps showing the Townsend deprivation quintile for each ED within the Trent region and an
inner-city area of Sheffield (striped areas signify missing data). (From 1991 Census: Digitised
Boundary Data (England and Wales); 1991 Census: Small Area Statistics (England and Wales).)

a deprivation score because of missing values in one or more of the census
variables, generally low counts and suppression thresholds built into the
census tables (Dale and Marsh, 1993). These EDs were given a deprivation
quintile value of 99. The remaining 10,470 EDs were equally assigned a
deprivation quintile on the basis of their Townsend score. A quintile value
of 5 indicated those EDs within the top 20% most deprived areas, and a
quintile value of 1 indicated those EDs within the top 20% most affluent,
relative to Trent.

Figure 8.2 shows the map of Trent region shaded into quintiles on the
basis of the Townsend deprivation score. Because of their relatively small
size and large number individual EDs are difficult to distinguish for the
whole of Trent. To show individual EDs more clearly, an area within the
south/center of Sheffield has been selected.

The maps of Sheffield center show that the more deprived areas are predominantly to the northeast of the map, within the wards of Castle, Manor,
Park, Sharrow, and Netherthorpe, which surround the south of the city center.

8.2.2.3   Standardized Long-Term Limiting Illness for People Aged
Less Than 75

Long-term limiting illness was also taken from the 1991 Census SAS. The
indirect standardization method was used, standardizing each ED by age
and sex to Trent region for all persons aged less than 75 years. The ED-based
population estimates used in the standardization were taken from the
Estimating with Confidence Project, which adjusted for the underenumera-
tion that occurred in the 1991 Census (Simpson et al., 1995). A value of
100 signifies that the observed number of persons with limiting long-term
illness under 75 years is equivalent to the number of persons expected,
taking into account the age-specific rates of Trent region overall. The

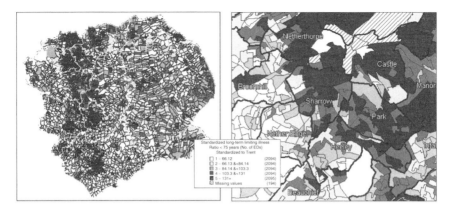

FIGURE 8.3
Maps showing SLTLI<75 quintiles for the EDs in the Trent region and an inner-city area of Sheffield (striped areas signify missing data). (From 1991 Census: Digitised Boundary Data (England and Wales); 1991 Census: Small Area Statistics (England and Wales).)

resulting SLTLI<75 values were then assigned to quintiles with the 20% lowest values assigned a quintile value of 1 and the highest 20% assigned a value of 5. The SLTLI<75 for 194 EDs could not be calculated because of confidentiality issues in the Census SAS tables (Dale and Marsh, 1993). These EDs were given a value of 99.

Figure 8.3 shows the SLTLI<75 quintiles for Trent region and for the selected area within Sheffield. The higher SLTLI<75 scores can again be seen predominantly within the northeast of the map, surrounding the city center to the south.

8.2.2.4   Adjacency Information

As well as each ED having a deprivation quintile and an SLTLI<75 value, each ED also has information about its neighboring EDs. The EDs were each assigned a number between 1 and 10,665. For each ED a list of neighboring ED numbers was recorded.

8.2.3   Storage of Information

All the information relating to each ED was stored on one space-separated text file. The file contained three parts. Part 1 held, on one line, the total number of EDs, the maximum number of neighboring EDs, and the number of variables. Part 2 held, for each ED, one line containing the ED number, ED name, the deprivation quintile, and the SLTLI<75 value. Part 3 held, for each ED, one line containing their ED number and the ED number for each neighboring (or adjacent) ED.

Table 8.1 shows an extract from the data file, showing part 1 and parts 2 and 3 for the ED 38PMFF03.

TABLE 8.1

Extract from the ED Information Data File

10,665	22	2 *(part1)*	
10,000	38PMFF03	4 4 *(part2)*	
10,000	9,998 9,999 10,001 10,002 10,003 10,004 0 0 0 0 0 0 0 0 0 0 0 0 0 0 0 0 *(part3)*		

Part 1 in Table 8.1 shows there were 10,665 EDs within the data file, a maximum of 22 neighboring EDs to any one central ED and two variables. Part 2 shows that the ED 38PMFF03 was numbered 10,000 and had a deprivation quintile of 4 and an SLTLI<75 quintile of 4. Part 3 shows the numbers of the six neighboring EDs. Because the maximum number of neighboring EDs was 22, the modified RASCAL program expected 22 numbers to follow each ED number in part 3. The ED 38PMFF03 had only six neighboring EDs, so 16 zeroes are included to ensure that the ED had the 22 expected values.

8.2.4   Queries

8.2.4.1   Query Patterns

Figure 8.4 shows the query patterns that were used to identify geographical patterns within the Trent region. These queries were developed to provide a range of pattern sizes and arrangement of deprived EDs of potential interest within the query pattern.

Query 1 is a fairly simple pattern looking for a central ED adjacent to three EDs, all with a deprivation quintile within the top 20% most deprived. Query 2 has a central ED adjacent to four EDs, all with deprivation quintiles within the top 20% most deprived and with the top 20% highest levels of SLTLI<75. Query 3 is looking for a pattern of EDs forming a chain of five, all with deprivation quintiles within the top 20% most deprived and with SLTLI<75 within the top 20% highest scores. Thus, although queries 2 and 3 both contain the same number of EDs, i.e., five, they represent very different shapes of patterns. For example, Query 2 could represent a tight cluster of deprived EDs and deprivation and poor health concentrated in a given area, whereas Query 3 could represent a chain of deprived EDs alongside, or bordering, a geographical feature, such as a road or river. Differentiating between clusters of deprivation and chains of deprivation in relation to geographical features in this way could be of value in understanding the local impact of deprivation and health for planning health-care and social-care services.

Query 4 is similar to Query 3 but seeks to identify chains of nine EDs. Query 5 is looking for a more complicated pattern of nine EDs all with deprivation quintiles within the top 20% most deprived and with the top 20% highest levels of SLTLI<75. Thus, similar to queries 2 and 3, both the queries 4 and 5 had the same number of nodes, i.e., nine, but represented different shapes of patterns that could be linked with geographical features.

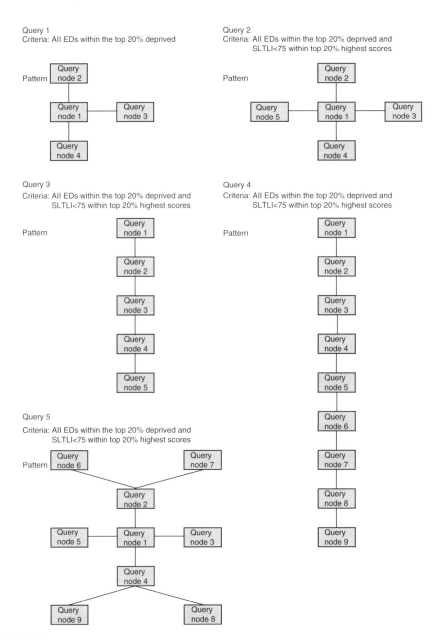

FIGURE 8.4
Diagrams showing query patterns and selection criteria.

8.2.4.2 Query Data File

The data files for each of the queries were set up in a similar way to that of the ED data file but with two extra parts. Part 1 held, on one line, the total number of query nodes, the maximum number of neighboring query nodes,

TABLE 8.2

Data File for Query 1

4	3	1		*(part 1)*
1	Q1	5		*(part 2)*
2	Q2	5		
3	Q3	5		
4	Q4	5		
1	2	3	4	*(part 3)*
2	1	0	0	
3	1	0	0	
4	1	0	0	
1	1			*(part 4)*
2	1			
3	1			
4	1			
1	0			*(part 5)*
2	0			
3	0			
4	0			

and the number of variables. Part 2 held, for each query node, one line containing the query node number, query node name, and deprivation quintile. Part 3 held, for each query node, one line containing the query node number and the query node number for each neighboring query node. Parts 4 and 5 allowed queries to be set up with ranges rather than absolute numbers. Part 4 held, for each query node, one line containing their query code number and a tolerance value percentage for the deprivation quintile. Part 5 held, for each query node one line containing their query code number and a tolerance direction for the deprivation tolerance value, which allowed tolerance values to be set around the deprivation quintile value, or set the tolerance value one way only, i.e., greater than or less than. The query data file for Query 1 is displayed in Table 8.2.

Part 1 of Table 8.2 states that there were four query nodes, a maximum of three connections, and one variable. Part 2 states that the four query nodes are called Q1, Q2, Q3, and Q4, with the query node numbers 1, 2, 3, and 4, respectively. All the query nodes have a deprivation quintile 5. Part 3 shows the connections within the pattern. It states that query node 1 is connected to query nodes 2–4, while query nodes 2–4 are only connected to query node 1. Part 4 states that all the query node deprivation values have a tolerance of 1%. In Part 5, all the EDs have a tolerance direction of 0 indicating that the tolerance is either side of the deprivation quintile, that is the deprivation quintile for each query node can be between 4.95 and 5.05. The query data files for query numbers 2–5 follow a similar pattern to the data file for Query 1.

The modified RASCAL program was used to run each of these queries against the Trent ED data file.

8.3   Results

Table 8.3 shows the number of EDs assigned to each of the deprivation quintiles and each of the SLTLI<75 quintiles. In total, 2094 EDs were given a top 20% deprivation quintile score and 2095 EDs were given a top 20% SLTLI<75 quintile score; 1341 EDs were assigned quintile 5 for both variables. Table 8.4 shows the results from running the RASCAL program for each of the five queries.

Query 1 identified 1527 EDs out of a possible 2094 top 20% deprived EDs matching at least one of the query nodes, and 1181 of these EDs were identified as matching query node 1. Figure 8.5 shows the EDs selected using Query 1 within the selected area of Sheffield. The solid colored areas show the EDs selected using Query 1 that matched query node 1, i.e., which make up the central ED for at least one geographical pattern. The striped ones are EDs that match query nodes 2–4 only, i.e., they are a neighboring ED for at least one geographical pattern but have not been identified as a central ED.

Query 2 identified 713 EDs matching at least one of the query nodes, out of a possible 1341 EDs within the top 20% deprived and top 20% SLTLI<75 highest scores. In these 713 EDs, 350 EDs identified form the central ED in at least one geographical pattern. Figure 8.6 shows a selection of the EDs identified within Trent region. The solid colored areas again show the EDs selected using Query 2 that made up a central ED for at least one pattern. The striped EDs matched query nodes 2–5 only, i.e., they are a neighboring ED for at least one geographical pattern but are not a central ED.

Query 3 identified 882 EDs forming part of a chain of five EDs all with both deprivation quintile and SLTLI<75 ratio within the top 20% highest scores for least one geographical pattern. Figure 8.7 shows the EDs selected for the area within another central area of Sheffield. In this case, all the selected EDs are shaded in a solid color, as the pattern of five linked EDs does not have a central ED to differentiate. The map shows that although

TABLE 8.3

Number of EDs by Townsend Quintile and SLTLI<75 Quintile

		SLTLI<75 Quintile						
		1	**2**	**3**	**4**	**5**	**99**	**Total EDs**
Deprivation Quintile	1	1,047	628	309	98	12	0	2,094
	2	672	652	512	216	42	0	2,094
	3	280	541	639	476	158	0	2,094
	4	84	236	491	741	542	0	2,094
	5	11	36	143	563	1,341	0	2,094
	99	0	1	0	0	0	194	195
	Total EDs	2,094	2,094	2,094	2,094	2,095	194	10,665

TABLE 8.4

Number of EDs Identified by the Queries 1–5

Query Number	Number of EDs Selected	Number of EDs Selected as Query Node 1
1	1527	1181
2	713	350
3	882	n/a
4	661	n/a
5	552	n/a

FIGURE 8.5

Map showing the results from Query 1 for the inner-city area of Sheffield. (From 1991 Census: Digitised Boundary Data (England and Wales).)

FIGURE 8.6

Map showing the results from Query 2 for the inner-city area of Sheffield. (From 1991 Census: Digitised Boundary Data (England and Wales).)

FIGURE 8.7
Map showing the results from Query 3 for an inner-city area of Sheffield. (From 1991 Census: Digitised Boundary Data (England and Wales).)

the defined pattern was five query nodes linked in a chain, the EDs selected are not necessarily forming a straight-line chain. What the pattern is actually identifying are any five linked EDs where each ED is linked to at least one or two EDs within the group of five in such a way that a chain can be formed.

Query 4 identified 661 EDs out of a possible 1341 EDs that form part of a chain of nine EDs, all with deprivation quintile and SLTLI<75 ratio within the top 20% highest scores, within at least one geographical pattern. Figure 8.8 shows the EDs selected for the area within the center of Sheffield. All the selected EDs are shaded within the map. This pattern of EDs,

FIGURE 8.8
Map showing the results from Query 4 for an inner-city area of Sheffield. (From 1991 Census: Digitised Boundary Data (England and Wales).)

FIGURE 8.9
Map showing the results from
Query 5 for the inner-city area of
Sheffield. (From 1991 Census:
Digitised Boundary Data (England
and Wales).)

forming a chain of nine, is a subset of the EDs identified in Query 3, which
form a link of five EDs.

Comparing Figures 8.7 and 8.8 shows that the shaded area within Firth
Park in Figure 8.7 is not shaded in Figure 8.8. This indicates that a chain of
five EDs could be identified within this area but no chain of nine EDs could
be identified.

Query 5 identified 552 EDs out of the possible 1341 EDs with both top 20%
deprivation and SLTLI<75 scores as matching at least one of the nine query
nodes in at least one geographical pattern. Figure 8.9 shows the selected EDs
for the area within Sheffield center. Just as with the results from queries 3
and 4, all the selected EDs are shaded in solid color and no differential has
been made between those EDs identified by query node 1.

The maps for queries 1–5 all show that the majority of EDs selected are
from the northeast of the map where the majority of more deprived EDs and
higher SLTLI<75 scores were found.

8.4   Discussion

The main aim of this study was to use graph-theoretical techniques to search
for increasingly complex geographical patterns in a database of public
health information. We were interested in identifying areas that had high
levels of deprivation and also deprived areas that had poor health. The
study has shown that the modified RASCAL program was successful in
identifying geographical patterns of EDs with these characteristics.

Overall, the attributes for which we were searching were fairly simple
and involved only one or two variables, namely Townsend index quintiles

and quintiles for long-term limiting illness. The deprived areas with poor health that were identified here represent areas that may be in need of additional health and social-care resources to meet the particular needs of the local population to improve its health and well-being. Because we were interested in identifying groups of deprived EDs and groups of deprived EDs with poor health, the same variable criteria were set for each query node within each query, namely the 20% most deprived and 20% with the highest SLTLI<75. The queries used in this study identified clusters and strings of EDs with similar attributes, but the MCS algorithm has also proved effective in identifying single deprived EDs that are adjacent to more affluent EDs (Bath et al., 2002b).

The queries developed in this study represent fairly simple attribute characteristics and do not exploit fully the capacity of the MCS algorithm, which can search among up to 20 attributes assigned to a node. Thus, the MCS algorithm could be used to search for patterns of EDs with much more complex sets of attributes and to identify areas with much more specific needs. For example, deprived areas that had extremely poor health, a relatively large population of older people, and high levels of long-term limiting illness and high mortality among the older people, may have particular needs. The ability to identify such areas could permit local health- and social-care providers, e.g., primary care trusts in England, to target and allocate resources more effectively for the local population. Current work is evaluating the MCS program for identifying areas of particular need.

The attributes that were used for the nodes were quintiles, so that areas of relative deprivation within the Trent region could be identified. Searching for areas of relative deprivation permits service planners and providers to identify those areas of greatest need in that area. However, searching among attributes that consist of the actual values is also possible and enables searching for absolute areas of deprivation, poor health, and so on.

In summary then, although the attributes used in the search queries in this study were relatively simple, the MCS algorithm is capable of searching for more complex sets of attributes. The MCS algorithm also allows complex shapes to be searched and identified and we discuss these now.

The query patterns 1 and 2 were of a very simple format and the results could have been identified using other available software such as Microsoft Access. The process to identify the patterns within Access would have simply involved: linking a file containing a list of all EDs and each of their adjacent EDs with a file containing the selection criteria for each ED; selecting only those EDs which matched the selection criteria; counting the number of adjacent EDs for each central ED; and selecting information from those EDs with a count of at least three or four, respectively. Queries 3, 4, and 5 however were more complicated patterns because the queries involve EDs that are not just directly adjacent to an ED but to EDs next but one or further apart, resulting in patterns that are far more complex for searching.

As was mentioned in the introduction, GIS software with ability to search topology could have, in theory, identified the patterns but it would have entailed writing separate complex computer programs for each individual pattern query. The advantage of the RASCAL graph theory algorithm is that it allows any pattern to be identified within the data simply by designing a simple query file containing the query pattern. However, the RASCAL program was less successful in eliminating the problems of time and data size that are experienced in the GIS software. The length of time it took to run the RASCAL program was related to the complexity of the query, with simple queries taking milliseconds and complex queries taking from between minutes to several days to run. Not only did the program identify all combinations of the EDs, but it also identified and retrieved all the possible permutations of the geographical patterns, generating extremely large result files. This has the potential to be a large problem when dealing with large geographical area split into many geographically small areas; it is, however, an inevitable consequence of the combinatorial nature of the isomorphism testing procedure.

The patterns selected were selected to show an increasing complexity and were not necessarily selected to demonstrate real life patterns that would be of interest for public health analysis. However, it is easy to visualize real-life complicated patterns similar to the ones selected in this paper; for example, identifying areas of high need for resource allocation where there would be interest in finding larger areas made up of small areas with the same variable composition as discussed above. The variety of shapes of patterns that were used for search queries in the study here may be of use within public health because of the distribution of EDs and their relationship to the geographical environment. For example, queries 4 and 5 both contained nine EDs, but had very different shapes. Query 4 contained the nine EDs in a chain, whereas Query 5 contained the EDs as a more compact structure. The EDs retrieved by the program for these queries, however, contained patterns of EDs common to both sets due to their adjacencies among the EDs, in addition to the ones specified in the query. Thus groups of EDs retrieved by Query 4 would also have been retrieved by Query 5, if in addition to the connections specified in Query 4, query nodes 6 and 7 had been connected to query node 2; query nodes 3–5 had been connected to query node 1; and query nodes 8 and 9 had been connected to query node 4.

If we were interested in using Query 4 to identify only chains of EDs, i.e., that were not shaped as clusters as in Query 5, then additional query structures could be constructed to identify retrieve those structures that had connections additional to those specified in Query 4. Retrieving chains or other very specific shapes of EDs might be helpful in identifying patterns associated with geographical features, for example a chain alongside a river or major road, or a cluster surrounding a nuclear power installation or landfill site. Current work is investigating and evaluating the usefulness of the MCS algorithm for identifying such exclusive patterns.

We have discussed the effectiveness of the program for identifying patterns of different shapes and containing different numbers of attributes and its potential for identifying clusters in public health. Public health surveillance is another area in which graph theory could potentially be used, as it involves more complex problems, such as identifying complex multidimensional patterns within large databases. Examples of such problems include examining health service use in relation to need, and access to health care in relation to socioeconomic differentials in health, and the techniques described here are clearly applicable to such challenging geographical and public health problems.

Acknowledgments

The authors acknowledge the Medical Research Council for funding this study under the Discipline-Hopping program. The authors thank Peter Fryers for providing the data on the enumeration districts, and Paul White and Paul Brindley for helpful discussions. Census output is Crown copyright and is reproduced with the permission of the Controller of HMSO and the Queen's Printer for Scotland. This work is based on data provided with the support of the ESRC and JISC and uses boundary material which is copyright of the Crown and the ED-Line consortium.

References

Alexander, F.E. and Cuzick, J., 1992, Methods for the assessment of disease clusters. In *Geographical and Environmental Epidemiology*, edited by Elliott, P., Cuzick, J., English, D., and Stern, R., pp. 238–250 (Oxford: Oxford University Press).

ArcInfo Version 8.2. Available from ESRI GIS and Mapping Software. Redlands, CA (http://www.esri.com/, accessed on 25 May 2002).

Arcview Version 3.2. Available from ESRI GIS and Mapping Software. Redlands, CA (http://www.esri.com/, accessed on 25 May 2002).

Bath, P.A., Craigs, C., Maheswaran, R., Raymond, J., and Willett, P., 2002a, Validation of graph-theoretical methods for pattern identification in public health datasets. *Health Informatics Journal* 8, 167–173.

Bath, P.A., Craigs, C., Maheswaran, R., Raymond, J., and Willett, P., 2002b, Use of graph theory for data mining in public health. Data Mining III. *Proceedings of the Third International Conference on Data Mining*, edited by Zanasi, A., Brebbia, C.A., Ebecken, N.F., and Melli, P., pp. 819–828 (Southampton: WIT Press).

Besag, J. and Newell, J., 1991, The detection of clusters in rare diseases. *Journal of the Royal Statistical Society Series A* 154, 143–155.

Dale, R. and Marsh, C., 1993, *The 1991 Census User's Guide* (London: HMSO).

Elliott, P., Wakefield, J.C., Best, N.G., and Briggs, D.J. (editors), 2000, *Spatial Epidemiology: Methods and Applications* (Oxford: Oxford University Press).

Knox, E.G., 1989, Detection of clusters. In *Methodology of Enquiries into Disease Clustering*, edited by Elliott, P., pp. 17–20 (London: Small Area Health Statistics Unit).

Kulldorff, M., 1999, Spatial scan statistics: models, calculations and applications. In *Scan Statistics and Applications*, edited by Glaz, J. and Balakrishnan, N., pp. 303–322 (Boston: Birkhauser).

Lawson, A., Biggeri, A., Bohning, D., Lesaffre, E., Viel, J.F., and Bertollini, R. (editors), 1999, *Disease Mapping and Risk Assessment for Public Health* (Chichester: Wiley).

Morris, R. and Carstairs, V., 1991, Which deprivation? A comparison of selected deprivation indexes. *Journal of Public Health Medicine* 13, 318–326.

Openshaw, S., Craft, A.W., Charlton, H., and Birch, J.M., 1988, Investigation of leukaemia clusters by use of a geographical analysis machine. *Lancet* I, 272–273.

Raymond, J.W. and Willett, P., 2002, Effectiveness of graph-based and fingerprint-based similarity measures for virtual screening of 2D chemical structure databases. *Journal of Computer-Aided Molecular Design* 16, 59–71.

Raymond, J.W., Gardiner, E.J., and Willett, P., 2002a, Heuristics for similarity searching of chemical graphs using a maximum common edge subgraph algorithm. *Journal of Chemical Information and Computer Sciences* 42, 305–316.

Raymond, J.W., Gardiner, E.J., and Willett, P., 2002b, RASCAL: calculation of graph similarity using maximum common edge subgraphs. *Computer Journal* 45, 631–644.

Simpson, S., Tye, R., and Diamond, I., 1995, What was the real population of local areas in 1991? *Working Paper 10. Estimating with Confidence Project* (Southampton: Department of Social Sciences, University of Southampton).

Townsend, P., Phillimore, P., and Beattie, A., 1988, *Health and Deprivation: Inequality and the North* (London: Croom Helm).

Willett, P., 1995, Searching for pharmacophoric patterns in databases of three-dimensional chemical structures. *Journal of Molecular Structure* 8, 290–303.

Willett, P., 1999, Matching of chemical and biological structures using subgraph and maximal common subgraph isomorphism algorithms. In *Rational Drug Design*, edited by Truhlar, D.G., Howe, W.J., Hopfinger, A.J., Blaney, J.D., and Dammkoehler, R., pp. 11–38 (New York: Springer).

Worboys, M.F., 1995, *GIS: A Computer Perspective* (London: Taylor and Francis).

9

Residential Property Utilization: Monitoring the Government Intensification Agenda

Peter Bibby

CONTENTS

9.1   Introduction

The Government is committed to promoting more sustainable patterns of development, by:

- concentrating most additional housing development within urban areas;
- making more efficient use of land by maximising the reuse of previously developed land and the conversion and reuse of existing buildings;

- assessing the capacity of urban areas to accommodate more housing;
- adopting a sequential approach to the allocation of land for housing development;
- managing the release of housing land; and
- reviewing existing allocations of housing land in plans, and planning permissions when they come up for renewal.

(Department of Environment, Transport and the Regions;
DETR, 2000c, para 21)

It seems peculiar to her suddenly that they should be living in this space: a hundred years ago it would have been a garment factory, where immigrants from eastern Europe stitched fabric into human shapes and practised getting their tongues around the muted diphthongs of English. This is what Lily loves about London, that every building, street, common and square has had different uses, that everything was once something else, that the present is only the past amended.

(Maggie O'Farrell, *My Lover's Lover*, London, Review 2002, p. 41)

9.1.1   Policy, Evidence, and GIS

In the opening years of the twenty-first century, planning policy in England and Wales was clearly directed to conserving undeveloped land and to the intensification of use of urban areas. DETR's Planning Policy Guidance Note 3 of 2000 (PPG3) encapsulated this emphasis. The term *intensification* denotes "a combination of changes in built form and activity" and focuses attention on the capacity of urban areas both to accommodate extra dwellings and to adapt to new economic roles. At the microscale, the term implies development of previously undeveloped pores within cities; the redevelopment of existing buildings and previously developed sites at higher densities; and the subdivision, conversion, and extension of existing buildings. All contribute to the intensification of use of existing buildings or sites and changes of use allowing increases in the numbers of people living in, or working in an area (Williams, 1999, p. 168). Policy has focused on amending the past in a manner which provides for more sustainable development and which celebrates—perhaps in the manner of O'Farrells's *Lily*—the values of urban living. Over the same period, across government, there was a reinvigorated interest in founding policy upon evidence. It therefore seems plausible that there might be some potential role for GIS (and indeed for Geographic Information Science (GISc)) in developing and monitoring policy for reshaping of the physical environment.

This chapter explores some of that potential. Its focus is on monitoring urban growth and the conservation of undeveloped land, on monitoring the mediating influence of urban land recycling, and on the reuse of existing buildings. It attempts to contribute to debate at three levels. Most immediately, it attempts to use GIS to draw some inferences about development patterns in England and Wales which might be pertinent to the assessment of policy. Second, it considers how particular techniques, including the use of natural language processing (NLP) with GIS, can contribute to the exploitation of data for policy purposes. Third and most fundamentally, it is concerned with the overall relationship between policy, evidence, and GIS and with the manner in which GIS use is and might be embedded within policy processes.

A prerequisite of addressing the first of these concerns is a broad understanding of aspects of relevant government policy in 2000 and immediately afterwards, while engagement with the third concern demands some explicit consideration of how the term *policy* itself is to be understood. The emphases of the 2000 revision of PPG3 reflect a commitment to regeneration and intensification, which suffuses popular planning thought and rests in turn on underlying concerns about sustainable urban living and broader notions of environmental sustainability. The 2000 revision of PPG3 must, therefore, be understood alongside a welter of other documents (including, for example, the urban and rural white papers of DETR, 2000a,b) and Prescott's (2003) statement on sustainable communities which depend upon the broader discourse of sustainable development. It must be emphasized, however, that other discursive currents influence present policy set out in the Communities and Local Government's Planning Policy Statement 3 (PPS3; CLG, 2006). CLG is the successor department to the Office of the Deputy Prime Minister (ODPM), DETR, Department of Transport, Local Government and the Regions (DTLR), and the Department of Environment (DoE).

The concept of policy pertinent to this chapter should neither be reduced to the text of PPG3 (or PPS3) nor bloated to include the sum of concerns about sustainability. In the tradition of Heclo, policy might be regarded as a "course of action or inaction" (Heclo, 1972, p. 85). The policy process might thus be seen as centering on the articulation of commitments intended to guide subsequent action. From this perspective, the prime significance of texts such as PPG3 is that they potentially allow such commitments to bind actors such as local authority planners who may be distant from central government policy making both in space and time. The policy process involves ensuring such attenuation, so that policy becomes a *"stance* which once articulated, contributes to the context within which a succession of future decisions will be made" (Hill, 1997, p. 7 ascribed to Friend et al., 1974, p. 40). The context reproduced by the policy process is sometimes referred to as the policy setting and includes an assumptive world of values, metaphors, and core narratives reflected in bureaucratic practices, operational definitions, and procedural rules.

Evidence is always used to support or supplant a story. Policy rests upon particular understandings of the nature of the world. Given the nature of policy, its relation to evidence is less straightforward than might first appear. Context denies the possibility of transparent empiricism, thereby complicating the role of GIS in monitoring its effectiveness. Sustainability, moreover, should perhaps be seen as an "essentially contested" concept in the spirit of Gallie (1955–1956). Without elaboration of a particular narrative, and of particular definitions, GIS, however useful, cannot provide a tool for distinguishing sustainable and unsustainable patterns of development. It, therefore, cannot somehow ground policy in evidence in an unproblematic manner. The evidence assembled using GIS is constrained by the data which it has been deemed worthwhile collecting and framed by particular narratives and images within the policy setting.

Understanding the potential of using GIS in policy monitoring involves appreciating the character of the traditional narratives. One such narrative provides an account of urbanization which focuses on the construction of dwellings, leading from the idea of exogenous household growth to expansion of the contiguous urban area and concomitant reduction in undeveloped land. The number of dwellings in Great Britain has increased by 80% in the last 50 years (Matheson and Babb, 2002, p. 163). The traditional narrative has moved with images such as "a Bristol a year," directly from increasing numbers of households to the expansion of the contiguous urban area, and this provides the imagery by which the press expresses the environmental consequences of household growth [see, for example, the transmutation of forecast changes in numbers of households into "twenty-seven huge new towns" (Daily Telegraph, 1996) or the invocation of "an area the size of Manchester" (Observer, 2003)]. They converge with images of urban growth, urban sprawl, and urban spread, which liken cities to organisms, demanding responses such as CPRE's *Sprawl Patrol*. Such images are reflected and supported by familiar cartographic devices, which record the expansion of particular towns over time, which may be replicated within GIS.

More recent narratives, however, qualify this story. Growth in numbers of households remains at the core. Although population growth has been modest in recent years, household growth—and hence urban growth—has continued (sustained by rising real incomes). This growth is to be understood in relation to changing lifestyle choices that show themselves statistically as continuing falls in average household size. Variants of the narrative typically question how new households or dwellings are to be accommodated, but not the sustainability of those social choices that allow household size to continue to fall (DoE, 1996). Through the 1990s policy discussion became increasingly concerned with the extent to which development might be concentrated on brownfield sites and hence mitigate pressure for urban expansion. This in turn prompted GIS development including both small-scale analytic work underpinning urbanization forecasts (Bibby and Shepherd, 1996) and development of a National Land Use Database (NLUD)—an inventory of brownfield sites.

In the absence of strong population growth, by 2000, household growth had come to coexist alongside crude housing surplus at national level (Matheson and Babb, 2002, p. 164). In particular cities and regions, problems of low demand for housing had come to assume prominence (e.g., Bramley et al., 2000) and these issues had risen high up the policy agenda. Narratives of urban growth thus came to interact with rather different narratives of local housing market collapse. These emphasized the rapid, extreme, and essentially arbitrary nature of local market adjustment as withdrawal of key actors (such as particular social landlords), vandalism against empty property, and outbreaks of social disorder might undermine the possibility of continued occupation. The specter of urban expansion running apace alongside the dereliction of redundant urban quarters had become evident.

Policy, moreover, must be concerned not only with substantive goals but also to the manner in which they are to be pursued. In a climate where evidence is used to legitimize policy, where there is a lack of confidence in forecasts, and where there is uncertainty over the performance of local housing markets, monitoring came and remains to the fore (in principle at least). The 2000 revision of PPG3 introduced a "plan, monitor, and manage" approach to planning for housing in preference to the previous regime— somewhat disparagingly dubbed "predict and provide" retrospectively (Prescott, 2000). This provides the context in which this particular series of GIS applications is set. It is very different to one in which housing demand—driven by population growth—would inevitably be met by the construction of family housing immediately recognizable by remote sensing and easily represented on large-scale maps.

9.2 Patterns of New Construction: Accommodating Housebuilding within Urban Areas

The introductory quotation from the 2000 revision of PPG3 (DETR, 2000c) focuses on three objectives: concentrating housebuilding on sites within urban areas, concentrating housebuilding on previously developed sites, and accommodating new dwellings within existing buildings. The remainder of this chapter treats each of these objectives in turn, using GIS to explore how far patterns of housebuilding in the 1990s proved consistent with the intentions set out in 2000 and exploring some of the issues arising. In so doing it must have regard to the closely linked intentions to

> avoid developments which make inefficient use of land (those of less than 30 dwellings per hectare net)
>
> encourage housing development which makes more efficient use of land (between 30 and 50 dwellings per hectare net)

and

> seek greater intensity of development at places with good public transport accessibility...such as city, town, district and local centres or around major nodes along good quality public transport corridors.
>
> (PPG3; DETR, 2000c, para 58)

The location of new development in relation to existing urban areas would appear to be an issue where there is a clear role for GIS and where the analytic issues are trivial. Effective monitoring might appear to depend simply on the availability of information on the location of new housing sites on the one hand and the boundaries of urban areas on the other recorded with sufficient precision and accuracy. Fortunately, Ordnance Survey (OS)—the national mapping agency—generate both sets of data. Since 1985 they have collected Land Use Change Statistics (LUCS) for what is now CLG as an adjunct to updating the national map base (Sellwood, 1987). This constitutes a tractable source of very fine-grained information about the location of new housebuilding (among other things). OS have also produced for CLG and its predecessors highly detailed boundaries of physical urban areas for use alongside Census statistics for 1981, 1991, and 2001 (for a discussion of these boundaries and their relation to other urban definitions, see Shepherd et al., 2002).

LUCS data refer to the land parcels shown on basic-scale maps (1:1250 in urban areas 1:2500 at the urban fringe and 1:10,000 in mountain and moorland areas). Where the use of any such parcel changes (on the basis of a 24-category classification) a LUCS record is created. It will include a 10-m grid reference for a representative point within the parcel, a one character code (e.g., R for residential) indicating the use before and another indicating the use after the change, an estimate of the year of change, an estimate of the area of the site, and (in the case of residential development) an estimate of the number of dwellings demolished and the number of units built. As shown in Table 9.1, these data indicate that in the years from 1990 to 2000 (inclusive), 1.45 million houses were built in England on 586 square kilometers of land (i.e., at an average density of 24.7 units to the hectare). It is important to note at the outset that the implied annual rate is historically low, although the scale of development is of the same *order of magnitude* as that required to meet household projections (e.g., DoE, 1995) or that suggested by the Barker (2004) review.

Digital boundaries of physical urban areas are generated for CLG by OS on the basis of a series of rules. The rules are used to aggregate parcels on the basis of their use and the distance between them. Any parcel on a basic-scale map is treated as being in either urban or rural *use*. The classification used is the same as that in LUCS, the individual uses being arranged into these two divisions. Parcels in urban use are then joined with their neighbors or other such parcels within 50 m to form *areas of urban land*. Open land totally surrounded by an area of urban land (such as Hampstead Heath or Richmond Park in London, or Sutton Park in Birmingham) is also treated as forming part of it. (Under the 1991 definition, a subset of these *areas of urban land* are deemed to be *urban areas*.)

Simply overlaying LUCS point data on the OS 1991 urban area polygons reveals that over the 1990s, in the order of 57% of new dwellings were accommodated within those urban areas (Table 9.1).* Although

---

* In the case of the boundaries produced by OS for use with the 1991 census, a distinction was made between areas of urban land and urban areas. An urban area for this purpose was defined as an area of urban land that impinged on four or more enumeration districts (the smallest units for which 1991 census data were released). This implied a variable lower limit to the population of urban areas (between 1000 and 2000 persons). The boundaries produced for the 2001 census encompassed a far larger group of settlements. For this study the term urban areas refers to physical settlements treated as urban areas in 1991 and with a 1991 population of 2000 or more.

TABLE 9.1

New Dwellings Built and Housing Land Developed, 1990–2000, England: Urban Areas (UAs) and Elsewhere

Year	Outside UAs			Inside UAs			Totals			% Inside UAs	
	Units	Hectares	Density	Units	Hectares	Density	Units	Hectares	Density	Units	Land
1990	51,516	3,337.9	15.4	110,046	3,841.0	28.7	161,562	7,178.9	22.5	68.1	53.5
1991	43,846	2,110.9	20.8	79,643	2,543.3	31.3	123,489	4,654.2	26.5	64.5	54.6
1992	52,126	2,582.8	20.2	78,714	2,626.8	30.0	130,840	5,209.6	25.1	60.2	50.4
1993	59,919	2,886.0	20.8	83,875	2,702.1	31.0	143,794	5,588.1	25.7	58.3	48.4
1994	70,735	3,500.0	20.2	81,493	2,772.4	29.4	152,228	6,272.4	24.3	53.5	44.2
1995	61,403	3,194.4	19.2	76,938	2,589.3	29.7	138,341	5,783.7	23.9	55.6	44.8
1996	61,967	3,029.5	20.5	66,095	2,101.5	31.5	128,062	5,130.9	25.0	51.6	41.0
1997	66,924	3,284.6	20.4	74,027	2,345.9	31.6	140,951	5,630.5	25.0	52.5	41.7
1998	65,839	3,118.9	21.1	69,252	2,239.3	30.9	135,091	5,358.2	25.2	51.3	41.8
1999	48,189	2,390.7	20.2	55,083	1,781.7	30.9	103,272	4,172.4	24.8	53.3	42.7
2000	45,100	2,124.3	21.2	45,703	1,533.9	29.8	90,803	3,658.2	24.8	50.3	41.9
Total(1)	627,564 H(RU1)	31,559.84 L(RU1)	19.9	820,869 H(UA1)	27,077.3 L(UA1)	30.3	1,448,433	58,637.1	24.7	56.7	46.2
Total(2)	570,784 H(RU2)	28,433.4 L(RU2)	20.1	877,401 H(UA2)	30,194.89 L(UA2)	29.1	1,448,185	58,628.3	24.7	60.6	51.5
Total(3)	564,516.3 H(RU3)	28,128.4 L(RU3)	20.1	883,773.7 H(UA3)	30,503.5 L(UA3)	29.0	1,448,290	58,631.9	24.7	61.0	52.0
2004 Definitions											
Total(4)	698,284 H(RU4)	34,752 L(RU4)	20.1	750,005.9 H(UA4)	23,880.2 L(UA4)	31.4	1,448,290	58,631.9	24.7	51.8	40.7

Note: The year-by-year values and Total(1) values have been calculated by treating LUCS data as points and overlaying them on urban area polygons. The values for Total(2) have been obtained by treating LUCS data as points and overlaying them on a 100-m grid derived from the urban area polygons. The values for Total(3) have been obtained by spreading LUCS data across a 100-m grid as described in Section 9.4 and overlaying the derived values on a 100-m grid derived from the urban area polygons.

there are no quantitative targets for the proportion of housebuilding to be accommodated within existing urban areas, it appears that these areas were able to absorb well in excess of 800,000 new dwellings in the period. The table appears to provide a substantial degree of comfort to those anxious to realize the government's goal of ensuring that by 2008, 60% of new housebuilding is accommodated on previously developed sites. (Note that this table says nothing about previously developed sites per se.) Those practitioners and commentators who remain profoundly skeptical of the realism of such targets might also find within Table 9.1 some justification for their position. They might question how long this pattern of development might be sustained, pointing out that while more than two-thirds (68%) of new dwellings appear to have been accommodated in urban areas in 1990, this proportion fell steadily through the decade, so that only half of all new dwellings were being accommodated in this way by 2000 (Figure 9.1). Moreover, it appears that less than half of all housebuilding land was found within the confines of urban areas as they had stood in 1991, and that this proportion too followed a distinct downward trend. This is consistent with the familiar view that with the passage of time it becomes progressively more difficult to identify sites within the urban area.

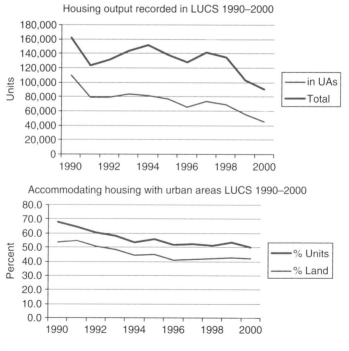

FIGURE 9.1
Housebuilding within urban areas in England between 1990 and 2000. (From LUCS. With permission.)

Unraveling these mixed messages and drawing out their implications demands a more thorough examination of the evidence, questioning the usual narratives more closely, and deploying GIS more creatively. Table 9.1 moves only a tiny step towards understanding how new dwellings have been accommodated or the extent to which they might be accommodated in urban areas in the future. The rest of this chapter attempts to move successively closer to definitions that are substantively meaningful in policy terms. This first definition of urban areas will be called UA1. The number of dwellings accommodated within the 1991 urban areas will be referred to as H(UA1), and the corresponding area of land developed L(UA1). Subsequent definitions of urban areas will be referred to as UA2 and so on, the general case being termed UAi (and the corresponding rural residual RUi). Development within UAi will be referred to here as *urban consolidation* (accommodating additional households within existing urban areas through either infilling of green pores or recycling of previously developed sites). In the 1990s, debate counterposed such urban consolidation against rural land conversion in the form of either *urban extensions* (UXi) or of *new settlements* (NSi) (e.g., Breheny et al., 1993). It is, of course, usually assumed that demand can be diverted between these different contexts and so it is impossible to understand the volume of new dwellings being accommodated in cities in isolation. As a next step we attempt to partition the total number of dwellings built over the 1990s, H(TO), into these components.

9.3 Accommodating Housebuilding: Urban Areas and Beyond

As the very idea of urban extensions embodies the metaphor of the city as polygon, elementary GIS operations should in principle allow for their direct measurement and for examination of their contribution to the housing land supply. Urban extension polygons might be defined as a subset of the difference polygons created by overlaying the urban area polygons defined by OS for use with the 2001 census with those for 1991 (defining UX1). New settlements (NS1) might be represented by urban area polygons not present in 1991 but found in 2001. Urban consolidation would occur in the polygons forming the intersection of the two sets (UA1). This simple geometric logic demands the recognition of two further types of circumstance which are more marginal to policy discourse. The first is represented by difference polygons referring to land considered urban in 1991 but not in 2001. These might be thought of as urban contraction polygons. The second comprises an outside remaining rural throughout also represented by a polygon (or in principle more than one). This last class of circumstance thus constitutes what might be termed as an exurban context (specifically XC1). The number of new dwellings accommodated in each of these contexts might in principle be assessed by overlaying LUCS point data on the polygons defined. Thus

$$H(TO) = H(UAi) + H(RUi)$$

or

$$H(TO) = H(UAi) + H(UXi) + H(NSi) + H(XCi)$$

and specifically

$$H(TO) = H(UA1) + H(UX1) + H(NS1) + H(XC1).$$

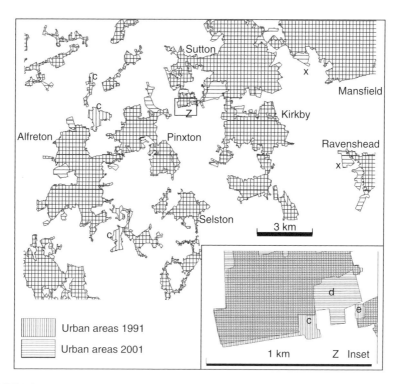

FIGURE 9.2

Urban area polygons in 1991 and 2001. Detail from part of the Nottinghamshire Coalfield.
Note: Because of the procedural rules used to define urban area polygons (see text), they are very convoluted. Comparison of polygons for 1991 with those for 2001 provides a clear indication of urban expansion (see for example areas of expansion such as those marked "X" on the western fringe of Ravenshead. Although they are not consistent with the notion that changes to urban use are fundamentally irreversible, areas of urban contraction are also found (such as those marked "c" above). While some of these appear to reflect change on the ground, other change appears to reflect differences of view. This seems particularly clear in the inset which shows the south-western limit of Sutton-in-Ashfield. Here differences in the western settlement margin appear to reflect a digitzing decision, and the minor contraction along the southern limit an arbitrary decision that in 1991 the A38 dual carriageway should be included within the urban area although it was excluded in 2001. The apparent contraction (c) seems to reflect a change of view, whereas the expansion (d) seems consistent with change on the ground, though the apparent contraction (e) seems to arise from another change of view.

In practice, such partitioning proves troublesome, both computationally and conceptually. Computational difficulties arise in overlaying two large sets of highly convoluted boundaries (Figure 9.2). The conceptual difficulties become apparent in examining the nature of change between the polygons representing the urban extent in 1991 and 2001 urban area polygons. Broadly speaking, such changes may arise

- Where building operations or changes of use imply an extension or contraction of the urban area on the basis of the definitional rules used by OS
- Where there is a difference in view of how the definitional rules should be applied
- Where there is a difference in view of the appropriate relationship between polygons and intensional settlements

The first source is, of course, the focus of immediate interest. As CLG and its predecessors have taken the view that rural to urban land conversion is essentially irreversible, it is not clear from a policy perspective how urban contraction polygons should be treated. In what follows, development that occurs within them is regarded as being within the urban area UA1. Differences of view become apparent when changes in the urban area polygons are examined in relation to (unchanging) detail of basic-scale maps. This second source of change is not necessarily conceptually difficult, but although potentially resoluble implies that measures of change based on the polygon datasets are not simply attributable to change in the built environment. Consistent with the expedient adopted with respect to urban contraction, areas viewed as urban in 1991 are treated here as remaining urban thereafter.

The third source of change raises more fundamental issues deriving from the general relationship between bounding and naming, which can only receive the most cursory treatment here (but see Jubien, 1993; Bibby and Shepherd, 2000; Bibby, 2005). The term *intensional settlement* is used here to refer to the places that those involved in the policy process talk about, or (more strictly) have in mind. (On intension, see Searle, 1995.) Despite the rationale for defining OS urban area polygons set out above, there is not a one-to-one correspondence between them and the named places assigned a unique identifier in the datasets. Many intensional settlements (e.g., Lincoln or Milton Keynes) are represented by more than one polygon. Although the polygons discussed approximate physical urban areas, unstated intensional definitions are in fact privileged. Conversely, a single contiguous area of urban land may be partitioned into several adjoining polygons. Hence although the physical definition of London does not extend to its administrative boundary in some places (e.g., LB Bromley) and extends beyond it in others (e.g., LB Hillingdon), boundaries between the boroughs are imposed.

The relationship between naming and bounding gives rise to a range of curiosa. Changes in view of the appropriate relation between places and polygons may, for example, alter apparent population sizes. Critically, the addition of a new urban area polygon cannot signal development of a new settlement without regard to the name or urban-area serial number attached to the place. Automated application of the city as polygon metaphor within GIS might potentially lead to interpretations too literal to be valuable for policy purposes (for example, placing too much emphasis on literal physical connection or disconnection and lacking real implications for likely travel patterns). Recourse to the intensional definitions mitigates this, providing a first illustration of how policy monitoring entails the management of metaphor.

For present purposes, it will be sufficient simply to flag each LUCS record with an indicator showing whether it fell within any urban area polygon: (a) in 1991 and (b) in 2001. This allows direct assessment of the role of urban consolidation H(UA1). Using lower case letters to express the role of particular contexts (e.g., UX1) relative to England as a whole (TO), it also allows assessment of proportion of new dwellings (h(UX1)) and of housebuilding land (l(UX1)) within implied urban extensions while abstracting from the details of the geometry of these areas. The identification of dwellings in new settlements is necessarily slightly more complex, as it must take into account not only of the matters just discussed, but also must recognize that polygons appearing for the first time in the 2001 dataset may demarcate preexisting settlements. It is therefore not possible to define in practice a new settlement context on the polygon logic (NS1 in principle). The approach taken, therefore, involved first identifying candidate new settlements (denoted by urban area polygons with codes appearing for the first time) and then overlaying these areas on concentrations of new housebuilding evident in LUCS. This was accomplished in fact by using a hectare grid representation of both the OS urban areas and the LUCS data, producing a definition of new settlement contexts (NS3) compatible with definitions of other contexts (UA3, UX3, and XC3) introduced below.

This implies an imperfect geometrical distinction between urban extension and new settlement, and so in the summary entries in Table 9.2, building in new settlement has been subsumed within UX1. It is immediately clear, however, that the contribution of new settlement to accommodating dwellings in the 1990s was minimal (NS3). Only five urban areas identified in the 2001 urban areas dataset but not in that for 1991 had concentrations of residential development in the 1990s. These were Cambourne in Cambridgeshire, Dickens Heath within Solihull District in the West Midlands, Whitley in Hampshire, Cotford St Luke near Taunton in Somerset (a new village on a former hospital site), and Hatton near Warwick, a village of mediaeval foundation with a former hospital site identified as a local growth point in the Warwick District Local Plan (Warwick District Council, 2003).* Using the grid method, it is

---

* Other locations identified in applying the procedure that strictly fail to meet the criteria are Millisons Wood (in the West Midlands), Dunkeswell near Honiton in Devon, Southfields in Essex Thameside (Thurrock), and Tanfield (a village abutting the urban area of Cheshunt in the Hertsmere District of Hertfordshire).

TABLE 9.2

New Dwellings Built and Housing Land Developed, 1990–2000, England: by Mode of Accommodation

	Units Built, 1991–2000			Land Developed, 1990–2000			Densities Achieved, 1990–2000		
	Outside UAs 2001	Inside UAs 2001	Total	Outside UAs 2001	Inside UAs 2001	Total	Outside UAs 2001	Inside UAs 2001	Total
Outside UAs 1991	205,505	422,059	627,564	14,248.67	17,311.17	31,559.84	14.4	24.4	19.9
Inside UAs 1991	5,256	815,613	820,869	261.16	26,816.13	27,077.29	20.1	30.4	30.3
Total	210,761	1,237,672	1,448,433	14,509.83	44,127.3	58,637.13	21.6	25.7	24.7
Percentages									
Outside UAs 1991	14.2	29.1	43.3	24.3	29.5	53.8			
Inside UAs 1991	0.4	56.3	56.7	0.4	45.7	46.2			
Total	14.6	85.4	100.0	24.7	75.3	100.0			

Descriptions
Exurban development — Urban extension
Reclassification — Urban intensification

	Units Built 1990–2000			Land Developed 1990–2000			Densities Achieved 1990–2000
Exurban development	205,505	14.2	XC1	14,248.67	24.3		14.4
Reclassification	5,256	0.4		261.16	0.4		20.1
Urban extension	422,059	29.1	UX1	17,311.17	29.5		24.4
Urban consolidation	815,613	56.3	UA1	26,816.13	45.7		30.4
	1,448,433	100.0			100.0		
Totals(1)							
Exurban development	205,505	14.2		14,248.7	24.3		14.4
Urban extension	422,059	29.1		17,311.2	29.5		24.4
Urban consolidation	820,869	56.7		27,077.3	46.2		30.3
Overall	1,448,433			58,637.2	100.0		

(continued)

TABLE 9.2 (continued)

New Dwellings Built and Housing Land Developed, 1990–2000, England: by Mode of Accommodation

	Units Built, 1991–2000			Land Developed, 1990–2000			Densities Achieved, 1990–2000		
	Outside UAs 2001	Inside UAs 2001	Total	Outside UAs 2001	Inside UAs 2001	Total	Outside UAs 2001	Inside UAs 2001	Total
Totals(2)									
Exurban development	170,910	11.8	XC2	11,789.8	20.1		14.5		
Urban extension	399,874	27.6	UX2	16,643.6	28.4		24.0		
Urban consolidation	877,401	60.6	UA2	30,194.8	51.5		29.1		
Overall	1,448,185	100.0		58,628.2	100.0		24.7		
Totals(3)									
Exurban development	185,628	12.8	XC3	12,597.9	21.5		14.7		
Urban extension	378,888	26.2	UX3	15,530.5	26.5		24.4		
Urban consolidation	883,774	61.0	UA3	30,503.5	52.0		29.0		
Overall	1,448,290	100.0		58,631.9	100.0		24.7		
2004 Rural Definition									
Exurban development	398,165	27.5	XC4	22,949.6	39.1		17.3		
Urban extension	300,120	20.7	UX4	11,802.1	20.1		25.4		
Urban consolidation	750,006	51.8	UA4	23,880.2	40.7		31.4		
Overall	1,448,291	100.0		58,631.9	100.0		24.7		

Note: The crosstabulated values and Total(1) values have been calculated by treating LUCS data as points and overlaying them on urban area polygons. The values for Total(2) have been obtained by treating LUCS data as points and overlaying them on a 100-m grid derived from the urban area polygons. The values for Total(3) have been obtained by spreading LUCS data across a 100-m as described in Section 9.4 and overlaying the derived values on a 100-m grid derived from the urban area polygons.

estimated that new settlements accommodated barely more than 2000 units over the period [or 0.2% of all dwellings; H(NS3) is 2207; h(NS3) is 0.2%].

Planning practitioners and analysts are unlikely to be surprised by the nature of these results (whatever the precise numerical values). Whatever definition might be adopted, enthusiasm for prospective new settlements waned over the decade, given the difficulties of overcoming risk on the one hand and public opposition on the other. It should also be clearly understood, however, that in contrast to the presumptions of more journalistic commentators, the 1990s witnessed the accommodation of more than 1.45 million new dwellings, without the construction of new settlements. Equally important in considering appropriate policy responses to household projections or to the recommendations of the Barker (2004) review, it becomes important to examine quite how such an apparently implausible volume of development has in fact been accommodated historically.

While the minimal role of new settlement should occasion no surprise, it is evident that new settlement and urban extension together accounted for barely more than one house in four [h(UX1) is 29%]. Despite the traditional narrative, the number of houses accommodated through urban consolidation was almost double that in new settlements and urban extensions. Besides urban consolidation, which accounts for 57% of new dwellings, there remains, however, a further component of accommodation, termed here *exurban* development. Although in GIS-based analyses it is not possible to ignore development in such contexts, it is relatively little discussed by policy makers and practitioners. Exurban development appears to account for some 200,000 dwellings in the 1990s [h(XC1) is 0.14]. Such development seems, moreover, to have proceeded at particularly low densities (14.2 dwellings to the hectare on average) and thus to account for a disproportionate share of all land developed for housing [l(XC1) is 0.243].

Examination of the potential for urban consolidation cannot be reduced simply to assessing the capacity of urban areas (on the one hand) and understanding (on the other) the competing attractiveness of sites within the urban area and those which would extend it. The scale and character of exurban development may imply a threat to urban consolidation and demands more sustained analysis. It therefore seems important to take care to see just what sort of development is involved—whether dispersed properties in sparsely populated areas, for example, or simply buildings very close to urban areas but deemed outside by imposition of a particular boundary. It will also be necessary to have regard to the overall level of demand.

9.4 Using Grids to Characterize Dispersal of Housebuilding

As a first step in visualizing the pattern of development, and in trying to understand the reasons for place-to-place variation in the relative importance of the modes of accommodation, it is convenient to transform the data to a fine (100 m) grid (each cell representing one hectare). Mapping the raw data points for LUCS is not helpful as some 223,538 observations underlie the summaries in Tables 9.1 and 9.2. Grid representation forms the basis for the remainder of the analyses in this chapter. Converting the 1991 and 2001 urban areas to hectare grid representations forms the basis for a revised definition of contexts UA2, UX2, XC2 allowing identification of urban consolidation, urban extension, and exurban development, respectively. Tables 9.1 and 9.2 therefore also show alongside estimates made on a point-in-polygon basis (UA1, UX1, etc.) variant estimates such as H(UA2), L(UX2), etc. calculated having converted the physical urban areas to a 100-m grid while still treating the LUCS data as points. The figures marked UA3, UX3, and so on are estimates calculated having converted both the physical urban area and the LUCS data to 100 m grids, with the LUCS measures spread as described above.

As LUCS data points refer to parcels of very different size, some preprocessing is required. While within LUCS the median recorded size of land parcels developed for residential use is 0.1 ha, 3.8% of points refer to parcels greater than 1 ha in extent, which account for 23.1% of units. Obviously, if a LUCS point refers to an area greater than 1 ha, it must overflow its cell. To offset this, areas of development in excess of 1 ha must be locally spread as the grid is created (implemented here using Prolog) although the configuration of the sites is unknown. In the discussion of earlier sections, where the relation between LUCS records and urban areas has been reduced to the point-in-polygon trope, this problem has simply been ignored.

Examination of Tables 9.1 and 9.2 shows that spreading out the LUCS data has only a modest effect on interpretation of how growth is accommodated [compare H(XC2) and H(XC3), for example]. Conversion of the urban area polygons does, however, have an impact on the overall picture. Table 9.1 shows that the proportion of housebuilding accommodated within the urban areas appears to rise from 57% [h(UA1)] to 61% [h(UA2)] as around 60,000 dwellings are reclassified. This stands as a warning of the potential sensitivity of any partitioning of modes of accommodation to the precise placement of the urban area boundary.

Figure 9.3a attempts to show variations in intensity of development at hectare cell level, and demonstrates the impossibility of grasping the pattern without some generalization. This can be achieved by using moving spatial averages, thereby commuting actual housing output in cell q [denoted here H(TO(q,0))] as shown in Figure 9.3a, to a tendency to develop over a particular radius r around cell q. The tendency to develop within 2000 m of cell q [denoted here H(TO(q,2000))] is shown in Figure 9.3b. When represented as a 2-km moving average in this way, the pattern of development in the 1990s becomes immediately obvious. Areas with limited or highly dispersed housing development are shown by the lightest shades (up to 1% of the land area being developed for housing over the period). In tracts with the deepest grays more than 30% of the area was developed for housing in the period.

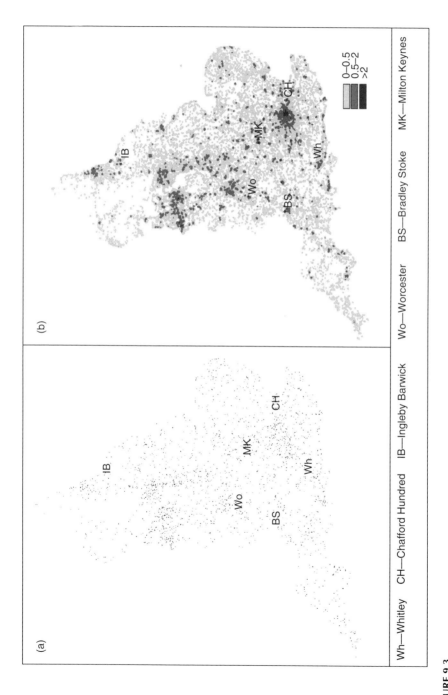

FIGURE 9.3
Intensity of residential development in England (1990–2000) by hectare cell.

Images such as Figure 9.3b provide the basis for a more intuitive grasp of the relative importance of different modes of accommodation. Figure 9.3b makes it a little easier to visualize the contribution of diffuse exurban housebuilding alongside the rather small number of major urban extensions [including Milton Keynes (MK), Bradley Stoke (BS) on Bristol's north fringe, Chafford Hundred (CH) in Kent Thameside, Worcester (Wo), or Ingleby Barwick (IB) near Stockton on Tees]. The new settlement at Whitley (Wh) in Hampshire is also evident, as are the largest areas of urban consolidation, e.g., London Docklands.

Applying spatial averaging at the 2-km scale to the individual accommodation modes (urban consolidation, urban extension, and exurban development) allows visualization of geographic variation in their contribution—see Figure 9.4a–c for H(UA3(2000)), H(UX3(2000)), and H(XC3(2000)), respectively. The minimal contribution of urban extension (Figure 9.4b) around London (more clearly evident in Figure 9.5a), despite the volume of units constructed (Figure 9.2c), must be understood primarily in relation to Green Belt policy. Figure 9.4b illustrates the somewhat larger contribution of urban extension to accommodation of new housebuilding in the Midlands, highlighting in particular the continuing expansion of Telford (a growth pole), and the expansion of Leicester (a city without a Green Belt) while once again suggesting the influence of Green Belt policy in limiting urban extension (around Birmingham, for example). More generally, the use of moving spatial averages calculated over different radii allows patterns to be analyzed and displayed at different scales, demonstrating the overall dispersion of development.

9.5   Using Grids to Explore Structural Effects and Market Relations

In order to understand these patterns, to assess whether they are in any way remarkable and most critically to begin to assess the limits to urban consolidation, a more analytical approach is required. The balance between urban consolidation and development elsewhere will depend in part on the capacity of urban areas to intensify (an issue which has been brought to the fore in policy terms), but also in part on the extent of development at competing sites at the urban fringe, or beyond. Departing a little from the dominant narrative, this section has regarded not only the geographic structure of a locality and planning policy considerations, but also the market for housing land. Actually representing markets within GIS is, however, a significant challenge. A market might (in the spirit of Cournot, 1838) be thought of as a conceptual space in which free communication ensures that identical goods command identical prices, but this begs the question of what is to be treated as identical and what is to be considered (geographically) unique. Moreover, the basic devices of market analysis are

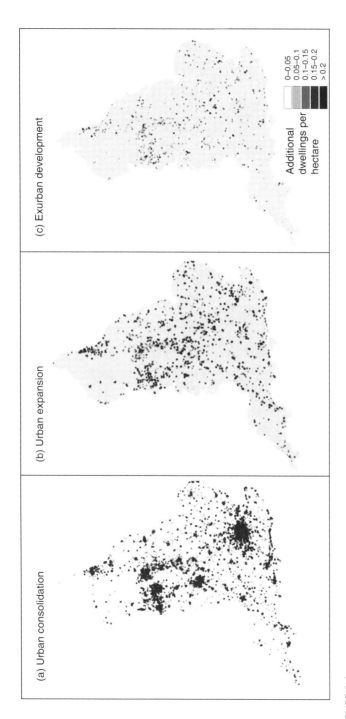

FIGURE 9.4
Components of accommodation, 1990–2000.

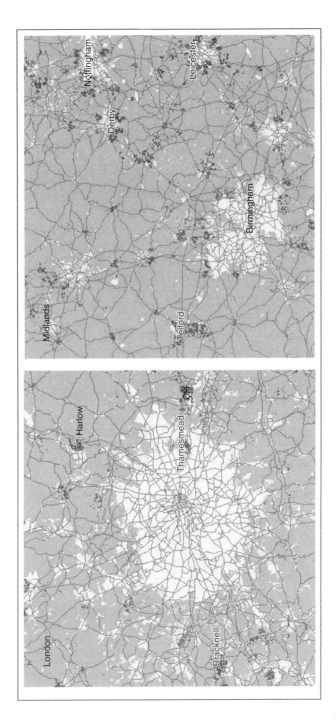

FIGURE 9.5
Dwelling units gained through urban expansion, 1990–2000 in London and the Midlands. (© Crown Copyright/database right 2007. An Ordnance Survey/EDINA supplied service.)

concerned with the realm of *intension*—quantities which actors might wish to provide or purchase at particular prices, whereas GIS are concerned necessarily with *extension;* that is, with dateable, placeable parts of the physical world (Bibby and Chowdry, 2001). So although the current framework of government policy refers to spatially delimited housing markets (ODPM, 2004), the structural metaphors of economics do not always easily fit with those of GIS. Resolving these contradictions would constitute a project of substantial practical and theoretical significance. On the relation between GISc and econometrics generally, see the work of Anselin and his collaborators (Anselin, 2000, 2001; Anselin et al., 2004.). This is beyond the scope of this chapter

To facilitate critical reflection on observed patterns, and to explore the relation between these two styles of thought (and their corresponding structural metaphors), this section pursues a much more modest goal. It explores what might be involved in assessing geographic variation in the degree of urban consolidation that might realistically be expected on the basis both of geographic structure and of market conditions. Founding expectations on evidence is clearly more complex than merely providing evidence of outcomes. The unit of analysis remains the hectare cell. Spatial averaging is used to create variables that generalize various quanta (e.g., housing output and house prices) over a 10-km radius around each cell, so that England is effectively analyzed as 13 million overlapping circles. The analysis appeals to a set of hypothetical circumstances, referred to as the 10-km radial model (10 KRM). Units constructed are assumed to vary in both plot-size and building footprint but to follow standard dwelling types and layouts, incidentally implying the development of residential enclaves of fundamentally similar character. They are assumed to occupy land of homogenous quality, but located alternately within the urban area as at 1991 (UA3) or outside it (RU3) [recall that $H(RUi) = H(UXi) + H(XCi)$]. Under these assumptions housebuilding thus implies the construction of suburbs, albeit that they may be discontinuous suburbs. Except in the presence of supply constraint, prices and quantities for each cell, as averaged over 10 km, are assumed to reflect market equilibrium. Rather than assuming geographically bounded markets a priori, conditions of demand and supply, and hence the position of the demand and supply curves, are assumed to change continuously from cell to cell. Elasticities are assumed to reflect more fundamental behavioral choices, conditioned by broader social values and hence to be constant across all cells. From this perspective, it would be the failure of these relations that would necessitate the identification of local markets.

The equilibria of the 10 KRM refer solely to two hypothetical homogenous goods: housing land and housing space. Actual outcomes will differ because of the varying relationship between the imaginary homogeneous good and units actually traded, and will depend on the specific characteristics of individual parcels and of their immediate environment. Nevertheless, 10 KRM should provide a benchmark identifying variability at the 10-km scale to which shorter wavelength variation might be added.

This model embodies a series of more specific assumptions:

(i)   That markets for housing land reach equilibrium in such a manner that within (overlapping) areas of 10-km radius around each hectare cell demand is equal to supply at the reigning price, except in the presence of supply constraints

(ii) That the demand for additional housing units in each cell stems from the formation of new households, and thus is a function of the distribution of existing households (no allowance being made for the formation of additional households headed by people living further afield)

(iii) That the demand for additional housing units is price and income inelastic, but that the demand for housing space per unit decreases with its price per square meter and increases with income

(iv) That the supply of urban land for housing is a function of the stock of urban land and increases with the price of land

(v)   That the supply of rural land is a function of the stock of land free of planning constraint and also increases with the price of land

(vi) That the demand for housing land is derived by reference to the demand for housing units and the demand for housing space

Following this logic, if supply is not constrained (e.g., by the planning system), the number of housing units constructed within each overlapping circle would be fixed by virtue of assumptions (ii) and (iii). Location within these circles would not, however, and together with the area of land to be developed for housing would depend on the intersection of a demand function based on assumption (vi) and a supply function based on assumptions (iv) and (v). On the assumption of identical character (and price), the equilibrium balance between development on urban and rural sites would be given by their contribution to supply at the relevant point on the overall supply curve (reflecting their different conditions of supply). The following paragraphs work through the assumptions of 10 KRM, attempting at each step to illustrate the relationships that seem to hold and to consider their implications.

Assumption (i) is not directly testable. It provides the logical link between extension and intension, allowing observable housing output across a circle of 10-km radius to indicate the demand for housing units and housing space at a given price—a hypothetical relation that is not directly observable. It also allows housing output to indicate supply in those circles where the flow of housebuilding land is not constrained.* Spatial averaging at the 10-km

---

* It is often assumed that the planning system constrains the residential land supply (or even that the supply of residential land is fixed). While it is true that England's area is roughly static, it does not follow that the flow of land that owners will wish to make available for a given use at a particular time is static. Neither is it self-evident that planning designations actually constrain supply (in this sense) below its free market level. This is a matter to be determined empirically (DoE, 1992).

scale is intended to acknowledge substitutability of sites and hence the possibility of diversion of demand within that radius.*

Assumption (ii), that demand follows the existing distribution of households and dwellings, appears vindicated. The stock of households has been estimated by using the grid references on the postcode address file (PAF; a virtually comprehensive register of postal addresses) to assign each address to its corresponding hectare cell.[†] There is a close relationship between the stock of dwellings within 10 km of a cell [$O(TO(q,10000))$]—the derived structural variable—and the number of additional units completed within 10 km of that cell [$H(TO(q,10000))$]. This accounts for 87.65% of the variability of the volume of new housebuilding. It appears that the number of new dwellings built over the 1990s was typically equivalent to 5.2% of the stock of dwellings in 2001 (implying a 5.47% increase relative to the starting stock).

Regression estimates based simply on geographic structure (ignoring price effects) highlight concentration in and around the major cities (Figure 9.6). The evident strength of this relationship should temper any tendency to posit a contrast between uniform decline in the North and rapid urban expansion of the Southeast. No immediately obvious North–South gradient is apparent in the regression residuals. It is clear, moreover, that demand around some northern cities (e.g., Liverpool, Manchester, and Leeds) is in fact higher than might be expected on the basis of the existing stock of households, though this is not true around others (e.g., Sheffield or Birmingham).

The findings underscore the necessity of having regard to both absolute and relative change in planning policy analysis (although this proves difficult in practice). The relationship displayed in Figure 9.6 underpins the absolute change in numbers of dwelling units. It highlights localities with substantial populations, and if the prime concern is with accommodating growth this should be predominant because the effect of place-to-place variation in the stock of households dwarfs place-to-place variation in rates of growth (departure from the regression parameter estimated as

---

* This device is intended to capture a situation of local spatial competition, but without imposing the housing market. The assumed spatial scale at which equilibrium is reached might be compared with the modal journey to work distance from the 2001 census (10.79 km).
[†] The generally close relationship between the distribution of households indicated by PAF [$O(TO(q,0))$] and that evident from the 2001 census is clear from Figure 9.13. This high level of matching confirms the potential value of PAF in monitoring change in land-use intensity. Figure 9.13b was constructed by converting an ArcView shape file representing Census Output Area boundaries to a 100-m grid with household density being calculated for each Output Area on the basis of its original geometry. Figure 9.13a was produced 2 years before from the 2001 Quarter II PAF. From the grid-analytic perspective of this chapter, it seems appropriate to think of statistical reporting units such as OAs, or units of administrative significance (such as postcode sectors) as averaging underlying data over an irregular area of typical but varying radius. The value of each cell q in Figure 9.13b might be thought of as $O(TO(q,c))$, c denoting the average radius of a Census Output Area.

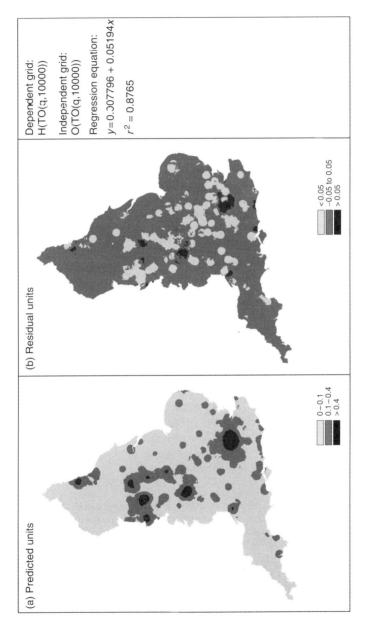

FIGURE 9.6
Additional residential units per hectare, 1990–2000.

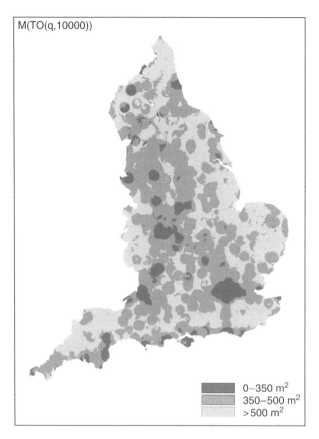

FIGURE 9.7
Inferred plot size (m²).

0.0078 in Figure 9.6). A focus on variation in rates of construction relative to the dwelling stock, by contrast, highlights a tract of land stretching from Devon to Lincolnshire referred to as Hall's Golden Belt (Figure 9.7; see also Bibby and Shepherd, 1991, 1996). It is important not to be misled by high rates of growth, however interesting they may be as an indicator of the leading edge of change. Crucially, areas such as the economically buoyant tract to the west of London, where Hampshire, Surrey, and Berkshire meet, possessed substantial stocks of existing households and experienced high growth rates. This underlies the pattern of positive residuals (Figure 9.6). Moreover, as 10 KRM takes no account of growth originating beyond this distance, other positive residuals inevitably highlight the principal policy-driven growth centers (such as Milton Keynes).

Assumption (iii) posits that overall demand for additional housing units does not vary with the price of units, but that the demand for housing space per unit decreases with its price per square meter and increases with income. This assumption thus redresses the exclusive emphasis on structural conditions in assumption (ii) by introducing price and income effects in a straightforward way.

The amount of inferred space sought per unit in cell q is estimated simply as the reciprocal of the density of new housing implied by LUCS, that is

$$M(TO(q,10000)) = 1/W(TO(q,10000))$$

where

$$W(TO(q,10000)) = H(TO(q,10000))/L(TO(q,10000))$$

while the price of housing space is estimated simply by dividing the appropriate estimate of dwelling price by the appropriate estimate of housing space:

$$P(M(TO(q,10000))) = P(H(TO(q,10000)))/M(TO(q,10000))$$

Perhaps the emphasis within policy circles on levering up densities,* and hence depressing space per unit, accounts to some degree for the very limited discussion of the (quite startling) geographic variations in the housing space measure, $M(TO(q,10000))$ as evident in Figure 9.7. Through the 1990s there was not only a sharp contrast between the density of new development in London and densities in the provinces, but also a very clear distinction between what might be termed the generalized urban realm (see Section 9.5 below), and elsewhere. This reflects and reinforces the conclusions drawn in relation to Table 9.1. Most obviously, the largest (inferred) plots were found in sparsely populated or remote areas, as defined below, such as North Devon, the Welsh Marches, north Northumberland, much of Lincolnshire, and parts of East Anglia, where densities achieved were only a quarter of those found in London. It also appears that densities may be typically lower in Hall's Golden Belt.

Estimation of price effects must rest on estimates of the price of all dwellings, as estimates of average prices for new dwellings are not distinguished in the published data. The available information derives from Her Majesty's Land Registry, which releases average transaction value by property type for postcode sectors (areas with typically 2655 households) on a quarterly basis. The overall average for each sector was initially assigned to all occupied hectare cells within it to estimate $P(O(TO(q,s)))$, where s refers to the average radius of a postcode sector. These values were then averaged over 10 km to estimate $[P(O(TO(q,10000)))]$ for each cell q.

The income measure is based on modelled estimates of average household income for electoral wards for 1998–1999 due to the Office of National Statistics (ONS, 2004). Analogously with house price data, the estimated values were initially assigned to all occupied hectare cells within the respective wards, to estimate $y(q,w)$, where w refers to the average radius

---

* Since the mid 1990s, U.K. policy has sought to increase density, depressing space per unit. This has been pursued particularly vigorously since publication of revised PPG3 in 2000 and Prescott's Birmingham statement in October 2002.

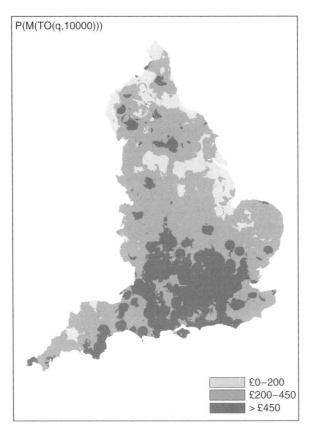

P(M(TO(q,10000)))

£0–200
£200–450
> £450

FIGURE 9.8
Price of housing space (£/m²).

of a ward. These values were then averaged over 10-km to estimate average household income for an area centered on each hectare cell [y(q,10000)].

Regression evidence provides some justification for assumption (iii). Under certain conditions, the observed relation between the amount of housing space and price must be interpretable as a demand function. This would seem to be the case where (a) the typical sizes of (inferred) plots sought by housebuilders and householders both fall as the price of land rises, and (b) the supply of land for residential development increases with its price per hectare, but does not respond directly to the number of units subsequently developed. On the basis that such a housing space demand function will take a loglinear form

$$\ln(D(M(TO(q,10000)))) = \alpha + \eta(\ln(P(M(TO(q,10000))))) + \beta(\ln(y(q,10000))) + \varepsilon$$

and estimating the price of housing space in each cell by reference to the price of semidetached property we obtain

$$\ln(D(M(TO(q,10000)))) = 5.1329 - 0.4538(\ln(P(M(TO(q,10000)))))$$
$$+0.5690(\ln(y(q,10000))) + \varepsilon \tag{9.1}$$

The estimate of the price-elasticity of the demand for housing space derived using 10 KRM is broadly consistent with those obtained using conventional approaches (e.g., Ermisch et al., 1996). The modelled demand for space is illustrated in Figure 9.9. Estimation of price-elasticity might be considered as a first step in assessing possible limits to intensification. The loglinear form of Equation (9.1) implies that elasticity is constant, i.e., that there is a constant relationship between a percentage change in the price of space and percentage change in the quantity of space sought. Thus the same absolute increase in price where price is very high (e.g., in central London) would imply a much smaller reduction in the quantity of space sought than it would where an average price prevailed. Consider a household with average income contemplating purchasing 333 m^2 (i.e., purchasing a dwelling in a development constructed at 30 units to the hectare; the lowest density now considered in PPG3 as commensurate with the efficient use of land), but facing the prospect of paying an extra £1000/m^2. On the basis of Equation (9.1), it would review its plans so as to seek only 203 m^2, i.e., a plot in a development constructed at 49.25 units to the hectare. If, on the other hand, that household anticipated densities of 50 units to the hectare, the upper guide to the efficient use of land, it should expect to have to pay £1560/m^2 and the prospect of paying an extra £1000/m^2 would have a less marked effect on its plans. If Equation (9.1) held, it would prompt a reduction in space sought to 160 m^2 increasing equilibrium densities, but only to 62.6 dwellings to the hectare. It might seem appropriate to consider on the basis of Equation (9.1), whether there is a point at which increases in price imply a negligible decrease in space sought (in absolute terms), thereby defining an effective maximum density and hence a limit to densification. It is perhaps surprising that estimates of price-elasticity based on 10 KRM and other approaches are similar given the degree of extrapolation involved, since the estimates of housing space in 10 KRM rest on whole houses recorded in LUCS and do not include measures of size for subdivided units which are evidently required to achieve the highest densities (in central London, for example). Moreover, one might expect spatial averaging at this scale to further reduce the possibility of estimating elasticities compatible with those based on individual survey data.

When due account is taken of variation in the price of housing space and of average incomes, expectations of the likely pattern of change in urban form and in urban consolidation differ somewhat from those based on structure alone. The principal effects are clear. The high price of housing space in London and along corridors stretching southwards to Brighton on the coast and westwards to Reading appears to have choked off demand. Income and price effects together combine to reduce expected density and increase anticipated landtake in Oxfordshire, Berkshire, and Buckinghamshire to the west of London, where pressure from large household numbers

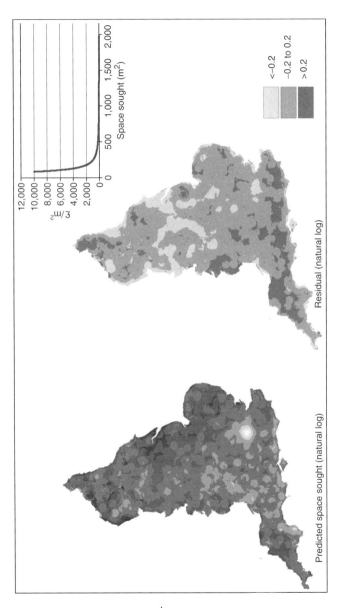

FIGURE 9.9
Demand for housing space in England, 1990–2000.

is compounded by high incomes. Nevertheless, the low price of space along the M62 corridor in Northern England (stretching from Liverpool, through Manchester and Leeds to Hull) does not appear to have encouraged increased consumption of housing space.

Equation (9.1), however, accounts for half the variance of the demand for housing space ($R^2 = 0.499856$). Thus, although the price of housing space [P(M(TO(q,10000)))], illustrated in Figure 9.8, and the level of household income exert influence on variations in density, they fail to account for them entirely. Even having taken account of observable variations both in the price of housing space and in incomes, it appears that over the 1990s there remained a continuing tendency for housing to be constructed at remarkably low densities in the rural domain. In analytic terms this brings into question the assumption of a single market for a homogenous good, raising the possibility that housing space in the countryside should be treated as a distinctly different commodity from space at the urban fringe (continuing to set aside the possibility that it might be necessary to define a set of geographically limited markets with entirely distinct demand and supply functions). In policy terms it raises the question of whether this exurban development prejudiced and might conceivably continue to prejudice the urban emphasis and the stress on intensification.

With assumption (iv), concern shifts from the demand for housing space to the supply of housebuilding land. It posits firstly a structural effect that the supply of urban housebuilding land depends upon the stock of urban land within 10 km. Second it posits a price effect. Figure 9.10 clearly confirms the structural effect. There is a very close relationship at the 10-km scale between the extent of urban land in 1991 [i.e., G(UA3(q,10000))] and the area of urban land developed for housing in the 1990s [S(L(UA3 (q,10000)))], accounting for 91.3% of the variability of the latter (Figure 9.10). Generally, therefore, it appears that the extent to which housebuilding has been accommodated on urban sites has varied in accordance with the extent of urban land. Across England, the flow of housebuilding land coming forward within urban areas was around 2.3 ha/km^2 of urban land over the period, or 0.2 ha/km^2 per annum.

Although the flow of urban land for housebuilding responds primarily to geographic structure, assumption (iv) also allows that the rate at which urban land enters the residential land supply should be expected to increase as the price of housing land increases.* Any such tendency is important as it potentially increases the scope for urban residential intensification (other things being equal). Exploring the validity of this assumption is difficult for two reasons. The first is the paucity of available housing land price data. The most detailed information available is due to the Valuation Office Agency, and relates to their local office areas. These data are very thin, however, and

---

* It is often assumed that supply of residential land is fixed. While it is true that England's area is roughly static, it does not follow that the flow of land that owners will wish to make available for a given use at a particular time is static.

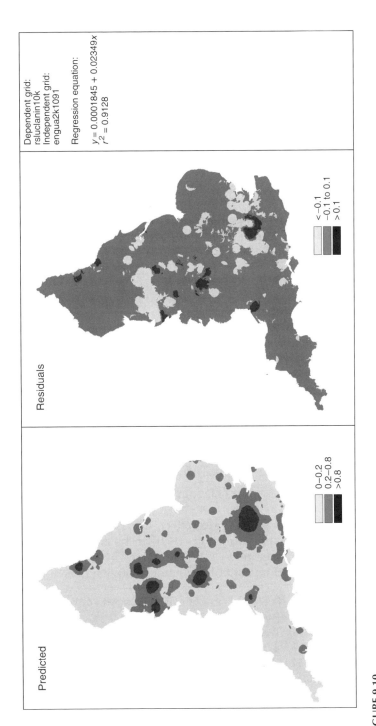

Dependent grid:
rsluclanin10k
Independent grid:
engua2k1091

Regression equation:

$y = 0.0001845 + 0.02349x$
$r^2 = 0.9128$

Residuals

< −0.1
−0.1 to 0.1
> 0.1

Predicted

0–0.2
0.2–0.8
> 0.8

FIGURE 9.10
Urban residential land supply, 1999–2000 (ha/km²).

show very little systematic variation apart from a distinction between Central London and elsewhere. For present purposes, two further estimates of residential land price consistent with 10 KRM were prepared, both using the principle of residual-land valuation. Residual valuation entails estimating land price [P(L(TO(q,10000)))] as the difference between the completed value of property developed and the total cost of its construction, that is

$$P(L(TO(q,10000))) = (P(O(TO(q,10000))) - C(H(TO(q,10000))))$$
$$\times H(TO(q,10000))$$

For present purposes, construction cost C(H(TO(q,10000))) has been assumed to be spatially invariant [i.e., simply C(H)], for a particular property type.* Two variant estimates of P(L(TO(q,10000))) were formed; the first with values based on overall average house prices [P(O(TO(q,10000)))], and the second with values based solely on prices for semidetached houses (Figure 9.11).

While regression provides some evidence of a land price effect, a satisfactory estimate of elasticity cannot be calculated. This is because of the second difficulty referred to above and one that is rarely discussed in the GIS literature. One might posit a relationship first to physical structure (i.e., the stock of urban land [G(UA3)]) and second to the price of housebuilding land of the form

$$\ln(X(L(UA3(q,10000)))) = \gamma + \phi(\ln(G(UA3(q,10000))))$$
$$+ \eta(\ln(P(L(TO(q,10000))))) + \varepsilon \qquad (9.2)$$

This could be estimated as

$$\ln(X(L(UA3(q,10000)))) = 1.0140 + 0.9118 \times \ln(G(UA3(q,10000)))$$
$$+ 0.1332 \times \ln(P(L(TO(q,10000)))) + \varepsilon \qquad (9.3)$$

Despite the positive sign on the price effect, the empirical Equation (9.3) cannot be treated as a supply function for urban housebuilding land [S(L(UA3(q,10000)))]. Any attempt to estimate a land price effect faces the problem that the demand and supply equations both have the form

$$\ln(X(L(UA3(q,10000)))) = a + b(\ln(P(L(TO(q,10000))))) + \varepsilon$$

and so any empirically derived relationship, such as Equation (9.3), is a weighted combination of the (unknown) demand (negative) and supply (positive) functions. This is an instance of the classic identification problem in economics. The parameter η underestimates the price-elasticity of supply.

---

* Costs have been estimated by floor areas derived from BCIS (2003) and construction costs per square meter from Surveyors Plus (2000).

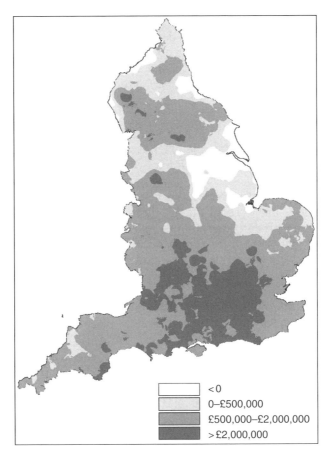

FIGURE 9.11
Modelled residual land value (£/ha).

It can, however, be concluded that the structural component dominates determination of the urban residential land, that place to place variation in the price of housing land is modest at the 10-km scale, and that there is a price effect levering in urban land for residential development.

The conditions of supply of land for housebuilding are utterly different outside the urban areas. Assumption (v) posited that the supply of rural land for housebuilding would be a function firstly of the extent of rural land (insofar as it is free of planning constraint) and secondly of the price of housebuilding land. Search for a structural effect without any reference to planning constraint demonstrates that there is virtually no relation between the extent of rural land and housing outturn on such land.

Of course, it might be argued that it is the extent of rural land that is accessible to concentrations of population and economic activity, which is pertinent to the supply of land for residential development rather than

the total extent of rural land, including remote moorland, for example.*
Here accessible rural land will be defined as that part of the generalized
urban realm (GU) outside the urban area, and context GU will be defined as
that set of hectare cells where the density of dwellings within 10 km exceeds
1 per hectare. This allows the specification of a supply function

$$S(RU3(q,10000)) = f(G(GU(q,10000)) - G(UA3(q,10000)))$$

but it is equally clear that any such structural effect is also negligible. Within
the generalized urban realm, the extent of accessible rural land accounts
for only 0.1% of variation in the flow of rural land developed for housing
(see the Idrisi scatter plot included in Figure 9.12a). Generally, the rate at
which housing land is supplied is very small relative to the extent of rural
land (0.3 ha/km^2 per decade) though the right-hand side of the plot illustrates the range of outcomes found in areas which are almost entirely
undeveloped. Large stocks of rural land may produce small supply flows,
but small stocks cannot produce large flows.

Such evidence might be differently construed within different narratives.
From an economistic perspective, this finding is far from surprising as the
flow of land to alternative uses should reflect opportunity costs rather than
the extent of stocks of land, albeit that place-to-place variation in the price of
development land seems relatively modest. Land-use planners, by contrast,
might hold that this results simply from failure to take account of the extent
of land actually allocated for housing in development plans. Although it is
not practically possible to establish the explicit pattern of residential land
allocations for England as a whole, it is possible to approximate the extent of
Green Belt designation within the 10 KRM framework and hence proxy
planning constraint. Green Belt policy, as set out in Planning Policy Guidance Note 2, is the most stringent form of constraint policy and the most
consistent nationally. It imposes a strong presumption against virtually all
built development. To test the land-use planners' perspective, one might
posit a very similar relation to that above:

$$S(L(RU3(q,10000))) = f(G(GU(q,10000)) - G(UA3(q,10000)) - G(GB(q,10000)))$$

using GB to designate a Green Belt context.[†] Once again, however, there is
no tractable relation between the stock of land (free of Green Belt constraint)
and the flow of land developed for housing (at the 10-km scale).

Articulated within the 10-KRM framework, LUCS data begin to provide
evidence of the nature of the effect of Green Belt constraint. Reflection on
Figure 9.12 serves as a starting point for an understanding of its relation to
urban consolidation. Figure 9.12a shows how the scale of the flow of rural
land being developed for housing in any 10-km circle varies with the

---

* Alternatively one might think that all land is of the same class but that demand for inaccessible land will be low.
[†] Data available for this study showing the limits of Green Belt relate to the late 1990s and are of
varying quality.

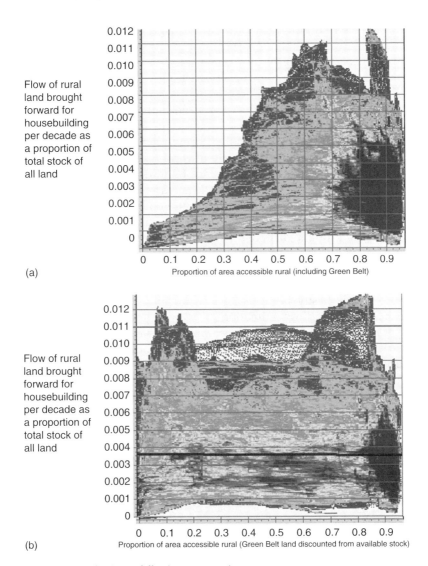

FIGURE 9.12 (See color insert following page 274.)
Relationship between the stock of accessible rural land and the flow of rural land for housebuilding.

corresponding stock of rural land. Figure 9.12b shows a similar relation but discounts Green Belt land. If the area within 10 km of a particular hectare cell were 20% urban and 50% Green Belt, the stock of potential development land would be treated as 0.8 (100%–20%) in Figure 9.12a, but 0.3 (100%–20%–50%) in Figure 9.12b. Towards the left of Figure 9.12b, a substantial scatter of points indicate circumstances where relatively high flows of housing land are associated with low stocks of unconstrained rural land.

There is no corresponding scatter to the left-hand side of Figure 9.12a. These circumstances cannot therefore be attributable to the extent of current urbanization or to the absence of accessible land. They must therefore be understood in relation to Green Belt constraint. They relate to cells at the center of 10-km circles which both encompassed substantial areas of land with Green Belt protection and experienced high rates of development on rural land, indicating the tendency, at this scale, for Green Belt designation to be associated with the diversion of development to unconstrained sites.

It therefore appears that the supply of rural land for housebuilding may more readily be understood by reference to an economistic narrative than that favored by planners. An economistic narrative might suggest moreover that constraint would imply upward pressure on land prices, thereby sending a signal for the release of further land for housebuilding. From such a perspective, one might dispense with attempts to examine structure and concentrate on responsiveness to price signals, simply positing that the supply of rural land would increase with price. It would be desirable to test whether such price effects are found, though this proves difficult in practice. This is firstly because the identification problem recurs with a vengeance, confounding estimation of supply and demand effects. Secondly, it is clearly possible that Green Belt policy constrains supply so that the market is not in equilibrium, and hence for some cells, observed price–quantity combinations fall on the demand curve but not on the supply curve. For all constrained cells, therefore, one can only be clear that the observed price–quantity combination will fall on the demand curve. Supply might potentially be constrained by restrictive land allocation, but cannot be increased beyond its market level. These estimation problems are only compounded by the relatively weak nature of the Green Belt data available for this study and the need to rely on estimates of residual values in the absence of residential land values. It therefore seems unwise to attempt to use the 10 KRM framework to attempt to unravel Green Belt price and quantity effects or to model the impact of the rural residential land supply on intensification.

While it is therefore not possible to address assumption (vi) satisfactorily, the attempt suggests some important conclusions. The analysis of demand for housing and housing space draws attention to the importance of existing concentrations of population, to the modifying effects of variations in income and to the price housing space, but ultimately highlights remarkably low densities in the rural domain. The analysis of supply, in the very gross terms that the 10 KRM permits, indicates first, that the flow of land entering the urban residential land supply is very much in step with the underlying urban structure, albeit modified by price effects. This obvious relationship is overlooked whenever the sourcing of housing land is treated simply as a matter of local policy choice (for example DoE, 1996 which takes no account of the structural limits to urban consolidation in areas with no substantial urbanization, e.g., Lincolnshire). Understanding the balance between the urban and rural components of the housing land supply is, however, far more difficult. The rural component does not respond to structure, and surely must be understood from an economistic perspective, though the paucity of reliable data with respect to prices and policy constraint militates against quantitative estimation of effects.

The foregoing analysis does not (and cannot) say anything about how the urban residential land supply at the 10-km scale comes to be in step with structure, thereby failing to address the search for sites at this scale—a major preoccupation of planning practitioners. The analysis does, however, point to two circumstances where social capacity to concentrate new dwellings in urban locations may be prejudiced or limited. The first of these might perhaps be thought of as involving an intensive margin (a point at which densification cannot be taken any further—an issue approached here through consideration of the price-elasticity of demand for housing space). Consideration of such a limit clearly plays a part in any analysis of the intensification agenda. Where demand for space is high, adaptive subdivision of buildings may form a particularly important component of the supply of dwellings. Monitoring must, therefore, concern itself with the utilization of existing buildings. The evidence of LUCS, however, deals (by virtue of its origin as a by-product of OS map revision) almost entirely with whole buildings, abstracting from subdivisions invisible on large-scale maps. Analysis of subdivision demands a different approach, and one possibility is considered further in Section 9.7.

The second circumstance might be thought of as an extensive margin beyond which development might be dispersed through the countryside, thereby threatening the achievement of policy goals. While this circumstance hardly seems to figure in discussions of urban intensification, much of the variation in the proportion of development secured on urban sites must be understood by reference to conditions beyond the urban area. In terms of the categories introduced above, development outside the urban areas takes place in two contexts: urban extension and exurban development. Urban extension has a very particular geography, subject to local policy allocating land for residential development as shown in Figure 9.4b and c. Although the flow of redevelopment opportunities on urban land accounts for the extent of urban consolidation [$H(UA3(q,10000))$], the absence of urban land does not necessitate exurban development [$H(XC3(q,10000))$]. In fact, use of regression demonstrates that there is a tendency for the proportion of development in exurban contexts (2000 exurban proportion) to increase as the endowment of urban land (10,000 urban proportion) increases. This component of development is overlooked (and indeed seems to be systematically ignored in the standard narrative) and while the framework introduced draws attention to the potential significance of exurban development, its essential character remains illusive. There remains a substantial problem in characterizing scattered development. Our traditional representational devices do not help; what constitutes a scatter depends on the granularity of the mesh on which it is represented; moreover within English the necessary distinctions do not seem to have been lexicalized.

Understanding the limits of possibilities for the intensification agenda depends on understanding more about the possibility for packing in the cities; and about patterns of development beyond (conscious, of course, of the significance of price effects in both cases). Case studies might help to provide an answer—focusing in the spirit of urban morphologists such as Whitehand (1992) on the subdivision of plots and of buildings in the cities, and possibly applying similar techniques to rural contexts, though there would remain problems of generalization. The remainder of this chapter attempts simply to identify and organize data which might provide leverage on change at these two margins, outlining an account which might be enriched by survey work, and providing a basis for constructing a sampling frame for case studies. While the analysis of subsequent sections is rudimentary, there is an element of innovation in the attempts to exploit existing address list data, and to move beyond cartographic tropes which at the microscale treat the building as an atom (failing to account for its further division into units of occupation) and at the broader scale treat cities as polygons.

9.6 Within the Urban Areas: Intensification of Units of Occupation 1998—Reconstructing a Grid Using PAF

To address the question of these limits on the effectiveness of the intensification agenda, it would ideally be desirable to consider all the ways in which additional households could be accommodated. This would entail consideration not only of the flow of newly constructed dwellings, but also of the flow of conversions, together with the assumed underlying patterns of demand and supply. This section approaches this problem in a manner which is in principle extremely simple. It involves constructing hectare grids from the individual postal addresses held on PAF to represent the density of units of occupation $[O(TO(q))]$ at different points in time (each conceived of as either a household or a dwelling or both). These grids should be expected to resemble both the distribution of households, as recorded in the 2001 census, and the urban areas defined by OS. The similarity between a grid based on PAF and one based on households recorded in the 2001 census is illustrated in Figure 9.13. Some of the changes recorded between successive versions of PAF should be expected to reflect housebuilding evident in LUCS, while other changes should capture aspects of intensification which cannot be assessed by reference to OS maps. Potentially comparison of successive grid representations of plot density $[C(TO(q))]$ and density of units of occupation holds the key to this, with NLP forming the basis for a detailed appreciation of the nature of change.

Given the amount of human resource devoted to delivery of the mail and given the imperative of maintaining PAF to satisfy business needs, it would seem reasonable to expect PAF to respond very rapidly to changes in the

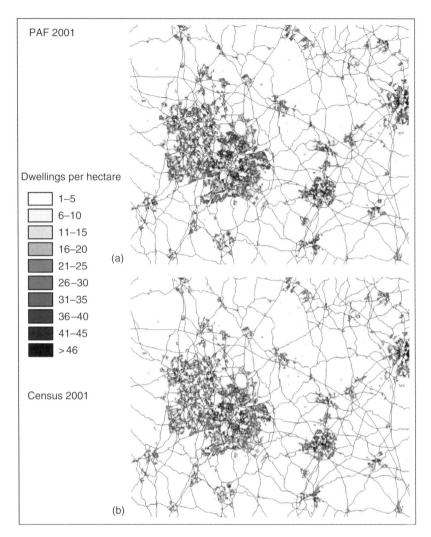

FIGURE 9.13
Distribution of households: comparison of census and the postcode address file. (© Crown Copyright/database right 2007. An Ordnance Survey/EDINA supplied service.)

distribution of units of occupation, suggesting the potential for this sort of approach. This section considers its feasibility further by comparing the pattern of property occupation for Quarter II 1998 with that for Quarter II 2001 (the time of the population census).* The practical difficulties of such a

---

* Obviously a longer run would be desirable, allowing comparison with LUCS. Nevertheless, 1998 is a convenient base year because it is the base-year used by U.K. government in the analysis of area-based social exclusion policy, and coincidentally it is the year selected by for the Countryside Agency's *Countryside Quality Counts* project (Countryside Agency, 2004).

conceptually simple task involve abstracting from all those sources of change that are not attributable to change in the underlying configuration of property or its occupation. For the period in question, this is not a trivial task. Initial attempts to compare grids for 1998 and 2001 produced quite implausible patterns of change, which cursory investigation showed were at least in part attributable to the apparent movement of unit postcodes induced by wholesale correction of grid references, as part of an initiative involving Royal Mail and OS.*

It is clearly necessary to abstract from these apparent changes. Where unit postcodes remained unchanged between 1998 and 2001, the more accurate grid references from 2001 were used to assign property to a grid for 1998. On this basis 1,206,815 unit postcodes were assigned, accounting for 21,605,582 properties. The extent of the improvements to spatial referencing was such that on average each property was displaced more than 208.49 m from the point it might have been considered to occupy on the 1998 version of PAF.

Complexities, however, arise where properties extant in 1998 were assigned new postcodes by Royal Mail during the period. New postcodes may be assigned in a variety of circumstances (aside from when property is newly built). There may be wholesale transference [in the case of Wirral, this involved re-post coding from L*n mkl* to CH*n mkl* (e.g., L45 6AB to CH45 6AB)]. In growing areas, Royal Mail may subdivide postcode sectors, producing postcodes that may be far from simple transformations of their predecessors. In all such cases, assignment to the 1998 grid was based on coordinate information on PAF for 2001, but this has entailed finding the appropriate 2001 postcode for each unit. The solution, which is extremely simple in principle, involves matching property addresses, identifying for each address the corresponding addresses in each unit postcode, and attempting to match them with the same addresses in 1998. The scale, however, is substantial: some 86,523 unit postcodes used to denote property within the study area in 1998 QII were not found in the 2001 QII. Deletion of postal descriptors without underlying change in buildings or property utilization implies a commensurate need to create new postcodes. Indeed, more than 114,000 new unit postcodes were introduced over the period. Matching involves structuring the addresses (and dealing with changes of abbreviation and of spelling and word-division). This was achieved by deploying a particular structure to represent the part–whole structure of properties and their addresses.

---

* The quality of grid referencing in the PAF was discussed in a classic paper by Gattrell (1989); see also Raper et al. (1992). During the period between the two cuts discussed here, the GRID-LINK initiative was introduced. Prior to this Royal Mail, who maintained PAF-stored coordinates for postcode centroids, from those held by Ordnance Survey. GRIDLINK established a process whereby the partners (Office of National Statistics, General Register Office Scotland, Royal Mail, Ordnance Survey of Northern Ireland, and Ordnance Survey) all used the same centroids leading to a marked increase in the positional accuracy of PAF references.

Although construction of the 1998 grid proves complex, subsequent identification of changes in the density of units of occupation is a trivial matter. Positive changes (i.e., increases in density of units of occupation between 1998 and 2001) result from the construction of new property or from the conversion of existing stock. Negative changes result from demolition or from losses attendant on conversion (as adjoining properties are converted into single units). Overall, it appears that the net increase over the period was 252,768 dwellings or 1.13% of the starting stock (i.e., 0.38% per annum). The annual net rate is less than the rate of new housebuilding per annum over the 1990s; but it takes account of demolitions and of the balance of both gains and losses due to conversions to and from residential property.

It is equally important, however, to appreciate the degree of place-to-place variation in the extent of net change and in the rate of intensification. Neighborhood functions can be used to capture the extent of variation at different scales. At the 10-km scale (consistent with the discussion of 10 KRM), stark contrasts emerge between different cities: net change reaches a peak in London where it is achieved by intensification rather than expansion. Net change appears negative by contrast in Birmingham—England's second city and the hub of the West Midlands conurbation, as it does in Tyneside (most acutely), Sheffield, and Liverpool. At this same scale it is notable that the three core cities of the East Midlands—Derby, Nottingham, and Leicester—all experienced a net increase in units of occupation. Far gentler smoothing—at the 800 m scale—reveals the very different patterns of accommodation in different cities. The centers of certain cities have been transformed and residential use has intensified. As indicated in Figure 9.14, these effects are very far from universal. There are, for example, very clear contrasts between central Nottingham which shows substantial residential intensification; and central Derby which shows hardly any (net gain evident at the 10-km scale being achieved by adding units of occupation nearer the margin). More generally, Figure 9.14 demonstrates the feasibility of exploring patterns of variation in cities, revealing for example the relatively complex pattern in the West Midlands conurbation with intensification of some of Birmingham's southern inner suburbs for example.

9.7 Within the Urban Areas: Intensification of Utilization of Existing Property

The preceding section has considered change in units of occupation, informally distinguishing those gains in fully developed areas achieved by increasing densities and those beyond achieved through expansion. While this allows some of the limitations imposed by the cartographic foundation of LUCS to be overcome, it does not allow explicit consideration of intensification of use of the existing building stock (e.g., through conversion). This section considers property utilization in more detail. Adaptive subdivision

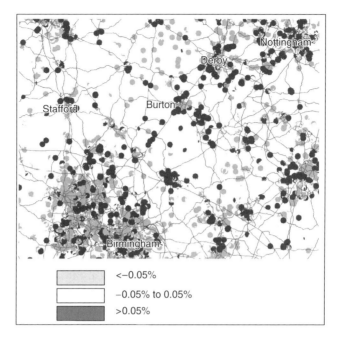

FIGURE 9.14
Rate of growth in stock of units occupation (1998–2001) in the English Midlands (800 m moving average). (© Crown Copyright/database right 2007. An Ordnance Survey/EDINA supplied service.)

of existing buildings embraces activities involving substantial capital expenditure (such as warehouse conversion), through to multiple occupation of houses, with a broad range of adjustments in between. In any investigation of the extent of property utilization, making a distinction between plot density and dwelling density and between buildings and units of occupation is critical in monitoring recent policy.

The disjunction between buildings (or plots) and units of occupation can be explored using NLP. Rather than observing physical objects such as houses, or inferring their existence from a digital map or LUCS records, the concern of this chapter lies in inferring the existence of social objects (units of occupation) from address data. This involves conceptualizing the nature of the objects that correspond to the words. Addresses do not directly reflect building form. Natural language descriptions instead relate directly to social objects (e.g., the Brown's home, Gladstone Garment Works, or Warden's Flat). Moreover, natural language accommodates changes in sets of social objects very flexibly (with terms such as new flat, Flat A, or the caravan). Property addresses might be thought of as referring to units of occupation with unknown physical limits. To the extent that it is possible to infer two sets of objects separately—plots $[C(TO(q))]$ on the one hand and units of occupation $[O(TO(q))]$ on the other, it becomes possible to assess intensity of use and the extent to which property subdivision contributes to intensification.

The work discussed in this section is complex and approaches the identification of units of occupation by developing a Prolog parser specifying the various levels of convention embodied in natural language addresses appearing in PAF. Only the sketchiest account can be provided here. Conventions specific to PAF overlie conventions applicable to addresses generally (for example, the break of geographic scale typically associated with a change of address line). These are in turn superimposed upon more general conventions of natural language (such as the nature of prepositional phrases, e.g., *rear of 42* or *over 15a*, and of noun phrases themselves, e.g., *Landslow Green Farm Cottage*—a cottage not a farm or a green). To serve their purpose, however, the rules which form the parser must also articulate ontology of plots and properties. The term property is here used to denote a whole—a conceptual structure comprising a plot, which may include one or more buildings and will include one or more units of occupation.

Addresses thus provide clues to the existence and character of the unit of occupation and of the property. Although little can usually be discerned about the building, street numbering will allow inferences to be drawn about plots (following the logic of the Conzenian tradition within urban morphology; see Whitehand, 1992), while the presence of terms such as flat, allows generation of a part–whole structure for each property. Property objects, therefore, are represented as a structure of parts bound together by relational processes and referenced by one or more natural language addresses from which their existence is inferred. The spatial arrangement of the parts, however, is not known, and so cannot be directly represented in GIS. Nevertheless, the point reference associated with each inferred plot may be used to assign it to a hectare cell. Thus on the basis of NLP, grids can be created representing the density of inferred plots and the density of inferred parts.

By considering natural language alone it thus proves possible to proxy the utilization of plots, allowing mapping and further analysis of the inferred property objects. This allows representation of variations in the intensity of property use underlying sharp distinction between dwelling densities in London and those elsewhere. Examination of conditions within London indicates that this arises out of the shifted relation between the density of plots [$C(TO(q))$] and the density of units of occupation [$O(TO(q))$], allowing the size of implied plots to fall below the minimum evidenced by LUCS. Figure 9.15 points up the contrast between the very modest variation in the density of inferred plots (presumably a reflection of original development densities), and the startling variation in the density of units of occupation, with a dense collar of subdivided property encircling the city's core. The pattern of property subdivision underlying dwelling densities in London should also be contrasted with much less intense subdivision in other cities such as Manchester, Liverpool, or Stoke-on-Trent (Figure 9.16), suggesting, of course, a clear relation to the variations in price of housing land and of housing space evident in Figure 9.8.

The methods introduced here allow an understanding of the intensity of development at a particular time and the manner in which it is achieved.

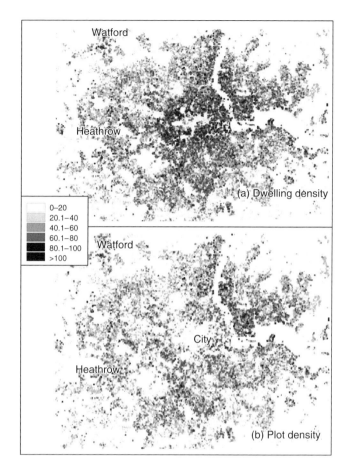

FIGURE 9.15
Density of occupation in Greater London in 1999 (per hectare).

They also allow analysis of the nature of change over time, decomposing the change evident in Section 9.6 into that due to increasing or decreasing subdivision. It is in fact possible to take this further and to analyze both the nature of the mix of uses and the pattern of change in uses associated with intensification and change—a use of NLP currently under development.

9.8 Constructing a Fine-Grained Settlement Geography to Identify Development Contexts

The second issue raised at the end of Section 9.5—the nature and extent of exurban development—is of a rather different character, demanding tools for description or visualization. To take this further involves attempting to describe very specific patterns in very general terms, but characterizing

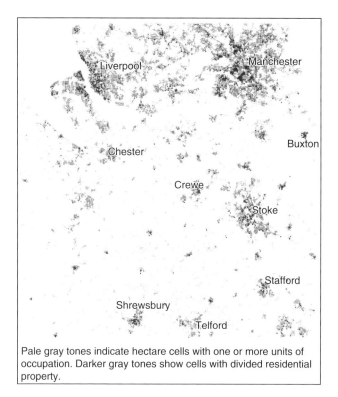

Pale gray tones indicate hectare cells with one or more units of occupation. Darker gray tones show cells with divided residential property.

FIGURE 9.16
Geographic variation in the incidence of subdivided residential property in the North Midlands in 2000.

dispersal meaningfully either statistically or in natural language proves difficult. In this section, this issue is approached by developing a finely differentiated representation of the settlement pattern summarizing both variations in density and settlement morphology and attempting to grasp the different sorts of geometric relation (e.g., in, adjoining, and near) to different types of settlement (villages, hamlets, towns, etc.). This allows a specification of the type of context in which each house comes to be built and hence assessment of the proportion of new development in particular development contexts such as town centers, small urban settlements, villages and their envelopes, and dispersed rural settlement in the sense traditionally used by geographers.

Grids based on the density of units of occupation evident within PAF can be used to characterize the settlement pattern, categorizing the contexts in which development occurs and monitoring change in the extent of the built-up area. Comparison of grids over time allows not only analysis of intensification, but also of urban expansion and of the relationship between the two. The key to such assessment is measuring density at different scales.

Density is a ratio, and the use of densities in geography tends to be troublesome, as researchers typically tend to focus on the (population) numerators to the exclusion of areal denominators. By calculating a density profile, however, that is to say by calculating a series of density measures with constant numerators but different denominators, a useful step can be taken towards measuring settlement pattern. To illustrate the approach, imagine that 32 dwellings stand within an individual hectare cell. This might either constitute an entire freestanding settlement—a small village—or be part of a larger one. In the former case, if density were recalculated over an area of four hectares (200×200 m) the estimate would fall to eight dwellings to the hectare, and recalculation over an area of 16 hectares (400×400 m) would imply a fall to two dwellings to the hectare (on the assumption that there are no further dwellings outside the village). More generally, if the properties form a freestanding village in open countryside, as the area over which density is calculated increases, the measured density will fall systematically. If the same 32 dwellings occupy a hectare cell near the center of a city, of course, density would not fall as the denominator increases in the same manner.

Different types of settlement thus have different density profiles and this provides the basis for a rule-based classification of settlement morphology. To construct profiles, it is convenient to consider density being measured over a series of radii (200, 400, 800, and 1600 m) around each cell (Figure 9.17). This is achieved by applying neighborhood functions to the one-hectare household grid. By calculating density at the 800 m scale and identifying cells with more than eight dwellings to the hectare, it is possible to approximate the urban areas defined by OS, and this forms the basis for Figure 9.18. A village, on the other hand, may be defined as having a profile with:

- A density of greater than 0.18 dwellings per hectare at the 800 m scale
- A density at least double that at the 400 m scale
- A density at the 200 m scale at least 1.5 times the density at the 400 m scale

To illustrate this, consider the example of Great Rissington in Gloucestershire, a village of about 360 dwellings. At the 800 m scale the typical density for a hectare cell in the village is 0.73 dwellings to the hectare; at the 400 m scale the corresponding density is 2.94 and at the 200 m scale it is 11.08 (Figure 9.18). Different profiles may be used to identify individual farmsteads, hamlets, villages, towns, and urban areas, together with settlement fringes, village envelopes, and peri-urban areas.* The classification of

---

* Such an approach has been used by the author to classify settlements in England and Wales for a consortium of government agencies including the Countryside Agency, the Department of Environment, Food and Rural Affairs (DEFRA), ODPM, the Office of National Statistics (ONS), and the Welsh Assembly Government (Bibby and Shepherd, 2004), while a similar approach has been used to characterize change shown in LUCS in work as part of the Countryside Agency's *Countryside Quality Counts* project.

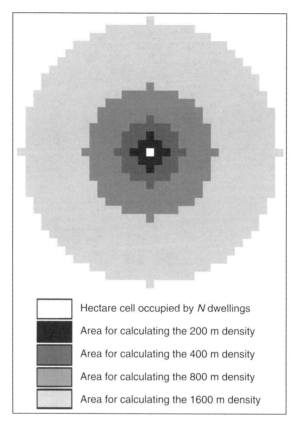

FIGURE 9.17
Moving windows for the calculation of density profiles (200–1600 m).

contexts may be further enhanced by using NLP techniques to distinguish varying kinds of dispersed settlement. In this manner it is possible to identify (historic) farmsteads as a first step.* Having done this, and having excluded those that lie in hectare cells already assigned to different morphological types on the basis of density profiles, it is possible to identify hamlets as clusters of 3–8 (historic) farmsteads following the traditions discussed by Roberts (1996).

It is thus possible to construct a morphology grid from the grid representing the density of units of occupation for any given period, which may be further refined through the application of NLP techniques. Then, just as it is

---

* A (historic) farmstead is here defined as a farmstead for which some evidence exists on PAF. It may be indicated by a business name or address (X Farm) or by a regional farm name (e.g., X Barton) or indirectly by a derivative name, e.g., X Farm Cottage or X Farmhouse. Pseudo farms with street addresses (e.g., Gabbotts Farm's a chain of butchers) were excluded.

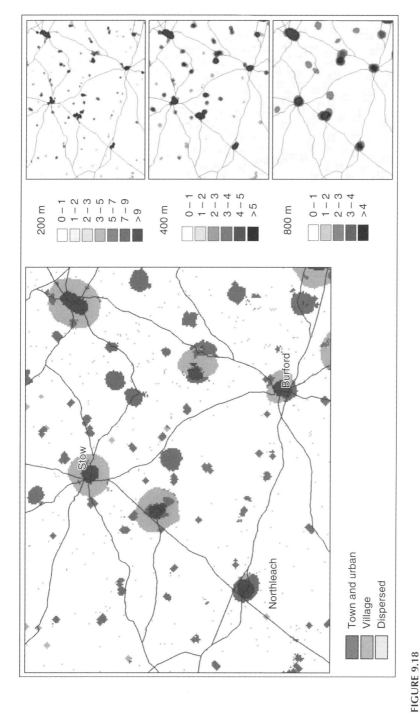

FIGURE 9.18

From density profiles to settlement form; Cotswolds, South Midlands, 2001. (© Crown Copyright/database right 2007. An Ordnance Survey/EDINA supplied service.)

possible by comparing the two underlying grids to look at shifts in intensification, it is also possible to assess changes in settlement form by comparing two morphology grids. Thus at the simplest, morphologies may be compared allowing a picture of growth of the contiguous urban area over the period 1998–2001 (as illustrated in Figure 9.19 which includes Cambourne, a very rare example of a new settlement, denoted by S), providing an alternative approach to assessing urban extension that does not depend on physical survey. It differs in that it explores changes

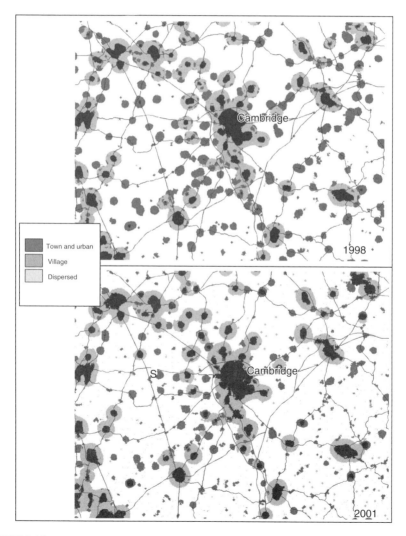

FIGURE 9.19
Cambridge area, 1998–2001: urban expansion or intensification. (© Crown Copyright/database right 2007. An Ordnance Survey/EDINA supplied service.)

in small-scale variations in the density of matter rather than comparing edges. Use of this approach highlights urban growth in Cambridgeshire for example, while confirming the almost unchanging extent of London (in contrast to the evidence of intensification in London discussed above). The grid framework set out here thus genuinely provides a method for simultaneously considering change in both morphology and density as an aid to assessing the effectiveness of intensification policy.

More significantly for our immediate purpose, recourse to density profiles allows assessment of the role of different types of context in accommodating new housebuilding. This demonstrates that in the 1990s only half of all exurban development (on definition, XC1) occurred within or involved the expansion of settlements that might be considered archetypically rural. Dwellings within villages account for about a quarter of new dwellings built in exurban contexts (26.4%) on this definition, and if the envelopes immediately around them were to be included, this proportion would rise to more than a third (35.7%) (Table 9.3). A further one-sixth of exurban development (15.2%) on definition XC1 occurred even further down the settlement hierarchy (the expansion of hamlets or construction of dwellings alongside isolated farmsteads). This development can be considered *dispersed*, in the sense it comprises small sites widely dispersed through and adjoining a wide scatter of small settlements.* To grasp its nature more clearly, it is probably easiest to consider a range of examples, though to characterize it satisfactorily may demand the articulation of a more nuanced set of natural language descriptions of context. Crucially, the density of these developments is very much lower (and thus the implied plot sizes very much greater) than those that typically pertain in "new urban enclaves" and it will be referred to here as countryside residential development.

By contrast, the remaining half of exurban residential development (on definition, XC1), occurred very close to the edge of larger settlements. It included some 62,000 houses outside the 2001 urban area boundaries, but which nevertheless fell within either the town category (18.8% of XC1) or the fringe category (14.5%) of the morphological classification, and together accounted for one-third of the volume of exurban development on this definition. The remaining 15.1% of all exurban development (on definition, XC1) was accounted for by piecemeal development, increasing density within what is termed here the peri-urban area, i.e., dispersed development physically separate from, but close to, towns and cities. It has a very strong statistical relationship to urban extension and is associated with growth points such as Milton Keynes, Telford, Ashford (Kent), Wokingham, and

---

* The type of sites discussed here might be thought of as "small sites, within and adjoining existing villages" similar to the type of sites which might be the subject of a "rural exceptions policy" designed to provide affordable housing to meet local needs as discussed in PPG3 Annex A (DETR, 2000c). It is not, however, being suggested that the development referred to here has resulted from such policies.

TABLE 9.3

New Dwellings Built and Housing Land Developed, 1990–2000, England; by Settlement Context

Context 2001	Inside ODPM Settlements Units	UA1 %	ODPM Settlement Extension Units	UX1 %	Outside ODPM Settlements Units	XC1 %	Total Units Built Units	%	Land Developed (ha)	Density Achieved (per ha)
				Point in Polygon Assignment						
Morphological Class										
No settlement	363	0.0	183	0.0	23,687	12.8	24,233	1.7	2,802.0	8.6
Isolated farmstead	17	0.0	2	0.0	2,812	1.5	2,831	0.2	284.1	10.0
Hamlet	18	0.0	5	0.0	2,350	1.3	2,373	0.2	242.5	9.8
Village	24,223	2.7	12,522	3.3	49,049	26.4	85,794	5.9	5,328.0	16.1
Village envelope	12,572	1.4	6,846	1.8	17,272	9.3	36,690	2.5	1,691.9	21.7
Peri-urban	7,947	0.9	4,758	1.3	27,982	15.1	40,688	2.8	2,811.1	14.5
Fringe	34,184	3.9	28,366	7.5	26,991	14.5	89,541	6.2	4,035.2	22.2
Town	76,288	8.6	37,721	10.0	34,977	18.8	148,987	10.3	6,319.9	23.6
Urban area (10,000+)	728,162	82.4	288,484	76.1	507	0.3	1,017,153	70.2	34,267.3	29.7
All contexts	*883,774*	*100*	*378,888*	*100*	*185,628*	*100*	*1,448,290*	*100*	*57,782*	*25.1*

Kent and Essex Thamesside. This quasi-exurban development may be considered *dispersed* in the sense that it was dispersed around the margin of substantial settlements rather than forming coherent extensions to the contiguous urban area. Significantly the density on these developments is on a par with reigning densities in urban extensions (or new urban enclaves) in contrast to countryside residential development.

Category XC1 is thus unsatisfactory, conflating two fundamentally different kinds of development whose geographic distributions are illustrated in Figure 9.20. Definition XC1 places too much emphasis on distinguishing the inside and outside of urban area polygons. On this point-in-polygon basis, 14.2% of all residential development in the 1990s appears to be exurban. While recalculating this proportion having converted the polygons to a hectare grid has only a modest effect on the estimate [reducing it to 12.8% (XC4)], discounting the 62,000 houses strictly outside OS polygons but categorized as town or fringe within the morphological classification reduces the estimated exurban component to 8.5% of the total. It appears

FIGURE 9.20 (See color insert following page 274.)
Distinguishing components of exurban development.

that the operational procedure adopted by OS necessarily creates hard urban edges at the level of representation even where they may be absent. The tendency to hard urban edges should be expected to be geographically variable; they are most likely to be found where there is both a highly restrictive planning regime and high demand pressure. Soft urban edges can be identified by the extent of quasi-exurban development (they are rare around London; Figure 9.20). In the present context, however, this is a relatively detailed aspect of form. Far more significant are the findings first that there is a substantial volume of housebuilding genuinely beyond the urban area (typically at low densities), but second that the volume of exurban development is much less than the city-as-polygon trope would suggest.

The effect of the use of density profiles to identify context has thus been to reduce the estimate of development in rural areas and to increase the estimate of urban extension from those based on point in polygon assignment, albeit that the evidence of LUCS still represents a challenge to the traditional narrative of urban expansion and to the assumption that rural areas contribute to meeting additional housing through the spread of urban areas. This underplays the role or rural locales in accommodating housing. In fact, the measures here provide only a modest estimate of housebuilding in rural England in the 1990s. In tracing through the method used by OS and ONS to amalgamate individual parcels of land "in urban use" to form urban areas, and in the series of subsequent analyses, the reader may have easily overlooked the arbitrary and extremely liberal definition of urban initially applied [one consistent with earlier projects in urban growth forecasting for government (Bibby and Shepherd, 1991, 1996)]. CLG itself now recommends that it is appropriate for most policy purposes to treat physical settlements (as defined by OS) as "urban" only where they have a population of 10,000 people or more (Shepherd et al., 2002), as opposed to the 2000 used here. Here there is a shift in the sense of the term urban from one which focuses on the use of an individual parcel of land to one which concerns itself with the size of a population cluster. Using the higher population cut-off, even deploying the grid-based definitions (as above) to filter out quasi-exurban development still implies that 342,000 of the houses built in the 1990s should be regarded as constituting exurban or rural development (24%). Urban definitions, like the areas to which they refer, are artificial.

9.9 Conclusions

This chapter has offered some evidence concerning the nature of recent patterns of residential development in England which were pertinent to policy aspirations over the period 2000–2006, and which remain relevant albeit in the more complex context provided by Planning Policy Statement 3 (CLG, 2006). It has also demonstrated a range of techniques that can be used to exploit a wider range of data sources to assist in policy monitoring. This section summarizes some of the main conclusions and reflects on aspects of the overall relationship between policy, evidence, and GIS.

9.9.1   Development Patterns and Policy Objectives

The evidence presented here goes some way to checking some common misconceptions and, by bringing together macroscopic and microscopic perspectives, demonstrates some facets of the relationship between policy targets at the national level and local development patterns. The chapter has illustrated the scale of annual housing output in England in a manner which shows what this has implied at the local level. Hopefully this serves as a reminder that the housebuilding targets that have occasioned such alarm are of the same *order of magnitude* as actual housing output in the 1990s and demonstrates the scale of accommodation possible incrementally without development of new settlements. The material in this chapter has demonstrated the generally substantial rates of urban land recycling. It has moreover demonstrated the tendency of these rates to reflect the geographically varying endowment of urban land rather than merely variations in market conditions and hence the degree of success of policy designed to concentrate new housebuilding within urban areas. It appears that policy objectives sometimes considered totally unrealistic are not far from being achieved.

These findings neither contradict our understanding that rates of housebuilding are historically low, nor provide grounds for complacency. Nevertheless a grasp of how new dwellings have actually been insinuated within the framework of existing settlement might allow more realistic debate about policy options and the respective roles of different levels of government. Accepting differential rates of growth between northern and southern England, it remains to insist on recognizing local diversity and the extremely strong relationship between the geography of new household formation and the geography of the population. From such a perspective, policy prescriptions might move away from ideas of regional shortages and surpluses and move to a finer level of spatial resolution. Nevertheless, urban expansion accounting for only a quarter of new housebuilding, this chapter has also pointed to the extent of development completely away from the urban area and the very low densities typical of development across the rural domain and hence the issue of urban intrusion.

9.9.2   Methods and Representations

This chapter has adopted a multiscale approach, concerning itself with how conclusions at one scale transmute to conclusions at another. Crossing scales to transform data into evidence involves recognizing the range of data sources and analytical tools that can be brought into play. Some of the techniques used here are commonplace in the GISc community; others are more novel. They have included standard geometric operations (point in polygon; boundary over boundary; and surface interpolation), relatively familiar statistical or econometric approaches (variants of the general linear model), and techniques from NLP and artificial intelligence (albeit that the nature of these approaches has not been considered in any depth). Critically, the patterns of urbanization, densification, extensification, and choice that can be demonstrated to inform policy discussions depend on the spatial scale at which they are represented, and this is ultimately limited by the manner in which data are collected. Thus the U.K. Sustainable Communities document, for example, draws out regional distinctions which, though valid, capture only a relatively small part of the variation discussed in this chapter. Microdata are important in unraveling patterns of change as they alone do not filter out variation at any spatial scale.

Crucially, the core datasets used in the chapter are microscopic: individual LUCS events and individual organization names and addresses from the PAF. Although the initial analyses relied upon assigning LUCS events (points) to urban areas (polygons), much of the work presented here has attempted to explore the possibility of using the same data in varying ways at different scales. At a conceptual level at least, LUCS points, PAF addresses, and hectare cells have been regarded as representing atoms from which objects of policy interest might be constructed, and it is this that has allowed microscopic and macroscopic perspectives to be brought together in this particular fashion. Thus the representation of wholes has depended explicitly on representation of their parts, with individual dwellings or units of occupation used to define the settlement geography in Section 9.8, for example. (Woods have been generated from trees.) More specifically, different analyses have depended on varying conceptualizations of the relations between wholes and their parts. The prime role of representational devices and spatial analytic functionality used has been to bring these parts together in different ways.

Much of the work presented here might be thought of as concerning itself with a sort of whole which within ontology is termed a mereological sum (e.g., Simons, 1987; Heller, 1990).* All England's new houses would be an example. Such uncomplicated (though possibly unfamiliar) individuals have parts (or atoms), but no specific form; the parts may overlap or be scattered. A mereological sum comprising so many dwellings (atoms) is identical whether the atoms are widely scattered across the countryside, concentrated in warehouse conversion projects, or built abutting say five existing urban areas in South West England. Many GIS users may be particularly reluctant to identify a scattered individual. Nevertheless, they might be more readily accepted and appreciated by users of grid-based GIS, where a count in a value attribute table might be thought of as characterizing an individual of

---

* Any combination of identified "atoms" may be thought of as a mereological sum constituting an individual (not just the general sum of all dwellings, for example). Indeed, mereological sums need not comprise atoms of the same kind. Generally, therefore, such individuals are far more numerous and far less structured than policy monitoring requires.

this type (referring to cells sharing a particular nominal value referring to a land-cover class, for example). A household forecast or a central government policy target might also be thought of as specifying only a mereological sum. The parts have no explicit configuration. Entries such as that in Table 9.2 referring to 398,165 units built in rural England on the 2004 definition are, of course, similarly underspecified.

This chapter has therefore sought in addition to specify rather more complex objects (i.e., at higher ontological levels, with more specific forms and causal powers). The analyses of Section 9.8 depend on specifying the circumstances (in terms of density in space and time) in which a mereological sum of LUCS events might be considered capable of flowing, hence operationalizing the familiar image of a residential development pipeline. Most of the analyses in this chapter concern themselves with configuration, at some degree of specificity, as well as mereology. In part the concern for the environment occasioned by household forecasts arises because the unspecified configuration of additional dwellings is assumed, and those assumptions are typically not informed by empirical examination of actual patterns of accommodation.

The spatial configuration of additional dwellings has been analyzed through a range of representational tropes. In the case of individual properties, inferences have been drawn about the configuration of wholes and parts on the basis of natural language. Throughout this chapter, neighborhood functions have been used within a grid framework to transmute objects into fields and back again, thereby defining further relations between wholes and parts (and allowing LUCS events to be commuted to flows). The notion of density has, moreover, been used to define settlement wholes from the dwellings that constitute them, albeit that the underspecification of the idea of density has been turned on its head by specifying a profile at a series of scales.

The work reported here has, however, also demanded consideration of far more complex part–whole relationships. Thus, the idea of a market in Cournot's classic formulation might be thought of as a whole in which asking-prices and bid-prices are the parts:

> Economists understand by the term Market, not any particular market place in which things are bought and sold, but the whole of any region in which buyers and sellers are in such free intercourse with one another that the prices of the same goods tend to equality easily and quickly.
>
> Cournot, A.A., *Research into the Mathematical Principles of the Theory of Wealth*, 1838.

Like the city as polygon, Cournot's region is a spatial metaphor that almost works. Whether this description actually implies a contiguous region of space–time is less clear. Market forces may cause similar housing in similar neighborhoods to command similar prices, although those houses and neighborhoods may be geographically dispersed. The demand and supply functions represent placeless intensional evaluations of placeable goods. They might even be considered as a mereological sum of intensional atoms. In this chapter, a rather different approach was taken to representing a market, relying simply on the assumption that averaging tendencies might operate at the 10-km scale, but without making the presumption of spatially closed regions.

9.9.3   Relation between Policy, Evidence, and GIS

Finally, the material of this chapter prompts some conclusions about the general relation among policy, evidence, and GIS, which cannot ignore these apparently arcane remarks about representation. The commonsensical and unexceptionable suggestion that policy should be based on evidence occludes all manner of questions about the relation among narrative, evidence, and representation. Because the overall pattern of urban development or the extent of urban intensification cannot be perceived in a glance or a series of glances, its appreciation is not a matter of percepts, but of concepts. It inevitably depends on representations. Although the images employed must have at least some degree of congruence with policy makers' direct perception of individual developments, GIS use must recognize and reproduce (though possibly in a modified form) widely-shared, credible, communicable representations which fit the categories used within policy discourse.

GIS use thus depends on and plays its part in reproducing the policy setting. In contrasting intensification with urban expansion, policy reflects and refines the "city as polygon" metaphor. GIS practices can either reinforce or check this tendency. Although this chapter has shown that this metaphor is inadequate for describing dispersed exurban development, it is privileged in policy discussion. Although household projections might only honestly be represented as a mereological sum, the polygonal template is used (unwittingly?) by journalists and deliberately by pressure groups, for example organizing dwellings (at the level of representation) into "twenty-seven huge new towns" in the manner discussed in the introduction. For policy makers faced with the conclusions of the Barker (2004) review, N dwellings may so easily become M towns. The leap between mereology and morphology is not inevitable, and GIS can be used, as in this chapter, to show that development has not taken (and need not take) the form feared. The differences in configuration obscured in this way have implications for environmental sustainability and for our understandings of how growth can and cannot be accommodated.

Representation within GIS is thus potentially contentious. Whether an individual is regarded as scattered depends upon the scale at which it is represented (any solid having spaces between its individual atoms) and so where policy discourse refers to urban spread and urban sprawl, graphics such as Figures 9.3 and 9.4 might easily be misrepresented or deliberately misused to suggest some form of development quite different to that which actually occurred and with quite different consequences. Unwittingly or quite deliberately such images may reinforce other metaphors for urbanization such as viruses, cancers, and tumors, which have been commonplace at least since William Cobbett (2001) penned his *Rural Rides* in the first decades of the nineteenth century. Furthermore, this is not simply a scientific problem of avoiding an ill-fitting spatial metaphor. In Section 9.8 above, it became clear that the appropriateness of the metaphor of the crisp urban edge was itself geographically variable and that this was a matter of public policy.* As settlements are artifacts, metaphors such as the urban edge can be forced to apply, whether by the application of design principles or by the application of rules such as Green Belt restrictions. Representations therefore presage physical outcomes and so GIS use is implicated not only in observing, but creating urban edges. Inability—or lack of will—to conceptualize or represent scatter within GIS not only reinforces the probability of unreflective shifts from mereology (N dwellings) to morphology (M towns) at the level of thought, but because this thought forms part of the discourse controlling urban form, prejudices choices about physical outcomes. Analogous issues (though more complex) arise in dealing with Cournot's region and policing the closure of spatial markets.

The analyses of this chapter have hopefully contributed something to an understanding of policy possibilities, but they are not completely concordant with policy concerns ruling immediately after 2000. In order to focus on urban areas and to consider some generic themes in GIS representation, no explicit attention has been given to the recycling of previously developed sites, which was a far more central policy concern. The percentage of new housing accommodated on previously developed sites is being a component of the quality-of-life barometer. On the contrary, while this chapter has explored possible ways of measuring urban intensification, it has also devoted substantial attention to exurban development which was of minimal policy concern and remains so (however much it might prejudice the achievement of central policy goals).

Whatever the reader's position on the effectiveness of policy, on the value of additional datasets incorporated, or on the relationship between policy and evidence, it is hoped that this chapter will have enhanced awareness of the range of analyses that are possible, given the richness of available data. It is clearly possible to undertake work which is national in scope, and multiscale in focus. Of course, the apparent richness of the data resource

---

* The hardness or softness of the edge of an urban area may be a matter of policy discussion. Thus a planning inspector appointed by central government to adjudicate in an appeal against refusal of planning permission by Newbury District Council concludes in his decision letter:

  However, there is one aspect of the proposal which does not in my view fully accord with the objective of EN5 to ensure a soft edge to the built area. This is in relation to the impact of the extension on Pincents Lane . . . The extension would involve the loss of much of the planting proposed under the scheme approved in September 1988 to be placed in the area to the east of the existing building. The gradual transition for those travelling down Pincents Lane from the rural area into the commercial area that this would have ensured would be lost.

depends on our will to locate it within a framework of meanings, our technical ability to exploit it, and our administrative ability to act upon it, but that is essentially the same story.

References

Anselin, L., 2000, GIS, spatial econometrics and social science research, *Journal of Geographical Systems* 2(1), 11–15.

Anselin, L., 2001, Spatial econometrics, In *A Companion to Theoretical Econometrics*, edited by Baltagi, B., pp. 310–330 (Oxford: Basil Blackwell).

Anselin, L., Florax, R.J.G.M., and Rey, S.J. (editors), 2004, *Advances in Spatial Econometrics: Methodology, Tools and Applications* (Heidelberg: Springer).

Barker, K., 2004, *Review of Housing Supply: Delivering Stability: Securing our Future Housing.* Final Report (London: HMSO).

BCIS Ltd, 2003, *Whole Life Cost of Social Housing, Phase 1*, Final Report (London: BCIS) (http://www.housingcorplibrary.org.uk/housingcorp.nsf/AllDocuments/52474 CEAE6829ADE80256DCC00530156/$FILE/Wholelife.pdf).

Bibby, P., 2005, GIS, worldmaking and natural language, In *Re-Presenting GIS*, edited by Fisher, P. and Unwin, D.J., pp. 55–84 (London: Wiley).

Bibby, P. and Chowdry, N., 2001, *Geographical Information Systems (GIS) and Housing Needs* (London: Housing Corporation Publications Section).

Bibby, P.R. and Shepherd, J.W., 1991, *Rates of urbanisation in England 1981–2001*, (Former) Department of the Environment Planning Research Programme (London: HMSO).

Bibby, P. and Shepherd, J., 1996, *Urbanization in England, Projections 1991–2016*, Department of the Environment Planning Research Programme (London: HMSO).

Bibby, P. and Shepherd, J., 2000, GIS, land use, and representation, *Environment and Planning B–Planning & Design* 27(4), 583–598.

Bibby, P. and Shepherd, J., 2004, *Developing a New Classification of Urban and Rural Areas for Policy Purposes—the Methodology* (London: HMSO) (http://www.statistics.gov.uk/geography/downloads/Rural_Urban_Methodology_Reportv2.pdf).

Bramley, G., Pawson, H., and Third, H., 2000, *Low Demand Housing and Unpopular Neighbourhoods* (London: DETR).

Breheny, M., Gent, T., and Lock, D., 1993, *Alternative Development Patterns: New Settlements* (London: HMSO).

Cobbett, W., 2001, *Rural Rides* (Harmondsworth: Penguin).

Communities and Local Government, 2006, *Planning Policy Statement 3 (PPS3): Housing* (London: Communities and Local Government).

Countryside Agency, 2004, *Countryside Quality Counts: Tracking Change in the English Countryside*, CRN85 (Cheltenham: Countryside Agency).

Cournot, A.A., 1838, *Researches into the Mathematical Principles of the Theory of Wealth*, translated by Bacon, N.T. (London: Macmillan/Augustus M. Kelly).

DETR, 2000a, *Our Towns and Cities: The Future Delivering an Urban Renaissance* (London: Stationery Office).

DETR, 2000b, *Our Countryside: The Future* (London: Stationery Office).

DETR, 2000c, *Planning Policy Guidance Note 3: Housing* (PPG3), Department of Environment, Transport and the Regions (London: Stationery Office).

DoE, 1992, *The Relationship between House Prices and Land Supply*, Gerald Eve, Chartered Surveyors with the Department of Land Economy, University of Cambridge (London: HMSO).

DoE, 1995, *Projections of Households in England to 2016* (London: HMSO).

DoE, 1996, *Household Growth: Where Shall We Live?* Cm 3471 (London: HMSO).

Ermisch, J.F., Findlay, J., and Gibb, K., 1996, The price elasticity of housing demand in Britain: issues of sample selection, *Journal of Housing Economics* 5, 64–86.

Friend, J., Power, J., and Yewlett, C., 1974, *Public Planning: The Intercorporate Dimension* (London: Tavistock).

Gallie, W.B., 1955–1956, Essentially contested concepts, *Proceedings of the Aristotelian Society* 56, 157–197.

Gattrell, A.C., 1989, On the spatial representation and accuracy of address-based data in the United Kingdom, *International Journal of GIS* 3(4), 335–348.

Heclo, H., 1972, Policy analysis (review article), *British Journal of Political Science* 2, 83–108.

Heller, M., 1990, *The Ontology of Physical Objects* (Cambridge: Cambridge University Press).

Hill, M., 1997, *The Policy Process in the Modern State*, third edition (Englewoood Cliffs, NJ: Prentice-Hall).

Jubien, M., 1993, *Ontology, Modality and the Fallacy of Reference* (Cambridge: Cambridge University Press).

Matheson, J. and Babb, P. (editors), 2002, *Social Trends 32* (London: Stationery Office).

Observer, 2003, *Prescott's homes plan will blight green land*, Sunday, February 2, 2003 (London: Guardian Newspapers).

Office of the Deputy Prime Minister; 2004, *Housing Market Assessment Manual* (London: Office of the Deputy Prime Minister).

Office of National Statistics, 2004, *Model-Based Estimates of Income for Wards in England and Wales, 1998/99: User Guide* (http://neighbourhood.statistics.gov.uk/information/income_estimates.pdf).

Prescott, 2000, *Ministerial Statement 7th March 2003 and DTLR News Release 2000/0164* (London: DTLR).

Prescott, 2003, Foreword, In *Sustainable Communities—Building for the Future* (London: Office of the Deputy Prime Minister).

Raper, J., Rhind, D., and Shepherd, J., 1992, *Postcodes: the New Geography* (London: Longman).

Roberts, B.K., 1996, *Landscapes of Settlement: From Prehistory to the Present* (London: Taylor & Francis).

Sellwood, R., 1987, *Statistics of Changes in Land Use: A New Series, Statistical News* 79, 11–16.

Searle, J.R., 1995, *The Construction of Social Reality* (London: Penguin).

Shepherd, J., Coombes, M., and Bibby, P., 2002, *A Review of Urban and Rural Area Definitions: Project Report* (London: Office of Deputy Prime Minister) (http://www.statistics.gov.uk/geography/downloads/Project%20Report_22%20AugONS.pdf).

Simons, P., 1987, *Parts: A Study in Ontology* (Oxford: Clarendon Press).

Surveyors Plus, 2000, *Estimating Social Housing UK*, Various Grades Construction Cost Briefing Building Costs May 2000 (http://www.surveyorsplus.co.uk/surveyorsplus/estimating/asp/estimatingdetail.asp?docid = 36).

Warwick District Council, 2003, *Warwick District Local Plan 1996–2011 First Deposit Version*, Leamington Spa, Warwick District Council (http://documents. warwickdc.gov.uk/stored-documents/WDC-Corporate-Documents/Local% 20Plan/Dep1Website/).

Weston, R., Oxley, M., and Golland, A., 2003, *A Methodology for Assessing the Economic Viability of Urban Housing Development* (Nottingham: Trent University).

Whitehand, J., 1992, *The Making of the Urban Landscape* (Oxford: Blackwell).

Williams, K., 1999, Urban intensification policies in England: problems and contradictions, *Land Use Policy* 16(3), 167–178.

Part II

Making Policy

Section I *Engaging with Policy-Makers*

10

Application of GIS to Support Land Administration Services in Ghana

Isaac Karikari, John Stillwell, and Steve Carver

CONTENTS

10.1 Introduction

This chapter reports on a research project whose objective was to understand the factors and processes that will underpin the successful adoption, implementation, and use of geographical information systems (GIS) by Ghana's land-sector agencies. In recent years, Ghana has formulated a land policy under the Land Administration Program (LAP) and made progress towards the development of a land information system (LIS). The country had previously received support from external donors in the transfer of GIS technology but with limited benefits for the administration of land by the relevant organizations. Thus, only slow progress has been made in automating the procedures for collecting and managing data on land ownership and transfer.

In response to the challenge for improvement, a customized application (prototype software system) has been developed to support some of the tasks involved in routine administration by the Accra Lands Commission Secretariat (LCS), the main institution responsible for managing public lands in Ghana. The technical structure of this prototype constructed using ArcView and Microsoft Access is the focus of this chapter. However, some attention is also paid to the institutional context and the roles of the agencies that operate in the sector.

10.2   Progress towards Land Policy and Land Information Systems

The main agencies operating in the land sector have not directly benefited from any specific project introduced by international donors. However, some land agencies have been involved in projects related to land and natural resources implemented by other sectors. Since these projects had land ownership and mapping components, it became imperative that the land-sector agencies like the LCS were involved. Such projects include the Urban II project and Ghana Environmental Resource Management Project (GERMP), both of which have now come to an end. In fact, the Government of Ghana had approved the GERMP in 1991, a project aimed at implementing the National Environmental Action Plan (NEAP) under the coordinating role of the Environmental Protection Agency (EPA). The Lands Commission, together with the Land Administration Research Centre (LARC) of the University of Science and Technology (UST), took responsibility for the land ownership aspects under the Land and Water Management Component of the project. Until the outset of GERMP in May 1993, no attempt had been made to computerize activities or information in the land sector. The commencement of the project was therefore welcomed in this respect and the LCS, like the other participatory agencies, inherited some GIS equipment from the project that is still available for use today.

In 1999, the Government introduced a land policy document (LPD) aimed at establishing and developing a LIS. It was noted that the basis of better management of information within the land sector might be brought about by analyzing and costing existing tasks, abandoning unnecessary procedures, and developing a better use of resources (UNCHS, 1990). One key objective was to reduce duplication in the storage of information and to replace registers, physically damaged through handling, with electronic versions. However, progress towards automation has been very slow in the agencies under the Ghanaian Ministry of Lands and Forestry. This can be attributed to inherent difficulties with the introduction of the technology itself and with problems of data conversion that confront organizations in this sector. Even though the Survey Department under GERMP used computers to perform land-survey computations, the development of GIS has remained very limited. The awareness of the potential of this technology appears to be growing yet the technical, economic, and institutional problems, even though outlined in the LPD, are yet to be addressed in practice. In 2001, however, the Government initiated the LAP to provide a platform to translate its national land policy into action. As expected, an up-to-date LIS that supports good management of land records is now to be constructed, and this provides the context for our research.

10.3   Need for Self-Determination

Africa still remains heavily dependent on donor support and development assistance from the West. GIS are regarded as a Western artifact and its introduction still requires assistance. This presents enormous challenges to countries such as Ghana. It is worth noting that the introduction of geotechnology in Africa by expatriate consultants has been positive in the main but has led to certain difficulties, particularly following the life span of GIS project implementation. Problems have included noncommitment or inaction by the adopting agencies, leading to nonmaintenance of equipment and therefore failure to incorporate Western GIS into their own systems. Even though Ghana is still subjected to the ebb and flow of donor thinking, it is imperative that the country seeks channels for more control over GIS project design and implementation. We therefore concur fully with the view that, if GIS are to be introduced successfully, then their development, modification, and control must begin with indigenes who have a deep understanding of the socioeconomic and political context of the situation as well as technical capabilities of GIS (Taylor, 1991). The reason for this position is straightforward. In Ghana, GERMP had been a notable precursor to the development of GIS in the land sector. At the end of the project, mainly designed and introduced by the World Bank, our investigations at the Accra LCS reveal that the adoption and assimilation of GIS had brought certain difficulties with it. After the consultants had departed, there appeared to be no one with the commitment to continue the process that had been started, and no plans were made to ensure optimum utilization of GIS equipment still available for use, i.e., that no integrated analytical tools are available in the Accra LCS.

The above observations suggest that there is a need for greater levels of self-determination and debate within Ghana in relation to the role of expatriate and indigenous experts and other areas of GIS diffusion policy. Cash et al. (1983) have noted that, since developing countries have distinctive environmental conditions that differ from developed countries, the formulation of effective strategies for developing computer-based information systems requires a good understanding of their special macro-external environment. Major gains can be obtained if indigenous experts with the requisite knowledge, understanding, and training are made, from the very beginning, to have greater control over awareness creation, design and implementation, and ex-post evaluation of GIS projects. We therefore argue that the approach to technology transfer needs to be strongly human-centered and less driven by technical prescriptions than the donors have frequently pursued and the governments of developing countries have allowed (Toulmin and Quan, 2000).

10.4   Land Types and the Role of Land Agencies in Ghana

Ghana has a complex land tenurial system in which state or public land is distinguished from customary or private land. Larbi (1995) provides a fuller discussion of these land types. Before 1983, only one agency (the LCS) was responsible for managing information on all such lands. Since this date, however, through the passing of legislation, the LCS has come to be restricted to managing mainly public/state lands. There are now five agencies with different and sometimes overlapping mandates and roles. The Office of Administrator of Stool Lands (OASL), for instance, now deals with the collection and disbursement of stool land revenue. Stool land is a type of customary land held in trust on behalf of subjects of the stool by a chief. This is despite the fact that the LCS still manages stool lands by the granting of concurrence (a form of certification to remove obnoxious covenants likely to be put in leases by such chiefs) on such stool land disposition/alienation. Ghana also has two forms of land registration—Deeds and Title—that are handled at the moment by both the LCS and the Land Title Registry. The country seeks to phase out the old deeds registration system and cover the whole country with title registration.

It is the view of Somevi (2001) that one factor responsible for the problems of land administration in Ghana has to do with the forces of politics and law accountable in the creation, shaping, and reshaping of institutions in the land sector. It is important to state that Ghana, like many developing countries, has tight budgetary constraints. There may be considerable advantages to building on existing institutions that are modified as necessary without setting up brand new ones with new operating budgets, new staff, and the time necessary to establish and entrench their legitimacy (Toulmin and Quan, 2000). In fact, ensuring effective coordination for all such agencies that are expected to be responsible for the GIS project is intrinsically complicated and must be well planned from the very beginning.

This implies that the diffusion of GIS into land-sector agencies will need some reengineering of the work processes of such agencies themselves. This will require a careful diagnosis in order to establish an overall strategy for change. Implementing this agreed strategy will require a selective approach. Initially, a few more critical problem areas would have to be tackled as comprehensive reform may be seen as too ambitious and unworkable. It will therefore mean that issues that seriously constrain the performance of these agencies, such as current manual work practices that are within their capacity to change, would have to be given priority. Public support for such a reform ought to be solicited.

It is well known that internal corruption is endemic in these agencies. The bureaucracy involved in land administration has remained unchecked and abuses such as extortion are entrenched. An empirical survey of selected respondents in September 2001 suggests that the reasons underpinning corruption in the Accra LCS, for instance, are, in order of priority: low wage levels, indiscipline, pressure from the public, sheer greed, and the very culture of the LCS itself (Karikari et al., 2002). Our observation corroborates evidence from research by Ghana's Centre for Democracy and Development (CDD) into corruption and other constraints on land administration in Ghana (CDD, 2000). Corrupt officials have seldom been sanctioned beyond being transferred to another department or sometimes to another regional office of the same department. It is accepted that civil servants should show due care and attention in their work and must be proven to be free from corruption. Failure to do so results in over-checking of work, duplication of effort, unnecessary costs, and excessive delays (UNCHS, 1990).

These issues emphasize that the introduction of GIS in Ghana's land-sector agencies must be put in proper perspective. As a technical tool, GIS cannot solve corruption in these agencies, and therefore this would have to be tackled in tandem. There must be zero tolerance for corruption. Paraphrasing Simpson (1976) who initially noted this about the system of land registration, we emphasize that GIS are not some sort of magical specific that would automatically produce good land-use and development although they may be invaluable aids to land reform. GIS are a means to an end, not an end in themselves, and GIS will require a commitment to change at the highest level (national reform). In fact, GIS are about people, not technology, and they are more than simply a project; they are a way of life. Technology provides the tools not the standards. A successful GIS requires not only strong public support but also strong management support (Burns et al., 1996). In the next section of this paper, we consider the Accra LCS in more detail and debate on some of the issues relating to it and GIS diffusion.

10.5   GIS Implementation: Institutional and Technical Considerations

Our research is based on the premise that there is a need for simultaneous attention to both institutional and technical processes. Our pilot project is part of an attempt to ensure successful implementation of GIS in the Accra LCS, in which one of the primary objectives has been to identify the range of factors relevant to GIS diffusion in this agency. This has been carried out using various

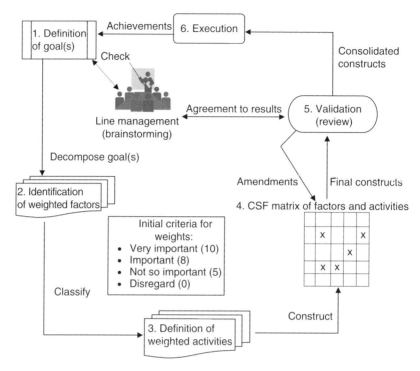

FIGURE 10.1
The CSF pilot study of performance needs. (*Source*: Modified from Wasmund, M., *IBM J.*, 32, 595, 1993.)

approaches including brainstorming sessions with staff at different levels, questionnaire surveys, and adoption of the critical success factor (CSF) method, following Wasmund (1993), to determine the most important factors considered to ensure that GIS adoption is successful (Figure 10.1).

The CSF methodology can be outlined using the framework shown in Figure 10.1 where the main components are shown. Double-ended arrows indicate processes that may be reviewed in transit, while single-ended ones indicate processes that are consulted by other processes. The processes are followed rigidly (from definition of goals to execution) and show the various criteria to be used. It will suffice to indicate the various steps and show how the objectives of each stage were achieved, namely:

- *Step 1*: We defined the GIS goals the LCS has set itself, and discussed and accepted them in a brainstorming session involving statements about purpose, scope, and time constraints through consensus building at middle or line management level. The rationale of the goal for GIS development is to build a system that delivers quickly and is error-free as far as possible and that generates the expected revenue with respect to land administration and management.

- *Step 2*: We identified and prioritized subgoals called factors using set criteria, discussed and added or deleted some factors as appropriate through brainstorming, while providing guidance in the whole exercise.
- *Step 3*: We then defined activities to support the subgoals in the brainstorming session.
- *Step 4*: We constructed the CSF matrix and tied in the activities and factors, removing all redundancies.
- *Step 5*: We reviewed and validated factors through intense discussion and reviewed the results with top management for acceptance through interview sessions (an iterative process).
- *Step 6*: We consolidated the constructs (accepted the factors and the activities) for execution.

A detailed analysis of the relative importance of the institutional factors pertaining in the LCS is available in Karikari et al. (2002). We will, however, discuss some of the elicited factors the respondents' thought would help in successful GIS implementation based on the CSF method.

What was the significance of these factors to the respondents? Respondents were of the opinion that to be able to have a successful GIS, the LCS must first and foremost aim at having a sound revenue base (current budget) on which to operate. This supports the view that future budgets must start with current resources (Wiley, 1997). Insistence on financial viability (based on the LCS's own resources) is an important means by which discipline would be imposed on the LCS while encouraging efficient management and use of available resources.

To be able to meet the overall goal set by the LCS on GIS implementation, it was agreed that the LCS must have well-trained staff needed to create the appropriate mind-set and establish the necessary technical skills. This supports the view that appropriate staff with requisite training in IT and GIS must be available and that members of staff who are lacking in the necessary technical and communicative skills are likely to mean failure (Edralin, 1990; Fox, 1991; Huxhold, 1991). Education and training, including in-house training and workshops, are therefore necessary ingredients to update knowledge. Training in IT and especially GIS may be sustained by external funding through sponsorship of courses abroad for a core group only who are expected to train others locally. Short refresher courses are needed for key staff with GIS knowledge to enable the acquisition, instruction, and communication of new ideas.

The existing hardware and software need to be upgraded or replaced regularly with all offices linked to the local area network (LAN), with each having computers for use so that work processes can be automated as far as possible. Closely linked to a good networking system, the group observed, was an effective maintenance program to maintain existing hardware and to upgrade software and make necessary backups. This presents

an enormous challenge to management as Ghanaians generally lack a maintenance culture.

Good leadership style and top-management support for GIS were required
for a lengthy time period. This supports the view that full management
support and involvement in the execution of GIS projects, with such projects
directed at intended GIS beneficiaries, are important requirements. For this
reason, inappropriate management styles will obscure direction and put
investment in GIS at risk. It is important that the right political leadership,
especially as it relates to the mode of appointment of the Executive Secretary
of the Lands Commission, be provided. All were in agreement that the post
ought to be advertised, and the selection procedure made more transparent.
Frequent and unnecessary changes of the top post must stop to enable
policies to be fully implemented. There is also the need for the LCS to be
given greater operational freedom (autonomy) to carry out its assigned task.

When it came to discussions on staff reductions, some respondents were
initially apprehensive. However, the majority were of the view that there
was need to remove underemployment, which already existed in the LCS as
part of the Civil Service legacy. Associated with this was the idea that
retained staff should be computer literate as far as feasible. Staff must
have some motivation to attain higher standards, especially in the area of
professionals who have to change their practice. This, of course, is related to
better remuneration since better salaries commensurate with job descriptions and performance are a necessary prerequisite for continuity in the job.
The idea is to stop brain drain after training. Discipline is required in a
system, such as the LCS where indiscipline thrives. From the very top to the
bottom, the chain of command and authority must be clear and respected.

Extensive consultation processes are, however, still needed to permit
effective engagements in the discussion of the proposed institutional options
for GIS implementation in the land-sector agencies. From the research undertaken and as indicated above, we conclude that the successful adoption of GIS
in Ghana's land-sector agencies will depend on satisfactory organizational
and institutional environments involving Ghanaians themselves. Moreover,
the successful initiation of GIS projects will depend on strong management
support (with GIS champions playing vital roles) and strong financial commitment from relevant sources. These circumstances would indicate the
readiness for GIS use, which would also depend on the political and economic situation of the country as well as sensitivity to the cultural norms of
the land-sector agencies, with donor countries playing catalytic roles.

In terms of the technical process, there is an initial need to study critically
the needs of the agencies involved in land administration, management, and
planning. This will help to bridge the gaps in conceptual understanding
between the potential GIS users (staff in the LCS and other land agencies)
and the GIS prototype developer. The information system design (ISD)
depends upon the development of concepts of performance and functional
needs for these land agencies. There must also be a sound basis for sharing
experiences between different African GIS laboratories. A mechanism to

facilitate a process of learning from best practice and experience has much value. Hastings and Clarke (1991) have indicated that it is possible, through bilateral and multilateral aid that is outside the umbrella of technical assistance, to ensure long-term vitality.

In the design involving the Accra LCS, the use of methods such as data flow analysis to elicit information about functional needs is paralleled by the development of prototype software for a range of routine land administrative purposes. The prototype requires careful explanation, documentation, and training if it is to be successfully implemented, if continuous funding (from within) is to be maintained, and if the pilot project is to lead to a sustainable implementation. Conventional wisdom requires that pilot projects be undertaken to test out the feasibility of GIS projects. While we share this view, we argue that a *selected features* prototype software (an operational model that includes some but not all of the features that the final system is expected to have) is developed initially, preferably by indigenous experts. This must precede implementation of a full-scale model of the system. Such prototype software will serve two purposes, namely, as a basis for deriving immediate benefits and as a basis for focusing donor support.

We have developed a prototype software application package called LANDADMIN for the Accra LCS that could easily be replicated elsewhere. The design of this software is technically appropriate, given the socioeconomic environment of the intended beneficiaries. Careful attention has been given to tasks normally performed by the LCS, with the view of having a user-friendly interface that is menu-driven and that properly represents the LCS work processes, while placing this within the requisite institutional arrangements of the Secretariat. This research, therefore, has been driven by the need to design a system that is to be optimally operational and that will be delivered to the LCS with the full participation of the users themselves from the outset. Fieldwork in July–September 2001 afforded the chance to trial the prototype software using ArcView to six selected respondents, as well as to investigate the way the Accra LCS was using Access as a database management tool. The GERMP was partly responsible for the choice of software available in the LCS, which includes ArcInfo (running on PC and Unix systems), ArcView, and Microsoft Access databases. The Accra LCS is at the moment using Access in its normal work processes albeit on a very minor scale (only for rent management). Suggestions from the present users (two at the end of September 2001) enabled the development of an Access interface within ArcView.

10.6   Prototype Software

The prototype software was developed using ArcView 3.2, the Avenue scripting language, and an Access database. The prototype has selected features that handle routine land-administration tasks such as the design of quick maps for field inspections and the generation of site plans to be included in leases in ArcView. It also provides lessees' general and billing information, rent demand notices and searches, personnel and secretariat information, clients address lists, rent positions, and rent demand notices on state land in Access.

Several authors and experts have suggested conceptual and practical design approaches arising out of the need to integrate spatial and nonspatial data in a reliable, easy-to-use, and cost-effective way. Black (1996) notes that the applications that claim to bring about such integration have not been without weaknesses and that a few have been problematic insofar as their usefulness is concerned. Lai and Wong (1996) have suggested (in relation to developing countries) that applications must conform to existing practices and their procedures simplified using expert systems and numerical models where appropriate. For land administration, as with other areas, there is need to provide interfaces that are user-friendly, very secure, and therefore both interactive and menu-driven, rather than command-driven. Screens must correspond to the mental picture of users' work processes and are therefore easy to comprehend. The bottom-line design issue is always practicality, particularly so in the Ghanaian context.

The conceptual framework developed by Pradhan and Tripathy (1994) offers promise for designing the prototype software for Ghana's land-sector agencies. These authors suggest alternative prototype architectures using a specialized database management system (DBMS) for implementing GIS. One suggested feature is what the authors call the multimedia database approach on an expert system that has the following features:

> [E]xtraction of geographic objects for map data and classification into points, lines or polygons; storage of the above in a relational database; unstructured background graphics and image retained in their individual databases; [and where a KBS is used], the KBS becomes the top layer of the system that serves as an intelligent interface between the user and the data.
>
> (Pradhan and Tripathy, 1994, p. 30)

Such a system maintains complete database integrity through the simple concept that the spatial location of a particular feature can be considered as just another attribute of that feature. Derived from this concept, both graphic and descriptive attributes are associated as a single record of a particular feature.

The use of this concept is influenced by the actual situation in the Accra LCS. The LCS has gone ahead in using Access (DBMS) separately with the aim of capturing attribute data and ensuring efficient management of its database with a little or no parallel development of a GIS. The proper use of their ArcView GIS will offer a better organization of the parcel data that the LCS uses and add the dimension of geo-referencing to the available data. At the moment, the two systems in the Accra LCS are not linked and therefore are not being used together.

An observation by Huxhold (1991) in the United States provides a very apt description of the Accra LCS situation. It was noted that one of the most accepted blunders about GIS is that nongraphics data from the organization's administrative databases must exist in the GIS with the base map information.

> The problem with this concept is that many local governments have already implemented information systems on existing computers to support their nongraphics data needs... Some of these systems may have been operating successfully for many years, and to abandon them in order to implement them on a new GIS would not only disrupt daily operations, but also create additional cost that may not be necessary.
>
> (Huxhold, 1991, p. 239)

Project designs, will therefore powerfully influence whether the information products generated will be well utilized or not.

Figure 10.2 shows the architecture of the prototype software based on the above principles. Conceptually, it takes the complete text database in Arc-View and stores them as attribute data in the Access database using the plot number as the unique identifier.* ArcView extracts geographic objects from the parcel data and classifies them into points, lines, or polygons and displays them as maps and so on, as specified by the user. The geometric relationships between points, lines, and areas in the structured layer are

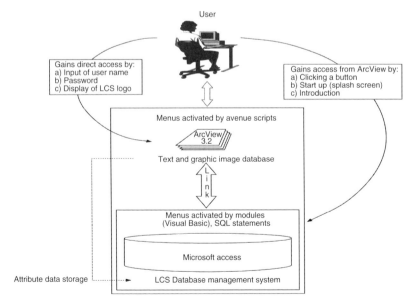

FIGURE 10.2
Detailed architecture of the LCS LANDADMIN prototype.

---

* In practice, however, the area and perimeter generated by ArcView is discarded in the Access database as they are of no practical use.

combined with other data, such as names and addresses, in the Access database to ensure efficient management of the database.

The communication medium between ArcView and the Access database is by executing system request commands through Avenue scripts in ArcView. This allows ArcView to be exited temporarily in order to perform functions and operations of Access. Since this is a link, minimizing the Access interface enables the user to interactively use both applications without realizing that they are not embedded. ArcView allows the input of login name and password, and displays the LCS logo. Access, when activated displays a splash screen and shows an introduction telling the user what to expect. This way, the right balance has been achieved between simplicity and complexity of project design. Scripts written using the Avenue programming language in ArcView 3.2 and a series of SQL statements and Visual Basic activate all menus. In this design, as already noted, the bottom-line design issue was practicality, as demonstrated below by some examples of tasks handled by the prototype software. The interfaces shown are improved versions following feedback given by respondents from the LCS.

10.6.1   ArcView 3.2 User Interfaces

Figure 10.3 depicts the ArcView Project window menu structure. Pull down menus need not be command driven but have been added to give some added boost for user interaction. Search reports are vital elements to land administration, and therefore, time was devoted to studying this process, which is used as an example in Figure 10.3. Two things are required for determining ownership status (search report), namely, site plans and attribute information on the plots. One would therefore require A (production of site plans through theme selection) or alternatively B (production of site plans by input of the plot number itself) plus the attribute information C. It is worth noting that of the six respondents, two preferred both A and B, three

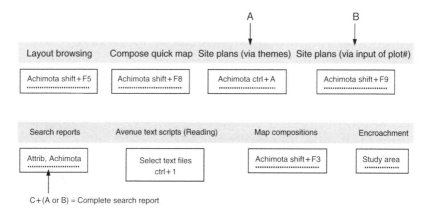

FIGURE 10.3
Overview schema of the menu structure in ArcView (Project window).

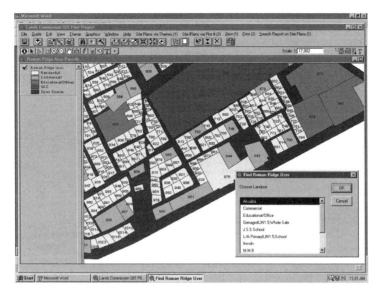

FIGURE 10.4
Site plans selection via Themes (on Roman Ridge Residential Area).

preferred A to B, and two preferred B to A. However, most were critical of the form that was output containing the attribute information (C) and suggested that it should look precisely the same as that produced by the manual system.

Figure 10.4 is a screenshot illustrating the selection of a land plot through theme selection (A) in ArcView. It opens two message boxes for selecting themes and reselecting associated plot numbers and pans the view to display the selected extent. The plot is highlighted and the end product is used as a site plan for a SEARCH report. An alternative process is by clicking an event script for a menu control that zooms into a select plot entered by the user by presenting a series of message boxes that gives feedback on the selection made. These processes prepare site plans for printing.

10.6.2 Microsoft Access Metadata Structure

There are three vital tables (T): (1) the AreaObjects table, (2) the AreaPosition table, and (3) the "TopicSelection" table that link the LCS main database and constitute the metadata structure (Figure 10.5). These tables are queried (Q) (as AreaLists and AreaTopicSelectionList) to produce the menu structure. Various modules, as noted (using the Visual Basic programming language), and SQL statements are used to generate forms, reports, among others.

10.6.3 Access Interfaces

ArcView is interfaced with Access. Launching into Access is by a simple Avenue script that uses the System.Execute command. The user activates it by clicking a button on the ArcView interface.

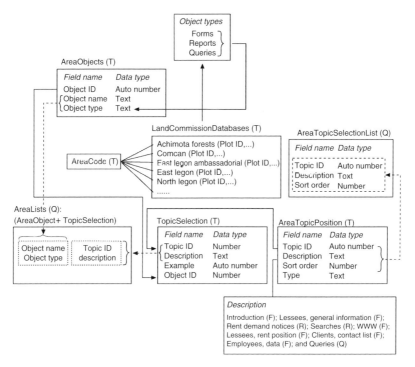

FIGURE 10.5
Structure for the LCS databases in Access.

Lessees' general information on Achimota Residential Area in Accra held within in the Access database. The user clicks a command button to view the detailed general or billing information of lessees. A dialog is presented allowing the user to keep open the selected information while new selections are examined.

The attribute information part of a search report is generated in Access. The report is the exact replica of that generated by the manual system and it is activated by the input of the plot number by the user. This report, together with the site plans, will constitute a complete search report as required by the land-administration system.

There is a link to the Internet from the Access interface. Provision is therefore made in Microsoft Access to link the LCS to the Internet (to the Ghana Home Page) where the LCS is expected to develop its own home page in future. Provision is also made for the LCS to have its own home page in Access. ArcView is the core of the prototype, and Access is launched from this interface, whereas entry to the Internet is via Access.

Providing support via the Internet or distributing metadata through web pages may appear not practical in Ghana where telephone services are erratic and electric power outages are the norm rather than the exception. Yet the philosophy is that agencies must start small and think big. Such innovations are relevant but must be pursued with circumspection.

10.7 Respondents' Reactions to the Initial Prototype Software

As already mentioned, six university graduates (of Lands Officer grade) were selected to provide some feedback on the first version of the ArcView-customized user interface. The group comprised those individuals more likely to use or supervise GIS applications in future and who were sufficiently well trained in land administration. They have the necessary influence on the operations of the Secretariat and will be expected to help impart knowledge gained to other users of lower grade. Each respondent was given an evaluation form that helped to structure his or her feedback. The form was so designed to allow respondents to provide comments on the layout and function of particular screens as they used the prototype. These comments were evaluated as a first step towards improving the software. Respondents spent between 1.5 and 2.5 hours testing the software by working through a questionnaire involving a series of applications.

When asked whether the software delivered will *help perform the selected tasks of the Accra LCS better*, all six respondents gave the affirmative. All six again indicated that the software would *help perform their task faster* than previously. Five *felt comfortable* using the prototype software with one feeling *very comfortable*. When asked which application respondents liked most, only one respondent gave the affirmative to all the features. One respondent each *preferred mostly*, application 1 (user name and password); application 2 (theme and plot selections as site plans); application 3 (plot selections by user input as site plans—alternative to application 2); application 5 (search report); and application 7 (quick maps for inspections). However, when respondents were asked to declare which applications they did not like, one was indifferent, one did not respond, and four indicated none.

The applications were not confusing to four respondents with two citing application 10 (lodging Avenue scripts as text files in directories) as confusing. Incidentally no comments were proffered. Two respondents wanted application 5 (search report) and application 6 (instant layout) to be presented in a better way. One was of the view that the search report could be presented in a form that would be easily and attractively readable. Another was of the opinion that the layouts should be in landscape form and the site plans should indicate the acreage, cadastral plan number, and the name of the area as a caption. Three thought the application could be improved with an equal number indicating their tacit acceptance of the applications as they stood. None rejected the applications.

Questions such as "What applications do you want added in future?" were added in the questionnaire to help in the expansion of the application to cover other areas. One view was that rent management in government residential areas was a priority and ought to feature in such prototype software. Others wanted the application to automatically calculate ground rents, to list all lessees in arrears for a given period, and to process documents so that an applicant can actually trace his/her document after it had been presented.

This exercise strengthened our belief that potential users must be involved in the design and implementation of GIS from the very beginning so they have a feeling of ownership of the system. Such views can be identified and accommodated by involving users in the design process itself. It is proven that system development that does not involve user participation has mixed adoption rates (Cavaye and Cragg, 1993). Systems built in reaction to needs expressed by users are adopted more quickly than systems built in the absence of expressed need. We note that change, as a consequence of introducing IT and GIS, cannot be effectively introduced if it is imposed or controlled (Campbell and Masser, 1995).

Land-use changes occur frequently, and it is recognized that the functionality of GIS need not be restricted to static displays. It is our intention that later versions of the software will facilitate the representation and modeling of phenomena such as encroachments on public land, a major issue in land management in Ghana, as a mechanism for advising government decision-making. Data on encroachments will need to be collected and combined within the GIS to show their spatial extent and their location in relation to other properties and infrastructure. The system would allow vulnerable structures to be identified and earmarked for immediate demolition. This type of development implies that GIS should be used more as an analytical tool in the land-sector agencies in future.

10.8 Observations

The successful implementation of GIS to support land administration in the land sector in Ghana is confronted with a series of challenges. This includes the need to provide frameworks within which GIS can evolve as a tool in an orderly way in the land-sector agencies; the need to find ways to democratize GIS in land administration and management systems and structures within Ghana; the need to generate designs that are innovative and practical so they will meet specific land sector needs; and the need to provide support infrastructure and services that will enable GIS to operate effectively and efficiently in the land sector.

It is important that organizations in Ghana's land sector contemplating GIS use consider the general definition of what GIS are (database, analysis, and visualization) in terms of what functions these systems may have within the specific agency and within its broader social context. The ability for GIS to do spatial analysis gives it geospatial prowess and distinguishes it from computer-assisted design (CAD) systems that are more suited for design and presentation. In the case of land-sector organizations, any definition that emphasizes the institutional aspect, yet deals with the appropriate technical aspects as well is seen as generic and therefore suitable for adoption.

Much reported research in Africa particularly, has either been anecdotally based (research not conducted under controlled conditions) or centered on retrospective single case studies of successful implementations. There is little by way of research to show the use of empirical methods in a more systematic study of GIS/LIS diffusion in Africa. This work, therefore, recommends a procedure of study aimed at examining prospective process of adoption, allowing new classes of users to learn the potential application of GIS in their work practices (Onsrud and Pinto, 1991).

There are fundamental similarities between GIS/LIS and information science (IS) or information technology (IT). To be able to develop an acceptable theoretical framework for GIS/LIS diffusion, the state-of-the-art research in IS or IT or even information management (IM) would have to be employed. There are different types of methodologies that lend themselves to GIS research. While the choice of methods may be critical, the review of literature suggests a suite of methods or an eclectic approach to the selection of these methods. No single method is most appropriate for studying a broad or complex research problem. Each method or combinations of methods have different assumptions, biases, and degrees of usefulness (Williamson et al., 1998). A common approach is to use several research methods that compensate for each other's weaknesses. Such methodologies should cover both performance and functional needs of the organization and must be applied under an acceptable conceptual framework.

There is general agreement on the fact that major problems to be overcome in improving land-information practices in developing countries are organizational, managerial, and human based (Zwart, 1990; Taylor, 1991). It is also clear that GIS technology is a Western artifact. Any wholesale diffusion of this tool therefore, without critical analysis and without considering the socioeconomic and institutional background of developing countries, may be counterproductive. Such critical analysis should include awareness creation among users and feasibility studies including a pilot project preceded by prototype software development. It should also include learning from the past and best experiences, development of appropriate systems required, project proposal, systems testing and evaluation, start-up, and, finally, systems operation and review tailored to specific agencies.

Finally, GIS installations in developing countries have mostly been donor designed and implemented exclusively by consultants. Governments must now demand and receive more control over project design and implementation. Ironically, the literature suggests that there is still lack of experience of experts from developing countries on the GIS/LIS diffusion processes and models for such countries. A suitable balance must be found. The development of new models and the refinement of existing ones to suit these countries are therefore necessary. While in many African countries there appears to be a hoarding of information, and therefore, little information exchange, effective project design should look at arrangements for data and information sharing. User needs, assessment, and training are also critical in the design process. Participation by NGOs and the private sector (as the engine of growth) in the use and production of geographic information must be encouraged.

10.9 Conclusions

Ghanaians have a strong attachment to land. The need to establish and develop GIS and networks among the related land agencies and to encourage internal cooperation and support in all aspects of land policy and administration engages the attention of Government. There is now a heightened awareness of the difficulties that often exist in trying to introduce GIS within land-sector organizations in Ghana. Our research relates to understanding not only the technical problems inherent in constructing GIS applications but also of the socioeconomic and political context within which GIS are to be adopted. This paper has outlined the development of the prototype software that will provide a system for much improved administration of routine land-management tasks in the LCS. This technical work on software development has been accompanied by fieldwork to identify the needs of the LCS as well as those factors deemed by existing staff to be critical if the prototype is to be accepted and full GIS implementation is to be successful thereafter.

References

Black, J.D., 1996, Fusing RDBMS and GIS. *GIS World* (http://www.gisworld.com).

Burns, T., Eddington, B., Grant, C., and Lloyd, I., 1996, Land titling experience in Asia. *Paper Presented at the International Conference on Land Tenure and Administration*, Orlando, FL, 23–26 November 1996.

Campbell, H. and Masser, I., 1995, *GIS and Organisations: How Effective Are GIS in Practice* (London: Taylor & Francis).

Cash, J.R., McFarlan, F.W., and McKenney, J.L., 1983, The issues facing senior executives, In *Corporate Information Systems Management: Text and Cases*, edited by Irwin, R.D., pp. 457–464 (New York: Irwin/McGraw-Hill).

Cavaye, A.L.M. and Cragg, P.B., 1993, Strategic information systems research: a review and research framework. *Journal of Strategic Information Systems* 2(2), 123–136.

Center for Democracy and Development, 2000, Corruption and other constraints on the land market and land administration in Ghana: a preliminary Investigation. *Survey Report Research Paper Number 4* (Accra: CDD).

Edralin, J., 1990, *Conference Report—International Conference in Geographical Information Systems: Application of Urban and Regional Planning* (Nagoya: UNCRD).

Fox, J.M., 1991, Spatial information for resource management in Asia: a review of institutional issues. *International Journal of Geographical Information Systems* 5(1), 59–72.

Hastings, D.A. and Clarke, D.M., 1991, GIS in Africa: problems, challenges and opportunities for co-operation. *International Journal of Geographical Information Systems* 5(1), 29–39.

Huxhold, W.F., 1991, *An Introduction to Urban GIS* (Oxford: Oxford University Press).

Karikari, I.B., Stillwell, J.C.H., and Carver, S., 2002, GIS application to support land administration services in Ghana: institutional and software developments. *Working Paper 02/02,* School of Geography, University of Leeds.

Lai, P.C. and Wong, M.K., 1996, Problems and prospects of GIS development in Asia, In *GIS in ASIA: Selected Papers of the Asia GIS/LIS AM/FM and Spatial Analysis Conference*, edited by Fung, T., et al., pp. 219–229.

Larbi, W.O., 1995, The urban land development process and urban land policies in Ghana. In *Our Common Estate* (London: The Royal Institution of Chartered Surveyors).

Onsrud, H.J. and Pinto, J.K., 1991, Diffusion of geographic innovations. *International Journal of Geographic Information Science* 5, 447–467.

Pradhan, A.K. and Tripathy, K., 1994, A conceptual framework of spatial and non-spatial data bases for the development of an efficient Geographic Information System. *GIS India* (January), 28–37.

Simpson, S.R., 1976, *Land Law and Registration* (London: Sweet and Maxwell).

Somevi, J.K., 2001, The effectiveness of institutions in land registration in Ghana. *Our Common Estate* (London: RICS Foundation).

Taylor, D.R.F., 1991, GIS and developing nations. In *Geographical Information Systems Vol. II: Applications*, edited by David, J.M., et al., pp. 71–83 (New York: Wiley).

Toulmin, C. and Quan, J. (editors), 2000, *Evolving Land Rights, Policy and Tenure in Africa* (London: DFID/IIED/NRI).

UNCHS (Habitat), 1990, *Guidelines for the Implementation of Land Registration and Land Information Systems in Developing Countries: with Special Reference to English Speaking Countries in Eastern, Central and Southern Africa* (Nairobi: UNCHS).

Wasmund, M., 1993, Implementing critical success factors in software reuse. *IBM Journal* 32(4), 595–611.

Wiley, L., 1997, Think evolution, not revolution, for effective GIS implementation. *GIS World Inc.* 10(4), 48–51.

Williamson, I.P., Chan, T.O., and Effenberg, W.W., 1998, Development of spatial data infrastructure—lessons from the Australian digital cadastral databases. *Geomatica* 52(2), 177–187.

Zwart, P., 1990, Land information management without computers. *Newsletter M. 5* FIG Commission 3 Working Group on Land Information Systems in Developing Countries, pp. 15–16.

11

Spatial Methodologies to Support Postwar Reconstruction

Sultan Z. Barakat, Adrijana Car, and Peter J. Halls

CONTENTS

11.1   Introduction

The Post-war Reconstruction and Development Unit (PRDU) of the University of York has been concerned with practical methodologies for community reconstruction after conflict for several years. Their work is predicated on experience derived from observation and participation with humanitarian aid agencies in the delivery of relief and in the encouraging and enabling of return and reconstruction. This experience, gained from work in a wide variety of theaters of conflict, demonstrates that postwar reconstruction (PWR) is a complex operation concerning individuals, communities, property, infrastructure, environment, and cultural heritage. It is the experience of the PRDU that an inclusive concept of reconstruction and development, which encompasses the key stages of relief, rehabilitation, and reconstruction leading to sustainability, is the most appropriate. This approach springs from the concept within the delivery of humanitarian aid of *do no harm*. Do no harm is concerned with the provision of emergency aid, food, clothing, medical care and shelter, and seeks to deliver this without prejudice to either the present situation or future reconstruction efforts. Initially conceived regarding the protection and emergency repair of cultural heritage, the concept underpins humanitarian aid.

Spatial Decision Support Systems (SDSS) "are explicitly designed to support a decision research process for complex spatial problems" (Densham, 1991, p. 403), and there can be few more complex spatial problems than rebuilding communities after war. Current methodologies in reconstruction are based on the creation of inter- and intracommunity focus groups, starting with a planning workshop, during which a serious attempt is made to identify and account for each community component's needs. These groups are inevitably multidisciplinary, and there are frequent problems relating to the need to interpret the understanding and language of one group of professionals to that of another and to the nonprofessional populace. Participatory planning GIS research appears to offer opportunities—especially where, in the postwar arena, communities remain dispersed owing to unsafe ground conditions or where individual groups of people cannot yet bring themselves to cooperate, face-to-face, with some other group or groups. The additional benefit of the potential to model possible scenarios in order to assess the likely impact and implications would be very valuable in helping to avoid mistakes which can frequently cost human lives.

This chapter reports work in progress to apply spatial methodologies for the design and implementation of decision-support systems for application in the task of community reconstruction after war. Although examples are cited from specific experience, we are seeking a general set of principles.

11.2   Postwar Reconstruction Process

While the relief effort is precisely that, rehabilitation is concerned with the support of afflicted communities. This includes encouragement and training such as to enable the people to organize themselves, to take control of their situation and environmental context, and to actively participate in the provision and distribution of aid and in the improvement of services, infrastructure, and conditions in general.

Reconstruction, the first step on the long-term recovery process, is concerned with the physical, the institutional, and the environmental aspects. There will be a need to jump-start the development process that has been interrupted and set back by the disaster. It is our experience that despite the hardship of survival during a protracted conflict, people continue to display a lively spirit of initiative and enterprise. People seek solutions to their immediate needs and an informal economy often operates within and across formal lines of demarcation involving local populace and combatants. Although formal administrative and international support is necessary to achieve large-scale reconstruction, at the local level recovery proceeds regardless of politics, international interventions, or formal planning—but is much slower and more painful than might otherwise be the case. Such local activities take advantage of whatever opportunities reductions in the conflict offer. By exploiting this local knowledge and expertise, local governance and aid agencies can enable the transition from emergency to reconstruction in more stable areas by the provision of incentives and, thus, strengthen moves toward peace. There is a symbiotic relationship between reconstruction and peace building: each is inextricably dependent on the other, and it is the experience of the PRDU that this phase can be an effective agent toward the achievement of peace. Clearly, unless care and a long-term view are taken, it is possible to take action at this stage which will adversely impact the subsequent phase of reconstruction. This final stage may be identified as beginning at the point at which the conflict can be defined as being over but has no discernable conclusion, gradually over time being subsumed into the normal planning activities of a sustainable community.

Planning depends upon the availability and access to up-to-date information and of effective means of communicating this information to all the actors involved. Information is the vital component of all decision-making as is the skill of prioritizing the available information in order to extract that pertinent to the case in point. It is also important that a system exists to maintain this information as and when the conditions change. One component of this information must be knowledge of the community dynamics, for example, how that community copes with disaster. Barakat and Deely (2001, p. 63) complain that communities' coping capacities tend to be ignored or underestimated and that there is little research or understanding of this vital community component. We will discuss this issue later.

Sparrow (2001) quotes Tenna Mengiste, formerly of the Ethiopian Red Cross, "relief will save the day, but when people are living on the verge of disaster it isn't enough to tell them to return to the status quo. If we can raise the quality of life, we may enable the most vulnerable not just to survive but to cope with, and prevent disasters." Sparrow reports that some places prone to natural disasters are becoming lawless and a threat to security, and charts political instability resulting from serious natural disasters from the volcanic eruption of the Aegean island of Thira in 1600 BC through the 464 BC rebellion in Sparta, triggered by an earthquake that left the city in ruins, to events in the present. While it is clearly simplistic to try to blame all conflict on natural occurrences, just as it is to blame despotic leaders, disaster resilience is dependent on sustainable livelihoods since income generation on the part of the populace boosts the chances of peace (Simms, 2001).

Simms (2001) outlines what he calls an ecology of disaster recovery. He sees instability, resulting from the currently changing environments of world trade and finance, and climate, such that aid and recovery efforts will increasingly be judged not by how quickly structures are rebuilt, only to be destroyed by the next disaster, but by how reconstruction contributes to the long-term resilience of communities (Simms, 2001, p. 35). Citing, by example, the biodiversity of the natural environment and noting that those healthiest parts are also the most diverse, Simms calls for solutions to include community involvement in the design and implementation of reconstruction, more local procurement of resources, and the introduction of new methods of assessment to ensure aid interventions sustain rather than undermine local economies. He notes that diversified local economies are stronger than monocultures and from his observation of the natural environment's diversity, argues for a diverse economy, built on small-scale enterprises that use resources sustainably and incorporate indigenous knowledge (Simms, 2001, p. 50). Any methodology must therefore be able to interact with the large numbers of local actors who will be involved.

For many years, public response to disasters affecting people has been to donate relief aid. Frequently, however, the aid has been delivered with but a poor and superficial prior analysis of what is required in the short term, let alone the requirements for a sustainable future. The way aid is delivered can prove counterproductive and continue to raise disturbing questions about whose needs are being served by aid—those of the donor agencies or their beneficiaries (Sparrow, 2001, p. 13). Sparrow quotes a Khurantatuth schoolteacher who, following receipt of aid after a cyclone had all but destroyed his village, said thoughtfully "The village was grateful, but a strategy of kum kum (little by little) would have been better. It was more than a family could consume." The quantity of relief delivered in a single operation had been so great that the villagers had not known what to do with it! Sayagues, concerned about the post-aid plight of flood victims in Venezuela, comments that the requirement is to put people at the center of disaster recovery and commitment to ensure that temporary solutions do not become permanent—aid dependency delays recovery (Sayagues, 2001, p. 86). The need is for careful targeting of relief aid in order to provide a secure platform from which those affected by the disaster can begin the slow process of recovery (Sparrow, 2001, p. 15).

Strategies and processes for reconstruction after conflict must be relevant to the political, social–relational, and environmental context in which they are to operate. Inappropriate or inadequately flexible processes or strategies can, however, fuel tensions and have the potential to lead to (re)new(ed) conflict (Stiefel, 1999). The period during, and immediately following, active conflict is one of rapid change. As the situation changes, information preciously gathered becomes outdated, people and resources become available, incapacitated, or move. The indicators used to measure local activity, economic, social, cultural, and environmental must be flexible and responsive to the changing situation. In addition to receipt of information from the locale, there must be a balancing flow of information back, to encourage and support the local people, or their goodwill, and the information source, will be short lived. In the progress from the anarchy of war to the security of peace people must be able to feel a part of the processes concerning them: participation is essential, not an optional extra. "Designs cannot simply be dropped from one environment to another and often the local workforce have no experience in working with the materials...sustainable change comes from within" (Jaquemet, 2001, p. 113). To achieve appropriate and sustainable designs demands close collaboration with the local people, to understand their environment and resources.

Sustainability must relate to the availability of resources, human, natural, physical, social, political, and economic. Skills obtained prior to the conflict may not be relevant to reconstruction, indeed may relate to outdated industry or industry for which the previous sources of raw materials or markets may no longer be accessible. New skills may need to be acquired. The reconstruction process must focus on the real needs and priorities of the community because they form the primary resource; it is essential that the local communities have a sense of ownership and responsibility toward the reconstruction projects (Barakat and Hoffman, 1995). One problem here is in managing the balance between external aid and the exploitation of local resources. Uncontrolled deforestation to yield timber products for rebuilding, for example, could lead to serious environmental problems that may threaten the survival of the community in the future. Care must be taken in the disposal of waste, in the location of new development and the impact of infrastructure reconstruction on natural phenomena, such as river flow and pollution. Hitherto, much of the focus of reconstruction has concerned the physical, urban environment yet harmony with the natural environment is critical for sustainability. In some cases, for example, following previously intensively worked agricultural land remaining fallow for several years may open up opportunities for a change to more sustainable and organic practices. In addition, movement away from intensive arable practices lessens the danger of groundwater pollution with consequent benefits for the urban population dependent upon subsurface water supplies.

Barakat and Deely (2001) contend that a rehabilitation methodology is required that is focused on the local populace and directed at enabling the development of a sustainable community. They set out potential conceptual, programming, and structural barriers to the process, which must be identified and evaluated, and recognize that the methodology must need adapt to certain of these barriers, in order to work around them. With each of the barriers enumerated, they outline likely benefits for acknowledging that type of barrier and the dangers of ignoring them—or, particularly in the case of structural barriers, to take account of the limitations of the reconstruction process and so inform progress in spite of that barrier. From this stance, they propose a methodology for community-led rehabilitation in terms of foundations and steps toward local ownership for sustainable recovery. They make the case that investing in rehabilitation can be a means of investing in peace, no matter how far off formal peace may appear. "The long term recovery of war-torn societies is dependent on increasing people's confidence in their future both at household and community levels. This is best achieved through initiating community-based rehabilitation programmes" (p. 80).

11.3 GIS in Postwar Reconstruction

Geographical information systems (GIS) are widely used as a tool in planning and modeling, town and country planning, retail development, environmental planning, and demographic analysis (Burrough and McDonnell, 1998; Heywood et al., 1998; Bibby and Shepherd, 1999; Cova, 1999; Larsen, 1999; Yeh, 1999). The PWR process is spatial in nature. The PWR problem is in reality a complex case of normal planning and decision support. Reconstruction after war is, like normal planning activities, a spatial problem. In the theater of conflict, GIS perhaps first became publicly used with regard to the Gulf War in 1990 when, in conjunction with satellite imagery, GIS tools were used to map and report pollution incidents in the Gulf and well fires on land. These incidents continued to have an impact on the environment long after the cessation of hostilities (USEPA, 1991). Since the actual war, GIS has been used as a modeling tool within the ongoing investigation into the health problems experienced by service personnel who served during this conflict, popularly termed Gulf War Syndrome. This work has included assessment of troops' exposure to atmospheric pollution from burning oil wells as well as other potential causes (Heller et al., 2000). GIS have been used for mapping war damage in Croatia during the early 1990s (Delać, 1994; Horvat, 1994; Kereković, 1994a,b). In Lebanon, following the beginnings of peace in 1990, a real-estate holding company, Solidere, was formed to bring together individual landowners, investors, and planning in order to facilitate the reconstruction of Beirut. Solidere, created by a law passed by the Lebanese parliament in 1991, was charged with implementing a master plan for the city, developed through a consultative process involving the Ministry of Public Works, the Beirut Municipal Council, and the Higher Commission of Urban Planning. This master plan concentrated on designations for the use of space in the city and a consequent need to update the city's cadastre, together with state information regarding properties and infrastructure and a GIS-based management system for this spatial information was adopted. Following the 1996 Israeli incursion into South Lebanon, a similar GIS-based information management and planning process was adopted for reconstruction planning (Ekmekji, 1997; Fry, 1997; Kabbara and Soubra, 1997; Kitmitto, 1999; République Libanaise, 1999; UNEP, 1999). GIS have been used recently in the planning of delivery of humanitarian aid in Kosovo as well as the delivery of aid and reconstruction planning following natural emergencies (Atkins, 2000; Korsey, 2000; Smith, 2000). Much of this work stems from the close relationship between the agencies delivering humanitarian aid in Kosovo and the UN peacekeeping force, KFOR. Most military operations use highly developed spatial planning models. Little, if any, of this GIS work was carried out in Kosovo, however. The operation led by the United States, for example, employed U.S. nationals located in mainland USA to perform the GIS-based analysis. Such remote use of such technology is understandable during the period of active conflict.

Despite these examples, the application of such tools does not appear to be widespread. One reason for this may have been perceptions of cost and hardware requirements: modern laptop computers now have the power to run the most sophisticated of commercial products and are relatively inexpensive and readily available. Additionally, there is a perception that a data-hungry technology such as GIS can never be satisfied in such situations. Kassa (2001), writing the Engineers for Disaster Relief (RedR) newsletter, states that "the main reason for its (GIS) limited use by emergency planners lies precisely with data, or the lack of it. Timely planning of response to humanitarian emergencies using GIS requires, amongst others, sufficient information about the nature and magnitude of the causal event, the affected area, as well as the population groups likely to be most affected." He goes on to recommend the use of remote sensing imagery as a source of suitable data to fill the gap and notes that this is less of a problem with slow onset events (such as conflict) than for sudden natural disasters. Another reason has been the concentration of information technology, and Internet usage, in a very small proportion of the world. Bolle (2001), citing UNDP sources, reports that the rich, OECD countries that make up just 14% of the world's population account for 82.5% of the Internet users. This divide results in a number of problems for the have-nots: barriers to knowledge, barriers to participation, and barriers to economic activity. In the developed world, we tend to take the availability of Internet resources for fact finding for granted yet those without the computing and communication infrastructures are effectively denied access.

There is some good IT news, however. Computing equipment currently has an active lifespan of a few, perhaps 18, months before being rendered obsolete by technological progress. Such equipment is only obsolete in that it is not as fast or is lacking in some novel feature when compared to its replacements; it remains completely serviceable for the vast majority of uses for which such computers are supplied and there is a growing transfer of such equipment to less technologically demanding parts of the world. Hermida (2002) describes how the Norwegian Ministry of Environment and the Dutch Ministry of Foreign Affairs have funded a network of such computers, donated by a private Dutch company, in Kosovo which has enabled the coming together of former combatants. This network, named Sharri.Net is enabling Kosovars to use the Internet to disseminate information about the environmental and reconstruction problems they face. The major issues are the reconstruction process and the environmental problems of mines, unexploded ordnance, and pollution. This facility has enabled ethnic Albanians and ethnic Serbs, former foes, to work together to solve their common problems. There are a number of agencies in the developed world undertaking the distribution of such computing equipments. Kassa (2001) concluded that "GIS presents an ideal platform for combining different datasets in order to model the impact of various humanitarian emergencies with the view to designing efficient responses…GIS will soon become a routine tool of humanitarian agencies for efficient and appropriate emergency response planning."

11.4   Use of Spatial Decision Support Systems in PWR

GIS and more especially SDSS are widely used in the planning process, in particular as an aid for urban planning. Moore (2000, p. 138), following Densham (1991), states that "Spatial Decision Support Systems are explicitly designed to support a decision support process for complex spatial problems. They provide a framework for integrating database management systems with analytical models, graphical display and tabular reporting capabilities." They incorporate and integrate the expert knowledge of decision makers, potentially from a number of relevant disciplines.

Decision Support Systems (DSS) are framework tools that enable the integration of information through database management systems, analytical models and graphical display to enable decision-makers better to explore the impacts of alternatives when making decisions. In particular, DSS are designed to handle problems that poorly defined, that are a mixture of quantitative and qualitative, and that are may be poorly or only semistructured (Yeh, 1999). The ability to handle problems that may be poorly structured or only partly defined and that comprise a mixture of qualitative and quantitative elements is so similar to the reconstruction process that further exploration is clearly justified, especially if it can be combined with some learning mechanism to benefit from the identification of faults. This leads us to see SDSS as a form of modeling tool which, drawing together the explanations of Cova (1999), Estes et al. (1992), Heywood et al. (1998), and Yeh (1999), we summarize in Figure 11.1.

There are primary flows between policies and the decision-making process, the one influencing the other and in turn, between the decision-making process and information and between the decision-making process and the problem (Figure 11.1). The information component comprises a number of facets, each of which relate to one or more sectors of the problem definition and which inform both the decision-making process and the solution model. The problem comprises a set of defining sectors, setting out the goals, objectives, alternative strategies, risks, evaluation and informed choice of strategy, and implementation and monitoring of that strategy. Following evaluation of the alternatives and risks involved, a strategy can be chosen for implementation or rejected and an alternative considered. Consideration of the implementation will inform the evaluation and, possibly, lead to the abandonment of that strategy or else the strategy

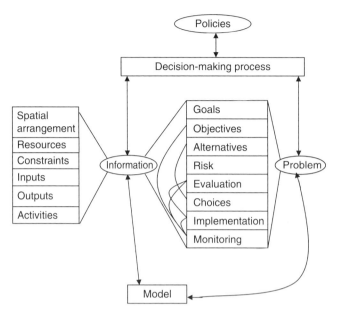

FIGURE 11.1
The decision support process for postwar reconstruction (PWR).

will be monitored throughout implementation for effectiveness and conformance with the objectives, feeding back to the evaluation for acceptance or rejection of the strategy. Throughout, the considerations are informed by the spatial arrangement of the components of the case study, by the resources available at any stage and the inputs required, by the way the activities involved interrelate to other parallel processes and by the availability of outputs from one process to facilitate another. The procedures may be used to model the whole of an activity, such as for regional planning, or to explore one component; all that changes are the formal definition of the problem, and the nature and scale of the associated information.

Halls and Barakat (1999) set out a five-phase process, which is similar to any complex planning operation:

1. Identify the problem
2. Plan for reconstruction
3. Implementation and management of the reconstruction plan
4. Evaluation and feedback
5. Maintenance and sustainability

These are to a large extent common sense, but frequently phase 2 is actioned before phase 1, or there is little provision for refinement in the reconstruction plan following changes on the ground. Phase 3 is often understated, with the emphasis confined to immediate achievement and little long-term thinking. The final phases, which should pervade the whole process enabling constant adjustment and refinement, are usually at best neglected. Halls and Barakat (1999) noted that despite the absolute requirement for a constant and effective information flow there is frequently a pervasive parochialism regarding information and that, prior to the conflict, there may have been little ethos of freedom of information or information sharing. Education, training, and public awareness hold the key to effective participation in the reconstruction effort: information must flow freely, be up-to-date, and be available where and when needed in as effective and timely a manner as possible.

11.5   Public Participation GIS

According to Barakat and Deely (2001), a reconstruction methodology must take into consideration the initial abilities of the target community to contribute to, and willingness to participate in, the rehabilitation of their community and must seek to develop their capacities such that they have the resources to continue the community sustainably. There are a number of factors involved in achieving such a target, including the attitude of agencies and the political will of local and international bodies. For example, agencies are often reluctant to transfer any real responsibility for programming, delivery or policy to local employees, communities, and officials, preferring instead to place expatriate workers in the afflicted communities. Simms (2001) reports that the resultant financial presence of aid interventions, agencies' demands for hotel and office space (for expatriate staff), food, fuel, and local employees can have a negative impact on the local economy as well as bring new money in. He notes how the spending power of the international aid workers and peacekeepers in East Timor is primarily benefiting Australian and Singaporean business, rather than stimulating growth in the local economy (p. 41). Barakat and Deely (2001, p. 63) state that "communities must be recognized as main partners as their participation ensures more effective management and relevant provision. Communities must be enabled to own, manage and support their rehabilitation."

Public participation in planning decision-making is a current goal in the changes being undertaken in Western nations, as they seek greater transparency and accountability in government. GIS are employed as one tool to facilitate public participation in decision-making. The Public Participation GIS (PPGIS) initiative within the Varenius project at the NCGIA (NCGIA, 1998) illustrates some of the work in this arena. The NCGIA initiative stems from an investigation into GIS and society, NCGIA Initiative 19 (NCGIA, 1998). The NCGIA is also engaged in an exploration of SDSS, under Initiative 17, Collaborative Spatial Decision

Making (NCGIA, 1996), which is closely related. The goals of these two projects include enhancing the availability of appropriate spatial analysis and decision modeling tools (Initiative 17) and an exploration of the role GIS might play in changing local politics and power relationships in decision-making. By the empowerment of local communities in decision-making and the exploitation of the value of sophisticated analysis, a means of promoting understanding of key issues (as opposed to the impact of such analyses in confusing and marginalizing individuals and groups) is achieved (Project Varenius). As these clearly have relevance to the issues of enabling and empowerment of people in the rebuilding of their communities during and after war, key points from the NCGIA Workshop on PPGIS are presented below.

Evans et al. (1999), Kingston (1998), Kingston et al. (1998, 1999), and Carver et al. (1999) describe work at the University of Leeds Computational Geography Group in the use of the WWW as a means of enabling individuals to participate in the planning process for their locality. In their pilot study for the town of Slaithwaite, they describe a mechanism by which members of the public can explore a planning context, interactively, using a dynamic map. Among the benefits they observed was the increase in information that it was possible to make available, over and above that possible using a traditional map. In addition, they reported that the two-way nature of the process enabled participants to view suggestions and alternatives proposed by others and to see how these fitted into the overall scheme. Carver et al. (1999) concluded that a WWW approach had an especial value for dispersed rural areas where it can be difficult to attend (meetings) at a particular time or place. In the context of PWR, such an approach has the additional benefit of enabling participation by, for example, inhabitants displaced—potentially to countries other than their homeland.

Kingston (1998) introduces a development of the public participation ladder, a series of steps from public right to know through to full public participation (Table 11.1). "In order to achieve (even) greater involvement

TABLE 11.1

The Public Participation Ladder

Public participation in final decision	Full	
Public participation in risk assessment and in recommending solutions		
Public participation in defining interests, actors and agenda		Public participation
Public right to object		
Informing the public		
Public right to know	Little	

Source: Adapted from Kingston, R. in Web based GIS for public participation decision making in the UK. *Paper Presented at the NCGIS PPGIS Meeting,* Santa Barbara, CA.

in (environmental) decision-making the public need to be provided with systems which allow them to create virtual spaces. Such systems should allow participants to explore the decision problem; experiment with choice alternatives; formulate one or more decision choices; and contribute to feedback and evaluation of the system" (Kingston, 1998, parentheses added). He goes on to elaborate these four steps, noting that exploration of the decision problem is an essential part of the users' learning process, which depends on direct and free access to all the information relating to the decision problem.

This approach to the analysis of participation, however, perpetuates the top–down nature of decision-making where the decision is made on behalf of the public, albeit with consultation. The usual developmental approach for decision-making has been to keep qualitative and quantitative data collection and management separate (Jordan, 1998). At a district or national level, studies often map socioeconomic indicators, commonly called indicators of development, although the people targeted for the development process are generally unaware that these indicators exist. Because such indicators are used for policy planning to identify both development priorities and geographic regions of activity, Jordan suggests that the developmental role of GIS often disempowers local people, because it involves a very low-level of public participation. But, Jordan says, "this is putting the technology before the people. Whilst it appears that GIS is being used for classic decision support purposes, the decision-making process itself is fundamentally flawed. There is little or no consultative process with communities."

Indeed, the affected people are most often simply told what is about to happen. This, we find, is close to what frequently happened with reconstruction aid: the people at the center of the community concerned merely become spectators of the reconstruction effort. Our experience in PWR shows that it is crucial for active involvement not only in making the decisions but also in carrying out the actions decided upon. We have witnessed too many situations where the local people have been relegated to the status of spectators, at best informed of the decisions made concerning them, frequently finding out only as others undertake the actions, as it were, on their behalf. This leads us to recast the participation ladder, as shown in Table 11.2.

TABLE 11.2

Public Participation Scale in Assisted Recovery

External support of local initiatives	Full	
Cooperative action		
Collaborative decision-making		Public Involvement
Consultation of local populace by external agencies		
Information dissemination regarding external agencies' intentions	Little	

The key difference between the public participation ladder of Kingston and ours is that as the level of public participation increases so the direct involvement of the external agencies decreases, to the point where they are providing support for initiatives entirely planned and implemented by the local people. By the term external, we understand both international agencies and also more local agencies where the decision-making is undertaken far from the scene of activity. Of course, in the early stages of PWR such detached planning may be inevitable: it is our experience that the reconstruction achieves greatest benefit where the local people are enabled to take over the decision-making, locally, as and when they have the resources. The role of the external agency becomes one of ensuring that these local resources become available.

In quoting "Tell me, I forget. Show me, I remember. Involve me, I understand." (from Moore and Davis, 1997), Howard (1998) makes an extremely apposite point in the context of reconstruction as well as for public empowerment. Spatial empowerment of the public requires their (the public's) utilization of appropriate technology together with the appropriate participation technique.

A critical part of learning, and of decision-making, is experimenting with alternative choices. The what-if approach forms a foundation to most GIS implementations, and the formulation of decision choices should aim to maximize consensus and minimize conflict. This is a difficult thing to achieve but is founded on compromise and one in which a neutral intermediary, such as a decision-making tool, may assist in encouraging evaluation from alternative points of view. Feedback and evaluation are essential components of any project: in decision-making, feedback to the users is essential to enable evaluation of how particular choices have arisen and keeps clients aware of the way they are formulating their choices. Having explored a case study of an implemented Web-based participatory planning system, Kingston concludes that participatory online systems will become an important tool in involving the public in planning decision-making and in formulating policy. He envisages such systems as providing all users, planners, and public alike, with appropriate mechanisms for exploration, experimentation, and formulation of decision alternatives (Kingston, 1998).

11.6 Deriving Ontology for SDSS for PWR

PWR is critically dependent upon knowledge sharing and information management and dissemination among and between individuals from a wide range of disciplines. While there are characteristics that are unique to the reconstruction process and the process is not only focused on the urban environment, there is a great similarity with the problems of urban management, and it is, therefore, worth exploring such tools.

Fonseca et al. (2000) describe the problem of dealing with the detail rich urban environment and with enabling the disparate disciplines involved to share their knowledge in order to better inform the management process. They note that there is a great deal of knowledge available, some of it already stored within management systems such as GIS, but that the lack of formal methods for the reuse of knowledge and data makes it difficult to gain benefit from the undoubted potential. Their proposal is to draw together the diverse ontologies involved in the urban environment, deriving computer software components from these ontologies, to provide a methodology for knowledge and data sharing. This approach is directly applicable to PWR for exactly the same reasons.

A frequent barrier to participation is the inaccessibility to the semantics of one discipline by another. Some have sought to address this in terms of public access, for example the work of Parker (1998), and by making use of the World Wide Web (Kingston, 1998; Parker, 1998; Kingston et al., 1999; Shiffer, 1999). Sometimes distinctive discipline languages have been developed as a means of professional security. In the context of PWR, which is an inherently cross-disciplinary activity, collaboration is necessary. To collaborate presumes the use of a common vocabulary as well as access. When international agencies become involved, these linguistic difficulties are increased by orders of magnitude—yet simple misunderstandings can be enough to rekindle the conflict itself. As our goal is to design and develop a dynamic model that can configure itself as the various actors define themselves and their recent thinking, a methodology that facilitates bridging across disciplines and language is essential.

In his discussion of an approach for handling knowledge and data portability and reuse, Gruber (1993a) notes that because different applications require different kinds of reasoning services or special purpose languages to express their knowledge there cannot be a common representation language or system. It would be inappropriate, for example, to require the engineer to express her knowledge in the language of the social worker but both have valid languages that are eminently suited to their respective disciplines. Gruber's solution is the identification of common language components, or ontologies that may be applied across multiple representation systems. In such a translational approach, the ontologies are specified in a standard, independent form and then translated into specific representation languages. For a worked example, Gruber selects reuse of a knowledge-based urban planning program/system. This is, of course, the domain of the (Spatial) DSS. Reusing such a system, or applying it to a new situation, entails the adaptation of the existing knowledge base to the new application domain or the creation of a new knowledge base from scratch. Detailed understanding of the kinds of information used as inputs or generated as outputs and of the kinds of domain knowledge applied is necessary. A common ontology can facilitate the sharing of knowledge (Gruber, 1993b) and so enable the participation of multiple actors by serving as a knowledge-based specification defining the vocabulary with which

Pockets of deprivation: Dalbeattie*

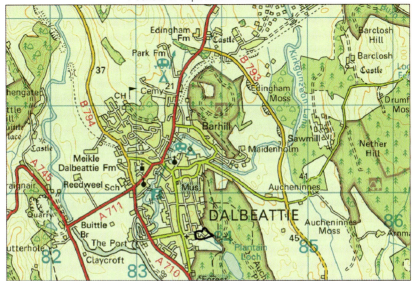

*Criteria for selection: OA Carstairs vs. PPS Carstairs ≥5; Population density ≤100; OA Carstairs ≥7.

(a)

Pockets of deprivation: South Kessock, Inverness*

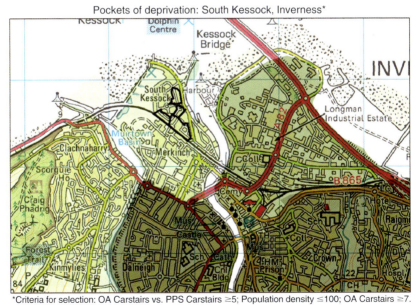

*Criteria for selection: OA Carstairs vs. PPS Carstairs ≥5; Population density ≤100; OA Carstairs ≥7.

(b)

FIGURE 5.4

Examples of pockets of deprivation located in (a) rural settlements, (b) urban areas

(*continued*)

Pockets of deprivation: Newbigging, near blairgowrie*

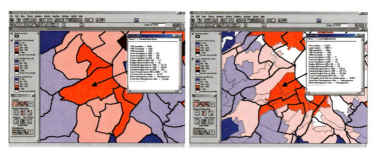

*Criteria for selection: OA Carstairs ≥7; Population density ≤100; OA–PPS Carstairs ≥7.

(c)

FIGURE 5.4 (continued)

(c) rural areas. (© Crown Copyright/database right 2007. An Ordnance Survey/EDINA supplied service.)

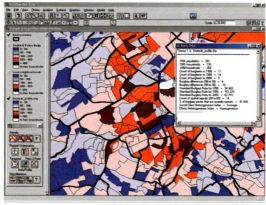

FIGURE 6.7

Comparison of standard burglary rates (SBR) for 2000 at varying spatial frameworks (police beats, neighborhood areas, and enumeration districts). (From 1991 Census: Digitised Boundary Data (England and Wales).)

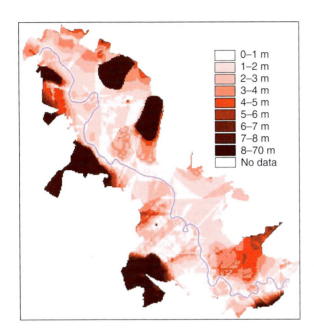

FIGURE 7.1
Modeled height above the river (HAR). (© Crown Copyright/database right 2007. An Ordnance Survey/EDINA supplied service.)

0–1 m
1–2 m
2–3 m
3–4 m
4–5 m
5–6 m
6–7 m
7–8 m
8–70 m
No data

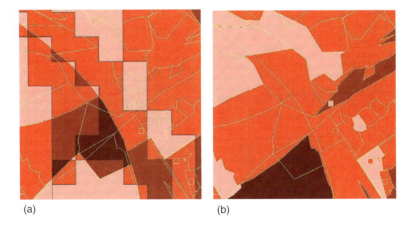

(a)   (b)

FIGURE 7.2
Associated vulnerability index with postcodes (a) before and (b) after dissolving (yellow lines indicate postcode boundaries). (© Crown Copyright/database right 2007. An Ordnance Survey/EDINA supplied service.)

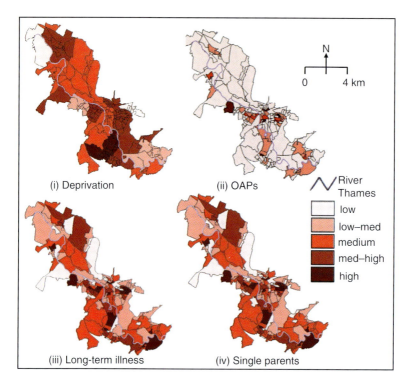

FIGURE 7.3
Output index for four of the separate domains that are combined to form an overall index of social vulnerability. (© Crown Copyright/database right 2007. An Ordnance Survey/EDINA supplied service.)

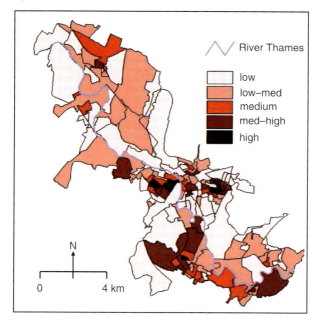

FIGURE 7.4
Index of social vulnerability. (© Crown Copyright/database right 2007. An Ordnance Survey/EDINA supplied service.)

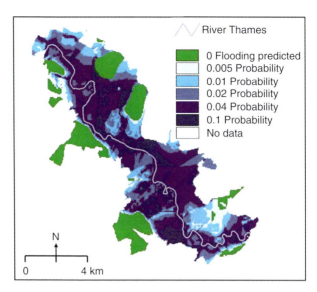

FIGURE 7.5
Index of flood probability. For each probability, any darker shading of blue is included in the full extent of the prediction. (© Crown Copyright/database right 2007. An Ordnance Survey/EDINA supplied service.)

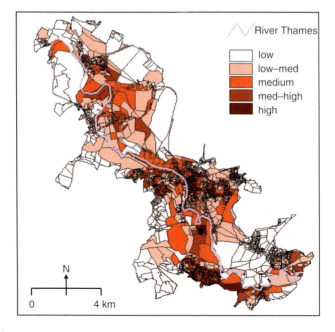

FIGURE 7.6
Combined index of flood vulnerability at the unit postcode resolution. A map viewer or GIS can be used to enlarge areas that are not clear at this scale. (© Crown Copyright/database right 2007. An Ordnance Survey/EDINA supplied service.)

FIGURE 7.7
Identifying vulnerable areas during a 25-year flood. To the left is a close-up of the coarser resolution map to the right. The green masks areas of the ISV that are safe and the Landline data helps see which particular properties are most vulnerable. Vulnerability is increased as the red shading becomes darker. The spatial units are not relevant for this demonstration. (© Crown Copyright/database right 2007. An Ordnance Survey/EDINA supplied service.)

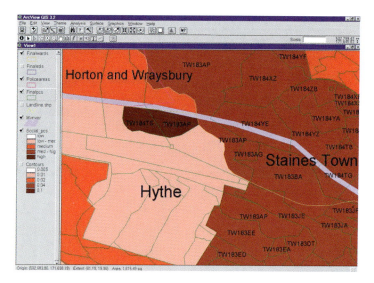

FIGURE 7.8
ArcView interface for flood-warning education (non-U.K. born displayed). The Environment Agency, local authorities, or police can easily identify enumeration districts or unit postcodes with particular flood-warning needs. The example given might result in flood-warning that does not rely on a full comprehension of the English language. The spatial units are not relevant for this demonstration. (© Crown Copyright/database right 2007. An Ordnance Survey/EDINA supplied service.)

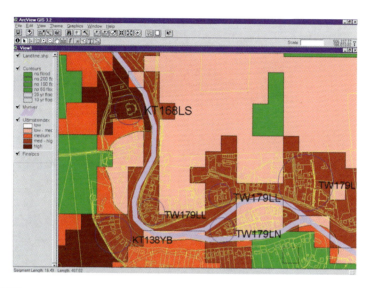

FIGURE 7.9

ArcView interface for emergency planning and response for a flood return-period identified as 25 years (green masking indicates regions of no flooding). Emergency services can easily identify the most vulnerable properties. The spatial units are not relevant for this demonstration. (© Crown Copyright/database right 2007. An Ordnance Survey/EDINA supplied service.)

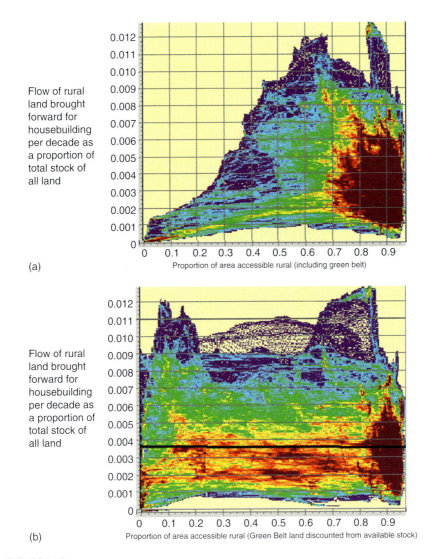

Flow of rural land brought forward for housebuilding per decade as a proportion of total stock of all land

Proportion of area accessible rural (including green belt)

(a)

Flow of rural land brought forward for housebuilding per decade as a proportion of total stock of all land

Proportion of area accessible rural (Green Belt land discounted from available stock)

(b)

FIGURE 9.12

Relationship between the stock of accessible rural land and the flow of rural land for house-building.

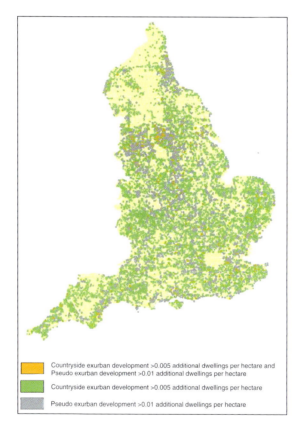

	Countryside exurban development >0.005 additional dwellings per hectare and Pseudo exurban development >0.01 additional dwellings per hectare
	Countryside exurban development >0.005 additional dwellings per hectare
	Pseudo exurban development >0.01 additional dwellings per hectare

FIGURE 9.20
Distinguishing components of exurban development.

FIGURE 12.3
Web-mapping interface (http://www.mepa.org.mt). (From Malta NPI.)

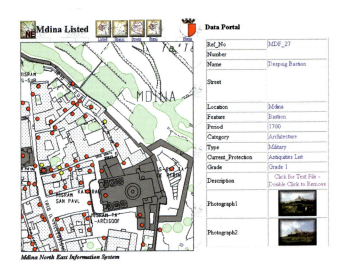

FIGURE 12.4
Interactive Image-Mapping system and data portal. (From Malta NPI.)

FIGURE 12.5
The 3D-mapping system: urban structures and topographic integration. (From Malta NPI.)

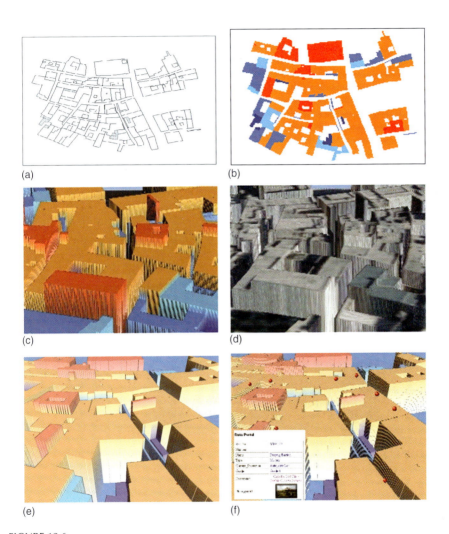

FIGURE 12.6

Spatial process employed in URML development. (a) The MapInfo stage: building heights; (b) conversion to a grid file through Vertical Mapper; (c) the pseudo 3D interface for polygon analysis; (d) integrating orthophotos with the 3D interface; (e) linking the NPI information nodes to the 3D interface and output to VRML; and (f) the final VRML output: the hot spots are linked to info in an immersive environment. (From Malta NPI.)

queries and assertions are exchanged between the parties involved, as demonstrated by Car (2001) in her work on SpeleoGIS of the Carinthian cave system.

Moore (2000) considers SDSS to be examples of spatial expert systems. "Expert systems are all about attempting to introduce human knowledge about problem solving into computer software. The general objective is to emulate the problem-solving capabilities of the human expert." (Openshaw and Openshaw, 1997, p. 108). A feature of the participatory systems described above is their role in collating knowledge from a wide variety of participants and of providing participants with the means to define or expand the rule base and so explore alternative scenarios. Indeed, any DSS should conform to Openshaw's definition of artificial intelligence (AI): "AI (then) is an attempt to mimic the cognitive and symbolic skills of humans using digital computers in the context of particular applications. This is not so much a matter of seeking to replicate skills that are already possessed by humans but of attempting to improve and amplify our intelligence in those applications where it is deficient and can benefit from it most" (Openshaw and Openshaw, 1997, p. 16).

Ellis (2000) goes further in his paper on dynamic SDSS and introduces concepts based on the application of cellular automata and multiagent objects. He is keen on the adaptation of rule-based systems for decision support but is concerned that the focus be properly maintained, concluding that, in the case of the use of intelligent agents, great care will be needed in order to define the identity and nature of individual agents. Rodrigues et al. (1998) have reported on the use of spatial agents in environmental planning. They quote Ferrand (1996) as introducing the use of agents for multicriterion Spatial Decision Support, with agent equilibrium indicating the arrival at a final planning decision. They then describe their own implementation and show how dynamic and proactive structure of the application facilitates the automatic creation and maintenance of the necessary information for simulation and integration and thus improving the decision-making process.

Harvey (2001) notes that "persons working with GIS wear different 'hats' depending upon the groups with which they associate. In actor network theories, these hats are different technological artefacts." He goes on to state that "the traces of relationships between people, institutions, and artifacts connected by agreements and exchanges" comprise an actor networks. While noting the similarity between Harvey's observation of GIS users and the many actors involved in PWR we also see a parallel between his actor networks and the modeling of population behavior using agents. Deadman and Schlager (2002) are concerned with modeling individual decision-making within populations and discuss a range of approaches prior to introducing an agent-based model. They state that "one of the strengths of agent-based simulation is the ease with which communication routines can be established between individual objects. Such simulations provide us with a tool to explore the effects of alternate communication routines on the behaviour of groups in a common dilemma" (p. 152). The various actors involved in PWR undoubtedly represent groups in a common dilemma. They conclude that "building effective simulations requires that the model of individual decision-making validly represent the real system that is the subject of the simulation" (p. 160), and set out requirements for testing the validity of the model. Some measure of the complexity of PWR is reflected in the problems of environmental management—especially balancing the requirements of conserving wildlife and supporting human recreation. Gimblett et al. (2002) describe an approach for modeling wild-land recreational use together with the inevitable conflicts of interest resulting from the interaction of people and wildlife. Their methodology is to use rule-driven intelligent agents operating in a GIS-like synthetic spatial world in order to simulate interactions and conflicts and find that they identify previously unrecognised encounters within interacting groups in addition to the expected conflicts. They conclude that "agent simulations seem to be an excellent method for modelling recreational encounters and ultimately conflicts" (p. 235).

Wooldridge and Jennings (1995) set out the basic theory of agents, comparing the formal theory to human activity. In doing this they explore agents in terms of intentional and representational logics and formalisms for representing belief, leading to behaviors driven by goals and desires. This leads to a variety of agent architectures, including, of relevance to this project, deliberative, or planning, agents, reactive agents and hybrids, and review implementation approaches. There is a lively research effort, with a growing literature concerning both theory and implementation, demonstrated by Davidsson (1996) and Gimblett (2002).

According to Jiang and Gimblett (2002), "The basic idea of agent-based approaches is that programs exhibit behaviours entirely described by their internal mechanisms. By linking an individual to a program, it is possible to simulate an artificial world inhabited by interacting processes. Thus it is possible to implement simulation by transposing a population of a real system to its artificial counterpart. Each member of population is represented as an agent who has built-in behaviours. Agent-based approaches provide a platform for modelling situations in which there are large numbers of individuals that can create complex behaviours." (p. 172). It seems as though they are almost thinking about the complex behaviors and interactions of the various disparate actors involved in PWR: for this reason it is appropriate to consider agent-based approaches for PWR.

11.7   Drawing Threads Together: Our Proposed Methodology

Gimblett (2002, pp. 11,12) states that "decision makers such as resource managers, environmental planners, and designers, faced with realistic environmental problems, would substantially benefit from simulation techniques that enabled them to explore alternatives and test ideas or theories, before expensive management plans are implemented. What is crucial, however, is that if decision makers have confidence in the use and results of agent-based simulations, the design of humanlike agents must be bounded by what can be synthesized from actual behaviour and grounded in this reality." When one is concerned with processes that may directly impact peoples' capacity to survive, such as with PWR, this concern is of great importance. Westerveldt (2002) is concerned about the integration of GIS and AI-based modeling and argues that "full use of the scientific models to affect land management decisions will not occur until the models are integrated as components of a simulation-based geographic modelling system" (p. 84). At the same time Deadman and Schlager (2002) are concerned about the role of the individual. They note that different approaches, models, are required depending upon the goal, for example "if one were interested in how and to what effect individuals learn about the consequences of a particular policy, then one should use a model of the individual that allows for learning . . ." (p. 144). They conclude their essay by stating that "building effective simulations requires that the model of individual decision-making validly represent the real system that is the subject of the simulation" (p. 160), and noting the need for replicative, predictive, and structural terms of validity.

In our discussion of the PWR problem, we noted that a requirement was to include the behaviors and capacities of the various actors, including the victims. We also noted that it is important to involve the victims and that any model must be capable of taking account of the developing capacities of the various groups as peace and reconstruction progress. The behavior of individual agents is defined by a rule-base pertaining to each agent (Moore, 2000) and we suggest that by drawing mainly on the work of Gruber (1993a,b), and Kuhn (2001), the actors may be given the opportunity to define themselves by parsing natural language descriptions to define a formal ontology for their activities. One of us (Car, 2001) has already demonstrated the efficacy of ontological parsing to obtain spatial information from natural language sources in order to build a spatial model of a complex environment. The use of a neutral interface, developed from the participatory GIS (PPGIS) work, together with the realization, as described by Hermida (2002), that the computing base may not be such a barrier, is a way of bringing the interested parties together, as well as to obtain the necessary descriptive rules to define each party. Of course, an iterative approach is needed: it will be necessary to permit parties to vary their behaviors in response to the actions of others. A modeling system, then, which enables the PWR actors to refine their responses by testing the likely reactions of others to the various options available to them, has many attractions—not the least being the possibility of alleviating antagonistic behaviors through a realization of greater benefits available when working together.

Most, if not all, GIS methodologies are predicated upon an unchanging spatial information base, or on managing the changes to the information base. In the PWR context, however, the spatial information base will necessarily need to be flexible in order to adapt as better information becomes available and to develop as the real world situation changes; a dynamic methodology will be essential.

In her work on hierarchical spatial reasoning, Car offers a methodology suitable to model each agent's activity (Car, 1997). Agents can be organized hierarchically. A concept for such an agent's hierarchy is determined by its nature (structural and/or functional hierarchy): a set of rules for reasoning among the hierarchical levels, i.e., agents, and a measure of goodness of the reasoning process. This hierarchy is a close match to the various levels of agent autonomy envisaged by authors such as Thalmann et al. (2000). Such a hierarchy offers a framework for dealing with the complexity of PWR and is consistent with the concepts of autonomous agents put forward by Davidsson (1996) and spatial agents in GIS as set out by Harvey (2001), Jiang and Gimblett (2002), and Gimblett et al. (2002).

11.8   Conclusions

In the complexity of PWR, of sustainable community rehabilitation, a tool or methodology that enables the combination of the skills of all the disciplines involved in such a way that all participants can benefit by applying knowledge from their partners should be especially applicable. Our suggestion is that, by drawing on the work of PPGIS, SDSS, AI and expert systems, a flexible and dynamic model of the PWR process is achievable such that individuals, groups, NGOs, official bodies, among others, can interact and explore the potential impact of the various potential actions in order to achieve continued consensus in deciding and implementing the best approach for a specific situation.

The outcomes of research and work presented in this paper are expected to show how this thinking is coming together in our proposal for a spatial decision support methodology for PWR that draws together these existing and developing methodologies in a novel way and which will offer important benefits to the participants.

References

Atkins, A., 2000, Project Kosovo: GIS and War Crimes Evidence. *Proceedings of the ESRI User Conference 2000* (http://www.esri.com/library/userconf/proc00/professional/papers/ PAP319/p319.htm, accessed on June 11, 2001).

Barakat, S. and Deely, S., 2001, Somalia: programming for sustainable health care. In *World Disasters Report Focus on Recover*, edited by Walter, J., pp. 59–81 (Geneva: International Federation of Red Cross and Red Crescent Societies).

Barakat, S. and Hoffman, B., 1995, Post-conflict reconstruction: key concepts, principal components and capabilities in post-conflict reconstruction strategies. *Paper Presented to the International Colloquium Organised by the United Nations Industrial Development Organisation (UNDO) and the Department for Development Support and Management Studies (DDSMS)*, Stadschlaining, 23–24 June.

Bibby, P. and Shepherd, J., 1999, GIS in analyses of land use and land cover data for policy formation. In *GIS: Principles, Techniques, Applications and Management* (2nd Edition), edited by Longley, P.A., Goodchild, M.F., Maguire, D.J., and Rhind, D.W., pp. 953–965 (Chichester: John Wiley & Sons).

Bolle, J., 2001, Can the Internet bridge the North–South divide? *Humanitarian Affairs Review* (Autumn), pp. 54–60.

Burrough, P.A. and McDonnell, R.A., 1998, *Principles of Geographic Information Systems* (Oxford: Oxford University Press).

Car, A., 1997, *Hierarchical Spatial Reasoning: Theoretical Consideration and Its Application to Modelling Wayfinding*. Published Ph.D. Thesis, Department of Geoinformation, Technical University Vienna, Vienna, Austria.

Car, A., 2001, Bringing Speleological Data into GIS. In *Proceedings of the 9th Annual Conference, GISRUK 2001*, edited by Kidner, D.B., David, B., and Higgs, G., pp. 41–43.

Carver, S., Evans, A., Kingston, R., and Turton, I., 1999, *Virtual Slaithwaite: A Web Based Public Participation "Planning for Real" System*. Centre for Computational Geography, University of Leeds.

Cova, T.J., 1999, GIS in emergency planning. In *Geographical Information Systems*, edited by Longley, P.A., Goodchild, M.F., Maguire, D.J., and Rhind, D.W., pp. 845–858 (Chichester: John Wiley & Sons).

Davidsson, P., 1996, *Autonomous Agents and the Concept of Concepts*. Published Ph.D. Thesis, Department of Computer Science, Lund University.

Deadman, P.J. and Schlager, E., 2002, Models of individual decision making in agent-based simulation of common-pool-resource management institutions. In *Integrating Geographic Information Systems and Agent-Based Modeling Techniques*, edited by Gimblett, H.R., pp. 137–169 (Oxford: OUP).

Delać, V., 1994, Reflection upon the reconstruction of Pakrac. Presentation to the *Settlement Reconstruction in Croatia* Workshop, Dubrovnik, September 1994.

Densham, P., 1991, Spatial decision support systems. In *Geographical Information Systems—Volume 1: Principles*, edited by Maguire, D., Goodchild, M., and Rhind, D., pp. 403–412 (London: Longman).

Ekmekji, J., 1997, Implementing GIS in the Lebanon—a case study. *Proceedings of the 1997 GISQatar Conference* (http://www.gisqatar.org.qa/conf97/links/i2.html, accessed on May 1, 2002).

Ellis, M.C., 2000, New developments in dynamic spatial decision support systems: an evaluation of current proposals. In *The Proceedings of the 4th International Conference on Integrating GIS and Environmental Modeling (GIS/EM4): Problems, Prospects and Research Needs 2000*.

Estes, J.E., Ehlers, M., Malingreau, J.-P., Noble, I.R., Raper, J., Sellman, A., Star, J.L., and Weber, J., 1992, *Advanced Data Acquisition and Analysis Technologies for Sustainable Development*. MAB Digest 12 (Paris: UNESCO).

Evans, A., Kingston, R., Carver, S., and Turton, I., 1999, Web-based GIS to enhance public democratic involvement. Geocomputation. *Conference Proceedings* (http://www.geog.leeds.ac.uk/research/papers/99-1/, accessed on June 11, 2001).

Ferrand, N., 1996, Modelling and supporting multi-actor spatial planning using multi-agents systems. In *Proceedings of the Third NCGIA Conference on GIS and Environmental Modelling, Santa Fe; USA*, January 21–25, 1996. Santa Barbara CA, NCGIA. CD version.

Fonseca, F.T., Egenhofer, M.J., Davis, Jr., C.A., and Borges, K.A.V., 2000, Ontologies and knowledge sharing in urban GIS. *CEUS* 24, 251–271.

Fry, C., 1997, Rebuilding Lebanon on digital foundations. *GIS Europe* 6(2), 24–27.

Gimblett, H.R., 2002, Integrating geographic information systems and agent-based modeling technologies for modeling and simulating social and ecological phenomena. In *Integrating Geographic Information Systems and Agent-Based Modeling Techniques*, edited by Gimblett, H.R., pp. 1–20 (Oxford: OUP).

Gimblett, H.R., Richards, M.T., and Itami, R.M., 2002, Simulating wildland recreational use and conflicting spatial interactions using rule-driven intelligent agents. In *Integrating Geographic Information Systems and Agent-Based Modeling Techniques*, edited by Gimblett, H.R., pp. 211–243 (Oxford: OUP).

Gruber, T.R., 1993a, A translation approach to portable ontologies. *Knowledge Acquisition* 5(2), 199–220.

Gruber, T.R., 1993b, *Toward Principles for the Design of Ontologies Used for Knowledge Sharing*, Technical Report KSL 93-04, Knowledge Systems Laboratory, Stanford University.

Halls, P.J. and Barakat, S.Z., 1999, Post war reconstruction: (spatial) challenges in rebuilding war-torn areas. In *GIS Challenges in Spatial Planning*, pp. 56–74 (Galway: National University of Ireland).

Harvey, F., 2001, Constructing GIS: actor networks of collaboration. *URISA Journal* 13 (1), 29–37.

Heller, J.M., Wortman, W.J., and Wier, J.C., 2000, Using modeled data and geographic information systems (GIS) technology for the investigation of Gulf war veterans' environmental exposure and illnesses. In *The Proccedings of the 4th International Conference on Integrating GIS and Environmental Modeling (GIS/EM4)* Banff Canada, September 2–8, 2000. Santa Barbara CA, NCGIA. (http://www.ncgia. ucsb.edu/conf/SANTA_FE_CD-ROM/main.html, accessed on June 6, 2001).

Hermida, A., 2002, *Internet unites Kosovo foes*. BBC News, Monday 22nd April 2002 (http://news.bbc.co.uk/hi/english/sci/tech/newsid_1939000/1939121.stm, accessed on April 24, 2002).

Heywood, I., Cornelius, S., and Carver, S., 1998, *An Introduction to Geographical Information Systems* (Harlow, Essex, UK: Pearson Education/Longmans).

Horvat, S., 1994, Renovation of private homes in Lipik. *Presentation to the Settlement Reconstruction in Croatia* Workshop, Dubrovnik, September 1994.

Howard, D., 1998, Geographic information technologies and community planning: spatial empowerment and public participation. *Paper Presented at the Empowerment, Marginalization an Public Participation GIS Meeting 1998*, Santa Barbara, CA, October 14–17, 1998. Santa Barbara CA, NCGIA. (http://www.ncgia. ucsb.edu/varenius/ppgis/papers/howard.html, accessed on June 15, 2001).

Jaquemet, I., 2001, Post-flood recovery in Viet Nam. In *World Disasters Report Focus on Recovery*, edited by Walter, J., pp. 103–123 (Geneva: International Federation of Red Cross and Red Crescent Societies).

Jiang, B. and Gimblett, H.R., 2002, An agent-based approach to environmental and urban systems within geographic information systems. In *Integrating Geographic Information Systems and Agent-Based Modeling Technique*, edited by Gimblett, H.R., pp. 171–189 (Oxford: OUP).

Jordan, G., 1998, A public participation GIS for community forestry user groups in Nepal: putting people before the technology. *Paper Presented at the Empowerment, Marginalization an Public Participation GIS Meeting 1998,* Santa Barbara, CA, October 14–17, 1998. Santa Barbara CA, NCGIA. (http://www.ncgia.ucsb.edu/varenius/ppgis/papers/jordan.pdf, accessed on June 15, 2001).

Kabbara, F. and Soubra, N., 1997, GIS in war damage assessment: case of South Lebanon. *Proceedings of the 1997 ESRI Users Conference,* San Diego CA, July 8–11, 1997. Redlands CA, ESRI. (http://gis.esri.com/library/userconf/proc97/proc97/to150/pap140/p140.htm, accessed on May 1, 2002).

Kassa, A., 2001, GIS and remote sensing: emerging tools for disaster management. *Engineers for Disaster Relief RedR Newsletter* 55, 6.

Kereković, D., 1994a, The Vukovar Project: from the ashes of war. *GIS Europe* 3(1), 30–32.

Kereković, D., 1994b, Project "proof of the identity of reconstruction project of the City of Vukovar", *CEUS* 18(2), 143–146.

Kingston, R., 1998, Web based GIS for public participation decision making in the UK. *Paper presented at the Empowerment, Marginalization and Public Participation GIS Meeting 1998,* Santa Barbara, October 14–17, 1998. Santa Barbara CA, NCGIA. (http://www.geog.leeds.ac.uk/research/ papers/98-5/, accessed on June 6, 2001).

Kingston, R., Caver, S., and Turton, I., 1998, Accessing GIS over the Web: an aid to public participation in environmental decision making. *Paper Presented at GISRUK98,* Edinburgh, Scotland, 31st March–2nd April, Edinburgh, University of Edinburgh.

Kingston, R., Carver, S., Evans, A., and Turton, I., 1999, A GIS for the public: enhancing participation in local decision making. *Paper Presented at GISRUK'99,* University of Southampton, 4–6 April 1999. Southampton, University of Southampton. (http://www.geog.leeds.ac.uk/research/papers/99-7/, accessed on June 7, 2001).

Kitmitto, K., 1999, *GIS Builds Beirut* (http://shakeeb.mcc.ac.uk/gisin.htm, accessed on May 1, 2002).

Korsey, S.G., 2000, GIS in the Kosovo ethnic conflict situation. The project Sentinel. *The Proceedings of the ESRI User Conference 2000,* San Diego CA, June 26–30, 2000. Redlands, CA, ESRI (http://www.esri.com/library/userconf/proc00/professional/papers/PAP929/p929.htm, accessed on June 11, 2001).

Kuhn, W., 2001, Ontologies in support of activities in geographic space. *IJGIS* 15(7), 613–631.

Larsen, L., 1999, GIS in environmental modeling and assessment. In *Geographical Information Systems,* edited by Longley, P.A., Goodchild, M.F., Maguire, D.J., and Rhind, D.W., pp. 999–1007 (Chichester: John Wiley & Sons).

Moore, T., 2000, Geospatial expert systems. In *Geocomputation,* edited by Openshaw, S. and Abrahart, R.J., pp. 127–159 (London: Taylor & Francis).

Moore, C.N. and Davis, D., 1997, *Participation Tools for Better Land-Use Planning: Techniques and Case Studies* (Sacramento: Centre for Livable Communities).

National Center for Geographic Information and Analysis (NCGIA), 1996, *NCGIA Initiative 17: Collaborative Decision-Making* (http://www.ncgia.ucsb.edu/research/i17/I-17_home.html, accessed on June 12, 2001).

National Center for Geographic Information and Analysis (NCGIA), 1998, *Empowerment, Marginalization and Public Participation GIS, a Project Varenius* Workshop. Santa Barbara, October 14–17, 1998. Santa Barbara CA, NCGIA. (http://www.ncgia.ucsb.edu/varenius/ppgis/ncgia.html, accessed on June 12, 2001).

Openshaw, S. and Openshaw, C., 1997, *Artificial Intelligence in Geography* (Chichester: John Wiley & Sons).

Parker, C.P., 1998, Living neighbourhood maps: the next wave of local community development. *Paper Presented at the Empowerment, Marginalization and Public Participation GIS Meeting 1998*, Santa Barbara, CA. October 14–17, 1998, Santa Barbara CA, NCGIA. (http://www.ncgia.ucsb.edu/varenius/ppgis/papers/parker.pdf, accessed on June 15, 2001).

République Libanaise, 1999, Programme Régional de Développement Économique et Social du Sud-Liban (Beyrouth: Le Haut Comité du Secours en collaboration avec Le Programme des Nations Unies pour le Développement) (In French).

Rodrigues, A., Grueau, C., Raper, J., and Neves, N., 1998, Environmental planning using spatial agents. In *Innovations in GIS 5*, edited by Carver, S., pp. 108–118 (London: Taylor & Francis).

Sayagues, M., 2001, Trapped in the gap—post-landslide Venezuela. In *World Disasters Report Focus on Recover*, edited by Walter, J., pp. 83–101 (Geneva: International Federation of Red Cross and Red Crescent Societies).

Shiffer, M.J., 1999, Managing public discourse: towards the augmentation of GIS with multimedia. In *Geographical Information Systems* (2nd Edition), edited by Longley, P.A., Goodchild, M.F., Maguire, D.J., and Rhind, D.W., pp. 723–732 (Chichester: John Wiley & Sons).

Simms, A., 2001, The ecology of post-disaster recovery. In *World Disasters Report Focus on Recovery*, edited by Walter, J., pp. 35–57 (Geneva: International Federation of Red Cross and Red Crescent Societies).

Smith, D.G., 2000, Kosovo: applying geographic information systems in an international humanitarian crisis. *Proceedings of the ESRI User Conference 2000*, San Diego CA, June 26–30, 2000. Redlands CA, ESRI. (http://www.esri.com/library/userconf/proc00/professional/papers/ PAP937/p937.htm, accessed on June 11, 2001).

Sparrow, J., 2001, Relief, recovery and root causes. In *World Disasters Report Focus on Recover*, edited by Walter, J., pp. 9–33 (Geneva: International Federation of Red Cross and Red Crescent Societies).

Stiefel, M., 1999, *Rebuilding After War. Learning from the War-torn Societies Project* (Geneva: WSP/PSIS).

Thalmann, D., Musse, S.R., and Kallmann, M., 2000, From individual human agents to crowds. *Informatik Informatique* 1, 6–11.

United States Environmental Protection Agency (USEPA), 1991, *Kuwait Oil Fires: Interagency Interim Report*, 2 April 1991.

UNEP, 1999, *Coastal Information System for Lebanon* (http://www.grid.unep.ch/activities/capacitybuilding/lebanon/, accessed on May 1, 2002).

Westerveldt, J.D., 2002, Geographic information systems and agent-based modeling. In *Integrating Geographic Information Systems and Agent-Based Modeling Techniques*, edited by Gimblett, H.R., pp. 83–103 (Oxford: OUP).

Wooldridge, M. and Jennings, N.R., 1995, Intelligent agents: theory and practice. *Knowledge Engineering Review* 10(2), 115–152.

Yeh, A.G.-O., 1999, Urban planning and GIS. In *Geographical Information Systems* (2nd Edition), edited by Longley, P.A., Goodchild, M.F., Maguire, D.J., and Rhind, D.W., pp. 877–888 (Chichester: John Wiley & Sons).

12

Malta NPI Project: Developing a Fully Accessible Information System

Malcolm Borg and Saviour Formosa

CONTENTS

12.1   Introduction

This chapter examines the use of GIS and Web-mapping technologies in the implementation of a heritage management system (HMS). Its use as a tool to warehouse, analyze, and publish data in an integrated and user-friendly format is examined. The paper is based on a methodology used spanning the process taken from the analog archives to a Web-mapping product.

The idea behind a digital national protective inventory (NPI) also referred to as an HMS, developed mainly on three fundamental ideals: accessibility, integration, and sustainability. In a country still synonymous with an obsession to hoard data, breaking through the accessibility barrier was seen as a major milestone, and also added the equally vital issues of integrating an information system and sustaining its upkeep.

The issue of accessibility was based on the slow process involved in monitoring and assessing change on Listed* buildings through the planning process. The analog system of existing data-capture sheets in volume form and a developing GIS network at the Planning Authority (PA) were seen as the right elements needed to develop an integrated system (Gatt and Stothers, 1996). Integration was also desirable in the evaluation process especially in valorizing assets that involved properties with different heritage aspects namely: archaeological, ecological, and cultural aspects. The project also considered sustainability as a fundamental issue both in relation to financial and human resources. These elements are encapsulated in the EU sustainable cities principle of ecosystems thinking (Borg, 1999a,b).

Once the integration process was completed, a dissemination phase was launched through the use of Web-mapping technologies and CD development that saw the previously inaccessible data converted to a simple information interface powered by HTML and JavaScript.

12.2 Setting the Framework: Integrated Heritage Management and Networking

The possibilities for establishing a combined national HMS have been pursued through a joint Planning Authority/Museums Department working group (Borg, 2000). The Heritage Act (2002) paves the way for the integration of the scheduling process, and for addressing issues in a holistic and integrated manner through the establishment of a joint Heritage Advisory Committee with representatives of the PA, the Environment Protection Department, and the Museums Department (Borg and Magro Conti, 2000).

Networking with the Environment Protection Department for monitoring, enforcement, and data gathering on natural habitats could also be strengthened. Such a system (required under the EU directives and the provisions of the United Nations Convention on Biological Diversity) could be further developed to include other environmental data with a geographical component, such as water and air quality data that could be more effectively linked with other spatial information for the elaboration of impacts, trends, and future scenarios. Recent changes such as the integration of the Environment Protection Department and the Planning Authority into one organization, MEPA, will help to align the process into a streamlined operating system.

---

* Listed properties, sites, and assets include those which are protected by the Antiquities Protection Act (1925), within the Urban Conservation Area boundaries, and in the National Protective Inventory.

12.3   Furthering the Network: Heritage Management and Sustainability

Sustainability and the integration of policy and action are key aims under-pinning much of EU legislation and policy. The conservation of architec-tural heritage and of urban or historic cores have a central place in the development of integrated environmental management strategies for sus-tainability. With regards to ecology and natural habitats the EU proposes standardization and sharing of data (through the EU Habitats Directive 92/43/EEC). The European Union and the Council of Europe provide the best guidance and parameters towards heritage management and conser-vation. Malta is already obliged to follow and implement certain conven-tions and charters because it has been in the forefront in signing these important documents.

In view of this, the Museums Department is envisaging the application of such an information system and the upgrading of the inventories not only for availability within the Department but also for broadening accessibility. Another option discussed was the possibility of networking since both inventories for sites and properties housed at the Planning Authority, and also the inventories of collections* can be integrated creating a hub of information on national heritage assets.

In October 2001, the Planning Authority in fact had made a proposal to integrate all systems in a drive to increase networking between the various entities responsible for heritage management. The partnership agreement signed by the Planning Authority, the Museums Department, the Malta Centre for Restoration, the Environment Protection Department, and the Restoration Unit (Works Division) for data access and sharing is a historical step which will see the building of a sustainable and integrated HMS which is GIS-based and covering the whole of the Islands' resources. The agree-ment for Developing an Integrated Heritage Management System for the Maltese Islands—Online Networking—was based on the following prin-ciples of partnership:

• To facilitate integrated heritage management
• To promote sustainability
• To promote networking
• To encourage wider access to environmental information

The integrated system will provide greater flexibility and dynamism in data compilation and processing, and in the development of inventories for

---

* Collections refers to all assets gathered, stored, or exhibited in all the state museums in the Maltese Islands and these may vary from archeological, natural history, documents, fine art, or intangible heritage.

natural and cultural heritage. Data accessibility will greatly assist in the processing of development applications, management of sites, preparation and review of local plans, formulation of development briefs, assessment of environmental impacts as well as the formulation of other policy documents by the Museums Department, the Environment Protection Department, and other Government entities.

Monitoring and enforcement will be integrated and coordinated since all entities will have standardized data and access to all aspects of the NPI. The HMS is essential for conservation initiatives, monitoring of environmental quality, and restoration/rehabilitation.

Effective sustainable management of resources requires more coordination and cooperation between government departments and other bodies. This HMS will integrate expertise and data to create a shared hub, avoiding overlaps, duplication of work, bureaucratic faltering, and data redundancy in a field where oftentimes impromptu intervention is necessary. Sharing of data also limits the recreation of further voluminous inventories, and the digital system will compress all this data and facilitate expansion. GIS are proving to be the tools that have halted the seemingly perpetual cycle of recreating the data wheel.

Networking between agencies is not only about the sharing of data. It is also important for management purposes. Streamlining and standardization are crucial in establishing modes of conduct, ethical approaches, and coordination on an official basis between the entities concerned. Legal boundaries will thus be limited by access and the sharing of a common field (Duff, 1999). It will also help clarify the remits and roles of the different entities. This, through specific access to data residing within a particular layer where a system of safeguards on layer access operates.

A broader aim of the partnership is to increase public accessibility to the data. Environmental education is assuming greater importance, and the network should have a role to play in facilitating this. Similarly NGOs and other bodies with an interest in the cultural and natural heritage would be able to access a wider range of environmental and other information, aiding them in their contributions to positive action for the management of this heritage and enabling them to play a greater role in the planning process. NGOs also have, within their membership, persons with expertise who can validly contribute to enhance the information base of the participating partners in their development of the HMS.

12.4 National Protective Inventory

The NPI will be of value when there is full availability to the broad range of business processes throughout the Planning Authority, spanning archival recording, development application, and policy making function. The Environmental Management Unit aided the development of the first phase of a geographical database system through the process of georeferencing storing of data and imagery scanning of each property. The GIS eventually provided a geographical link to all the property information within the same dataset, while the interface also allowed this data to be correlated with various other spatial datasets within the authority. The system maximizes the availability of the NPI enabling a broad spectrum of operational staff to access the NPI online.

The ultimate objective in this digitizing process is to increase availability of the system within the authority and establish a platform/inventory that could be later distributed/sold or act as a hub on a national level. The standardization of the system will be in line with Council of Europe's report on a computerized heritage documentation center and Recommendation No. R(95)3 of the Committee of Ministers of Member States. This HMS will improve the scheduling process, monitoring of scheduled areas, zoning within urban conservation areas, delineation of UCA perimeters, and enforcement.

12.5 Process

The NPI system has been developed since 1967. There were successive attempts in scheduling methodology. Although the system always followed criteria set by the Council of Europe (IECH), it was only after 1987 that the Malta Town Planning Services Division resumed the exercise of filling in the data-capture cards. In 1991 an expert from the Council of Europe reported on the situation. In 1992 a report was presented to the Council of Europe on *Technical assistance for a computerized heritage documentation center in Malta*; however, this system was never applied.

The data-capture cards in use at the time were unfortunately rarely compiled according to the standards proposed in the manual for the use of the research team assistants, prepared in August 1990, and the result of this exercise leaves much to be desired. All the cards are partially empty (20%–50%) and significant data has been left out, e.g., architectural history, typological data, basic bibliography, legal data, ownership, and proposed utilization.

The initial stages of the data-mining process required a considerable amount of scanning and data entry in order to capture the available data cards, many of which were in handwritten format.

Thus the digitization process provided a base for the different data formats to be compared and integrated into one main information system, aptly named the HMS. Where problems were encountered, a system upgrade would also take into consideration the revision of these cards. This process may have provided a platform for the digitizing of each card; however, the data that was not compiled still has to be filled in.

Considering that the analog phase of the project has been 35 years in the making, the digitization process was smooth to say the least, albeit having to

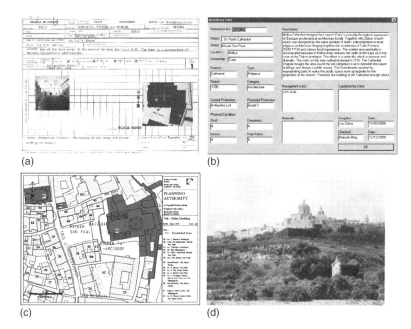

(a)   (b)

(c)   (d)

FIGURES 12.1

Information sources: card system (analog and digital), scheduling-base-mapping, and imagery. (a) The analog card, scanning process; (b) the digital card, necessitated an extensive input process; (c) the PA scheduling maps—linking the information at the PA to the NPI; and (d) the imagery— linking of the analog/digital cards to the real world-links to aid recognizance. (From Malta NPI.)

overcome a number of flaws and lacunae. One such significant flaw in the system adopted to-date is the approach, which was not systematic, and critical areas have been left out, as in the case of Mdina, Valletta, Victoria, and Rabat. Even areas of Cottonera have not been covered till date though initiated in 2007. Where data existed, this was cataloged for the eventual development into a GIS template as shown in Figure 12.1b.

12.6   Heritage Management and the NPI

The decision to develop a GIS-based system over a database came about through a logistical issue and circumstantial developments. Over the past eight years, the Planning Authority has been developing various cultural and natural HMS for use in its day-to-day planning work, including local plan formulation, development control, and scheduling of property. In addition to housing the National Mapping Agency, the organization has developed an extensive GIS spanning all area related to land use and land cover. This bed of information was the natural springboard for the siting of the NPI from which the HMS was developed. In view of this, currently the

TABLE 12.1

Status of the NPI up to December 2001

Data cards (hard copy)	
Cultural sites	91 volumes containing 9631 cards
Archaeological sites	11 volumes containing approximately 1000 cards
Digital cards	
Cultural sites	2100 cards
Natural resource surveys (habitats)	Complete coverage (data available in digitized format except for Gozo—work in progress)
Natural sites	13% of areas surveyed included in NPI
	12.9% of Maltese Islands scheduled

PA houses the most advanced local systems in the compilation of digital heritage inventories covering the whole of the Islands for both natural and cultural heritage.

With the system all set up, today, a stocktaking exercise can tell that the current situation is looking healthy and more information is being created to aid the planning process. A key component in this is the regeneration of the NPI, updated by the Heritage Inventory update, also undertaken by the PA. The current status of the NPI comprises both analog and digital versions as shown in Table 12.1.

12.7 GIS Factor: Frontiers in Data Dissemination

The GIS allows users to analyze data spread over space and time. Since most information on daily activities are based on a specific spatial area such as a street, council, house, clinic, or playground, it would be ideal to have a system that provides the users with ready-made data at their fingertips. Even further, users could create their own requests and receive specific results. GIS provides one of the answers to such a scenario and is currently being introduced in a variety of organizations in the Maltese Islands.

Apart from limitations based on the availability of specialized resources in spatial data creation, the main stumbling block concerns the actual dissemination of data to a wider audience. In addition to the organizational-level data hoarding issues mentioned elsewhere in the document, other limitations to access involved the specialist level of expertise needed to access data through a high-end information system such as a full-blown GIS. Until recently, GIS developments had restricted access to a few specialists in areas mainly related to the physical world and its permutations into the economic world. GIS has come a long way beyond that time, and it is now opportune to look at the social aspects that make up a system closely relating to the real world. This has been achieved both through high-end system-use as well as through the possibility of disseminating the data to a wider non-GIS audience. In effect GIS may not need to feature since it could be a hidden system that users may not be aware of.

In line with developments in other countries, the Maltese GIS community strives to grant access of a number of information systems to the general public using online facilities as a medium to convey data both on the physical and social aspects. Going through a three-phased process, a series of Internet-based GIS projects were created, from the initial creation of an image-mapping system through a GIS–client system and finally to a full WebGIS. These projects, depicted in a format that users easily become familiar with, can be accessed from any location, when required without the constraints of owning proprietary software for map generation and data analysis (Carver et al., 1997; Kemp, 1997; Rowley, 1999). The Web is relevant for GIS data owners since data can be stored where users can find it, in a format that they can read irrespective of platform or system they are using, as well as using a common and free browser; an ideal high-tech tool made accessible irrespective of the client means (Schon, 1996).

As GIS are spatial tools that integrate mapping and data analysis functions of the diverse information layers into one integrated system, data from the PA and any other participating organizations can be processed through the layering option. Users can access and integrate the data to produce unique information outputs, both as maps and databases, something that is quite innovative locally but may yet help to bring together the different organizations currently using systems that are not able to communicate or still have ownership barriers to trust upon them. As an open system, which can take-up more data, the HMS may be connected to other digitized networks as in the case of the land registry, facilitating the investigation of protected property in view of the landowners' status.

This base led the developers of the HMS to look into the possibility of using dissemination tools such as compact disk (CD) and the Web. Standalone CDs were seen as a concurrent option in order to reach those persons who do not have a Web connection, while the Web version was seen as the best alternative to the main GIS. The HTML/JavaScript version was developed for both Intranet and Internet use and subsequently launched.

12.8   Looking for a Comprehensive System

The process of analog card digitization was further developed into a GIS through the linkage of the digitized cards to a vector system using MapInfo and in-house MapBasic programs (Figure 12.2). Points and polygons were used, mainly points for the location of dwellings through an address-point database system and polygons for areas such as archeological sites. The latter would require a buffering process for eventual inspections should a development application fall within that same boundary. This is

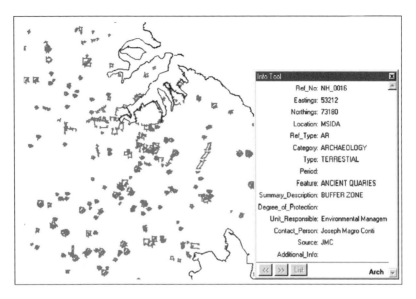

FIGURE 12.2
GIS—heritage management system. (From Malta NPI.)

especially important in a country such as the Maltese Islands, a virtual archeological site.

The first phase of a geographical database system has been developed, which allows for the georeferencing and storage of data and images of each property in the inventory (Figures 12.1 and 12.2). The immediate result of this exercise was the availability of data-capture cards on a GIS that has produced a powerful search system comparable to those available in more advanced European countries.

The issues pertaining to the NPI concentrated on the gathering of data and integrating the whole into a seamless GIS. Integrating analog and digital materials proved a major hurdle on time-constraints, personnel, and hardware. The later phases of this system have been greatly speeded up owing to new software being developed as well as the availability of faster hardware.

The information system now includes the information on:

a. Unique reference number

b. x–y coordinate location using a UTM non-Earth meter projection

c. Local council location

d. Reference type

e. Category and type of property

f. Period of development

g. Feature type

h. Summary description

    i. Degree of protection

    j. Unit responsible, contact person, and source of information

    k. Additional info

12.9   Access

Granting access to the general users necessitated the implementation of a series of steps primarily based on the use of a dedicated GIS package, namely MapInfo and a proprietary MapBasic program called NPI.mbx specifically created for the HMS. However, the GIS was developed for use within the PA, and internal users only could access the information through the organization-wide GISViewer. Together with the development planning information system which houses all applications from 1993 to date, the NPI could now be shown against constraints for new development. This system proved to be a success; however, the need to go further was felt as the data needed to be disseminated on a wider level—the public level. This involved discussion on the type of data to be disseminated, which imagery could be produced as well as the type of access that users could have.

    The debate at this stage concentrated on the medium to be used. The best option would have been to go for a high-end WebGIS where users could access the GIS directly from their home through the employment of a dedicated map-server (Green, 1996; Laver, 1997; McGill, 1997; Plewe, 1997; Harder, 1998; Greenwood, 1999). This thin-client fat-server system is an ideal scenario though it required the setting up of an organization-wide map-server that would also comprise of development application maps, ecology maps, constraints maps, and a host of other layers residing in the PA systems. At this stage, the Web-server had yet to be developed. It was eventually launched at the end of 2001. Also, being a high-end system with a complex interface, the map-server could create some problems for users who may not be so comfortable with technology. Thus the next option was to go for a simple low-cost, low-end thin-client–thin-server option through the use of HTML and JavaScript, a process called Image-Mapping (Formosa, 2000). It was used with success in the development of a Web site for Census of Population and Housing (http://www.mepa.org.mt/Census/index.htm), and in this case proved to be the best option for the dissemination issue as a CD could be subsequently developed based on the same setup, something that was not possible with a dynamic Web-server.

12.10   Conversion

The launching of the second phase of the project necessitated that the spatial entities and attribute designations be integrated with the digitized-card

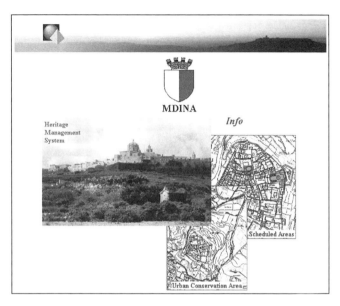

FIGURE 12.3 (See color insert following page 274.)
Web-mapping interface (http://www.mepa.org.mt). (From Malta NPI.)

material enabling the developers to create an Image-Mapping system. This phase was based on the creation of a template through the use of the MapInfo (MapInfo Corp.) add-on WebPublisher (DataView Solutions). The latter software generated an HTML frame setup that was incorporated within the main site framework. Creating the site necessitated a degree of HTML authoring. The prototype was based on the NPI of the town of Pieta and later the city of Mdina (Figures 12.3 and 12.4).

Once the prototype was rectified for errors, feedback on ease-of-use was received and encompassed into the system. The site has since been uploaded into an Intranet and the Internet (http://www.mepa.org.mt) as well as being promoted through the publication of a number of CDs, satisfying the major user-base.

The Image-Mapping system provides the user with a graphic interface that allows for preprepared maps of the town in question, which maps (Figure 12.4) highlight the scheduled property, clicking on which activates the data portal relevant to that property. Further clicking within the data portal activates text boxes and imagery with details on the property.

As new technology was introduced within the PA, the Pieta and the subsequent Mdina HMS were seen as the main information systems that could be included in the new Web-server. Plans are in line to convert this system into a full WebGIS as an add-on layer to the Internet-based map-server developed since the Image-Mapping launching. This technology is a massive improvement of the Image-Mapping format since it caters for dynamic updating from the main MapInfo information systems,

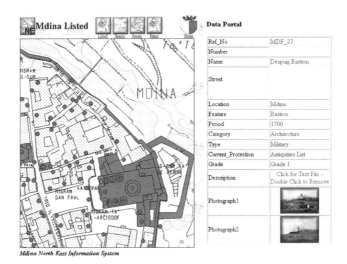

FIGURE 12.4 (See color insert following page 274.)
Interactive Image-Mapping system and data portal. (From Malta NPI.)

which in the Image-Mapping option is a laborious task and needs programmer intervention. The map-server option is seen as the natural heir to the Image-Mapping system. However, even this technology is being dated with such developments as XML and GML. GML2 is even better than XML as it deals with spatial objects and GML3 is envisaged to deal with other issues such as querying and topology, among others.

12.11   Future of NPI

Plans are in line for the expansion of the boundaries of the NPI GIS. The next phase for the NPI is to develop a full 3D-modeling information system. This aims to create a model of the city of Mdina in 3D allowing the GIS specialist to analyze the planning developments in real-time 3D, create a digital mirror of the town and analyze the impact of development (Tufte, 1997). In effect, this calls for the creation of development control models through the 3D plotting of planning applications to determine the effects on the fabric prior to erection (Brail and Klosterman, 2001). In addition, a 3D model would allow users to build archaeological and heritage systems in 3D through the conversion of 2D vector layers residing in the MapInfo HMS. The software envisaged for this role includes Vertical Mapper, Virtual Frontier (both from Northwood GeoScience Ltd), Bryce (Corel Corp.), and VRML development tools such as Pavan (InfoTech Enterprises). Models of building are envisaged to be created through a number of technologies such as property photogrammetry, laser scan, creating unique

FIGURE 12.5 (See color insert following page 274.)
The 3D-mapping system: urban structures and topographic integration. (From Malta NPI.)

dxf models through software such as ACAD (Autodesk) and Canoma (Metacreations Ltd), which are then converted to the Web format. The idea is to connect and integrate the tools for planning and conservation in this 3D model for monitoring, enforcement, evaluation, and assessment purposes. Figure 12.5 depicts the ancient city of Mdina as is being developed in a 3D project that would eventually be integrated into the NPI information system.

The way forward is therefore an integrated 3D HMS, using the urban fabric setup of the city, for conservation parameters, policies, and enforcement issues in planning legislation. Dissemination of the 3D model data will be based on the use of VRML as a conveyor-format for Intranet and Internet use (Liggett et al., 1995; Moore et al., 1997; Raper et al., 1997). The VRML model will allow users to access the city model as well as any related planning, architectural, archaeological, and sociocultural information as well as a virtual tour of the city and buildings spanning the wider needs of development control, heritage management tourism, museums, and the general public.

The process to build the VRML will be based on the same initial MapInfo–Vertical Mapper process but will be based on the generation of an immersive building and topographic model that will be integrated with the information accessible through the Image-Map and map-server options. Figure 12.6 interprets the process that will be developed for this stage of the model.

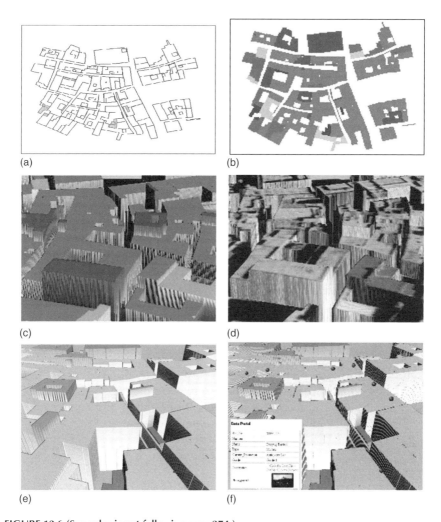

FIGURE 12.6 (See color insert following page 274.)
Spatial process employed in URML development. (a) The MapInfo stage: building heights;
(b) conversion to a grid file through Vertical Mapper; (c) the pseudo 3D interface for polygon
analysis; (d) integrating orthophotos with the 3D interface; (e) linking the NPI information
nodes to the 3D interface and output to VRML; and (f) the final VRML output: the hot spots are
linked to info in an immersive environment. (From Malta NPI.)

The resultant information system would be based on a layered approach
where users can access data that is available in an immersive clickable
scenario through direct linking to 3D models of buildings within historic
cities. In addition, the system would incorporate links to multimedia,
imagery, walk-throughs, HMS data, and access to a dynamic array of live
information systems.

12.12   Conclusions

With the completion of the NPI GIS, the dissemination of the data has become a primary focus. The NPI GIS has been developed into an online interactive system that allows users to access data from any computer without the need for users to purchase expensive and highly specialized software. The Pieta pilot project (a town adjacent to the Grand Harbour city of Valletta) proved a success and the NPI Web-Map was launched through the PA Web site. Subsequent development of an online Mdina NPI further proved its popularity that a complementary project of a stand-alone Mdina CD was launched. There is considerable scope for enhancement particularly with such developments as 3D mapping and VRML. With the recent launching of full-online WebGIS, users can access live data through the map-server installed in the PA Web site. This system allows users to view information layers in conjunction with other maps, permits, orthophotos, and other land-use data. The NPI could be integrated into the map-server allowing users the full access to the PA Web site as well as other information that may be uploaded, inclusive of information from other departments. This would eventually lead to a nationwide integrated system.

This document encourages the development of both the current Web-mapping system as well as the map-server option and in turn suggests that the best way forward would be the development of a 3D information system to the eventual use of an immersive VRML information system. While the latter three developments allow users to interact with the different land-use data beyond the NPI, the former allows the parties to disseminate the data as a unique system and through other media such as CDs. There is, then, a great potential for the sharing and dissemination of heritage data online. Data on scheduling and listing may be accessed by NGOs with heritage interests and by the public, thus complying with EU access standards. Developing the system further to include sites of environmental importance would be of wider benefit, not least in facilitating education on environmental issues.

References

Borg, M., 1999a, Protecting the cultural and natural heritage of the world, *International Conference on World Heritage Sites*, Tokyo, Japan.

Borg, M., 1999b, The way forward for the cultural heritage sector—integrated conservation—development planning issues and cooperation, *National Conference on Heritage Issues organized by the Museums Department*, Malta.

Borg, M. and Magro Conti, J., 2000, *Draft Urban Conservation Topic Paper* (Media Review Section, Planning Authority), Malta.

Borg, M., 2000, Technology as a means of networking—GIS and conservation, *Joint Seminar Museums Department and Planning Authority*, Malta.

Brail, R.K. and Klosterman, R.E. (editors), 2001, *Planning Support Systems: Integrating Geographic Information Systems, Models, and Visualization Tools* (Redlands, CA: Environmental Systems Research Institute).

Carver, S., Blake, M., Turton, I., and Duke-Williams, O., 1997, Open spatial decision-making: evaluating the potential of the World Wide Web, In *Innovations in GIS 4: Selected Papers from the Fourth National Conference on GIS Research*, edited by Kemp, Z. (London: Taylor & Francis).

Duff, A., 1999, Accessing the spatial data warehouse, In *AGI 1999 Conference Proceedings*, (San Francisco: Miller Freeman).

Formosa, S., 2000, *Coming of Age: Investigating the Conception of a Census Web-Mapping Service in the Maltese Islands*, Unpublished MSc Dissertation, University of Huddersfield, Huddersfield, U.K.

Gatt, M. and Stothers, N., 1996, The implementation and application of GIS in the Planning Authority of Malta, In *Geographical Information, Second Joint European Conference and Exhibition on Geographical Information*, Barcelona, Spain.

Green, D.R., 1996, Between the desktop and the deep blue sea, *Mapping Awareness* 10(3), 19–22.

Greenwood, C., 1999, Internet mapping made simple, *Mapping Awareness* 12(7), 31–32 (GeoTec Media, Huntingdon, U.K., http://www.geoplace.com).

Harder, C., 1998, *Serving Maps on the Internet: Geographic Information on the World Wide Web* (Redlands, CA: Environmental Systems Research Institute).

Kemp, Z. (editor), 1997, Innovations in GIS 4, In *Selected Papers from the Fourth National Conference on GIS Research* (London: Taylor & Francis).

Laver, J., 1997, GIS/vector mapping via the Internet: system security, the key to electronic-commerce, In *Geographic Information—Exploiting the Benefits, AGI 97* (London: The Association for Geographic Information/Miller Freeman).

Liggett, R., Friedman, S., and Jepson, W., 1995, Interactive design/decision making in a virtual urban world, In *Visual Simulation and GIS* (http://www.aud.ucla.edu/~robin/ESRI/p308.html).

Malta, Government of, 2002, *Cultural Heritage Act*, Act N. VI of 2002, Valletta.

McGill, M., 1997, State-of-the art network-centric GIS: why and where Internet and Intranet GIS applications are successful, In *Geographic Information—Exploiting the Benefits, AGI 97* (London: The Association for Geographic Information/Miller Freeman).

Moore, K.M., Wood, J.D., and Dykes, J.A., 1997, *Using Java to Interact with Geo-Referenced VRML within a Virtual Field Course*, ICA Visualization Commission, Gavle, Sweden (http://www.geog.le.ac.uk/mek/usingjava.html).

Plewe, B., 1997, *GIS Online: Information Retrieval, Mapping and the Internet* (Santa Fe, NM: OnWord Press).

Raper, T., McCarthy, T., and Williams, N., 1997, Integration of real-time GIS data with Web-based virtual worlds, In *Geographic Information—Exploiting the Benefits, AGI 97* (London: The Association for Geographic Information/Miller Freeman).

Rowley, J., 1999, Raising the standards for Web mapping, *GEOEurope* 8(11), 24–25 (http://www.geoplace.com).

Schon, D., 1996, High Technology and Low-Income Communities: Prospects for the Positive Use of Advanced Information Technology, *Colloquium on Advanced Technology, Low-Income Communities and the City* (London: MIT, http://sap.mit.edu/projects/colloquium/book.html).

Tufte, E.R., 1997, *The Visual Display of Quantitative Information* (Cheshire, CT: Graphics Press).

13

A GIS-Based Methodology to Support the Development of Local Recycling Strategies

Andrew Lovett, Julian Parfitt, and Gilla Sünnenberg

CONTENTS

13.1 Introduction

Following the publication of the Waste Strategy 2000 and its associated performance targets (DETR and National Assembly for Wales, 2000), there is presently particular interest among local authorities in improving their household waste recycling rates. The government has set a national household waste and composting target of 25% by 2005–2006, to be met through the cumulative effect of individual statutory targets for each local authority (Cabinet Office Strategy Unit, 2002). Many authorities have introduced kerbside collections of recyclables or established facilities at Civic Amenity (CA) sites, smaller neighborhood recycling centers, or supermarkets where households can deposit materials such as glass, paper, or cans. The form and extent of such recycling infrastructure provision varies greatly between local authorities (Parfitt et al., 2001; Community Recycling Network, 2002), and it is not straightforward to assess the effectiveness of different initiatives. Partly this is because recycling rates also tend to differ according to the socioeconomic characteristics of the population involved (e.g., Tucker et al., 1998; Perrin and Barton, 2001; Barr, 2002; EnCams, 2002), but there are additional difficulties arising from the need to examine the interaction between the recycling scheme and use of other household waste stream outlets such as conventional house-to-house refuse vehicle collections. A further complication is that operational weight data often relates to a diversity of zones (including collection rounds or catchment areas of different sizes around "bring" sites), so making the task of reconstructing the household waste stream and linking such information to defined populations a challenging one.

Geographical information systems (GIS) can make a significant contribution to the problems of evaluating recycling strategies because of their ability to integrate data from operational entities such as CA sites or refuse collection rounds with details of the population resident in particular locations (Lovett, 2000). This capacity to link data sources based on geographical position, coupled with the means to calculate additional information such as accessibility measures or catchment boundaries, in turn permits the reconstruction of waste stream characteristics at the local scale. To illustrate the potential for enhancing waste management and planning through the use of GIS, a research project was carried out in one local authority (South Norfolk Council) to link subdistrict operational data (from both collection round and site-based sources) with population details. This database was then used to predict the impact of several possible new schemes for improving recycling performance and assess the extent to which they would help the authority meet their recycling targets for 2005–2006.

13.2   Research Methodology

South Norfolk Council covers a predominantly rural area of 908 km^2 extending from the southern suburbs of Norwich to the county border with Suffolk (Figure 13.1). In the 2001 census, the resident population was recorded at 110,710, an increase of some 8% since 1991 (Office for National Statistics, 2003). When the research began in 1999, the district used black plastic sacks for refuse collections and had no kerbside recycling schemes but a high density of bring sites through the provision of a network of minirecycling centers and facilities at CA sites or local supermarkets. South Norfolk was chosen as a focus for the study on several grounds that included the presence of a consistent and well-maintained waste statistics database (Parfitt, 1997), previous involvement in other GIS projects (e.g., Lovett et al., 2003a), and proximity to the research team based in Norwich.

The three main phases of the research are summarized in Figure 13.2. In the first phase, a GIS database was compiled that included details of recycling facility locations (Figure 13.1) and a road network with attributes such as speed estimates to allow the calculation of car travel times (e.g., Lovett et al., 2004). Boundaries of refuse collections were defined by generating Thiessen polygons around addresses in the street listings used by contractors, and then dissolving the dividing lines between areas served on the same round day. Overall, there were 40 round days in the district

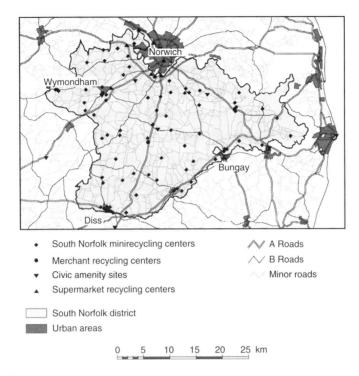

FIGURE 13.1
Recycling facilities in South Norfolk, spring 2000. (© Crown copyright/database right 2007. An Ordnance Survey/EDINA supplied service.)

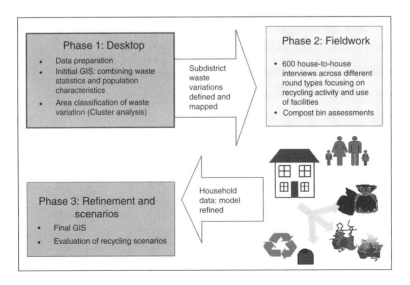

FIGURE 13.2
The overall research methodology.

(eight rounds each weekday), but only 13 of these (32.5%) represented a single contiguous polygon and there were many irregularly shaped boundaries. Figure 13.3 illustrates a typical situation using the example of Round 5 in the southeast of the district. Operational factors (particularly vehicle capacities) were the main reason for the complex configuration of round days. For instance, some villages were served when a vehicle was traveling to/from a landfill site, while in other cases full loads were generated by collecting from part of a town followed by a section of the surrounding rural hinterland.

Given the nature of the round-day boundaries it was difficult to reliably estimate the populations of these areas from even the finest resolution census statistics (i.e., enumeration districts) and there was also the problem that the most recent details (from 1991) were somewhat outdated. As an

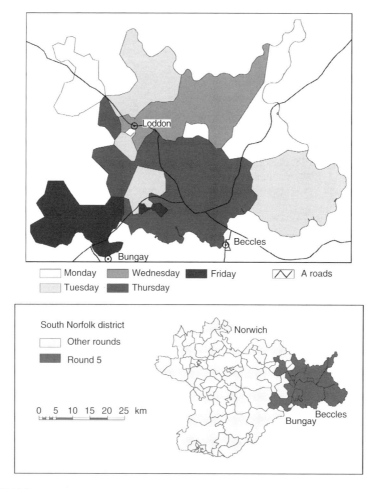

FIGURE 13.3
Boundaries of daily refuse collection areas in Round 5. (© Crown Copyright/database right 2007. An Ordnance Survey/EDINA supplied service.)

alternative source, South Norfolk Council provided data from their Council Tax register as of August 1999. The Council Tax is a compulsory local taxation system based on residential property values and the register is updated annually for each local authority (Eastham, 1993). Several studies have identified associations between the propensity of households to participate in recycling activities and the Council Tax valuations of their properties (Mansell, 2001; Lovett et al., 2003b).

There were 47,474 properties on the South Norfolk register, of which 47,291 (99.6%) had useable postcodes. By applying geocoding and point-in-polygon techniques to the data (household totals by unit postcode and Council Tax band) it was straightforward to estimate the numbers of households in each round day, while the proportion of properties in different valuation bands also provided a simple measure of socioeconomic status. Figure 13.4 maps the percentage of households in bands A and B (the lowest value properties) for each round day and indicates a general pattern of more affluent areas around the southern fringe of Norwich.

Details of the refuse weights (in metric tons) collected from each round-day during the period March 1999–February 2000 were supplied by South Norfolk Council. Other data on the weights of materials collected at

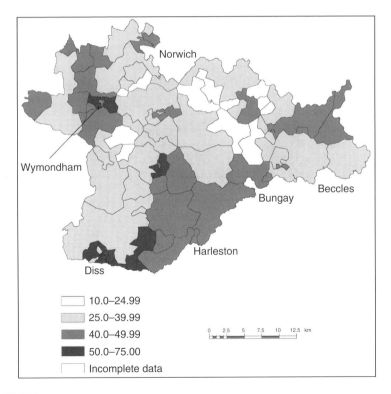

FIGURE 13.4
Percentage of households in Council Tax bands A and B. (© Crown Copyright/database right 2007. An Ordnance Survey/EDINA supplied service.)

different types of bring facilities (e.g., minirecycling centers or CA sites) were obtained from the relevant authorities. This included information from sites outside the district that were known to be used by South Norfolk residents. All these sets of operational records encompassed at least a 12-month period which helped to take account of known seasonal variations in collection weights and so provide a more representative picture of overall trends. An areal interpolation method was then used to apportion the site weight data among refuse collection round days. This involved first defining sets of catchment areas (based on shortest travel time) for each type of bring site. Figure 13.5 shows an isochrone (travel time) map for CA sites and Figure 13.6 the resulting catchment boundaries. The travel-time map highlights the influence of main roads (such as the A11, A140, and A146) radiating out of Norwich, while the catchment areas indicate that in several parts of South Norfolk (actually encompassing over 35% of residents), the nearest CA site was outside the district.

Overlay techniques were employed to calculate the numbers of households in each refuse collection round that fell within different bring-site catchments. Separate analyses were carried out for minirecycling centers and CA sites. From the resulting allocation tables it was straightforward to

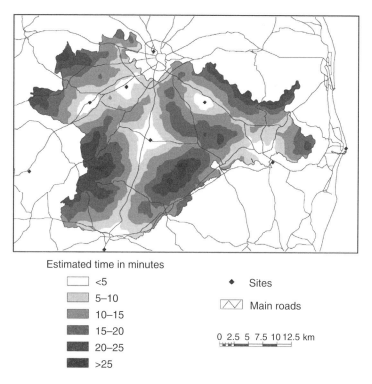

Estimated time in minutes

- [ ] <5
- 5–10
- 10–15
- 15–20
- 20–25
- >25

♦ Sites

▱ Main roads

0 2.5 5 7.5 10 12.5 km

FIGURE 13.5
Estimated travel time by car to nearest CA site. (© Crown Copyright/database right 2007. An Ordnance Survey/EDINA supplied service.)

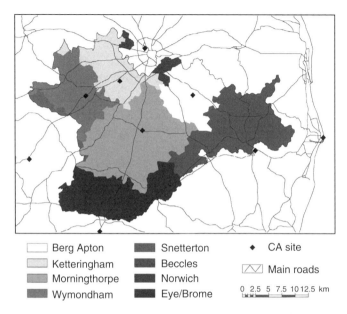

☐ Berg Apton	▨ Snetterton	♦ CA site
▨ Ketteringham	▨ Beccles	▨ Main roads
▨ Morningthorpe	▨ Norwich	
▨ Wymondham	▨ Eye/Brome	0 2.5 5 7.5 10 12.5 km

FIGURE 13.6
Predicted catchment areas (based on shortest travel time) of CA sites. (© Crown Copyright/database right 2007. An Ordnance Survey/EDINA supplied service.)

distribute the site weight data among collection rounds on the basis of household numbers and so reconstruct the average weekly elements of the domestic waste stream on a local area level.

Statistical analyses were conducted to examine associations between socioeconomic variables (e.g., percentage of households in Council Tax bands A and B) and the relative use of different outlets for household waste across the 40 round days. A hierarchical cluster analysis was then performed which divided the round days into five distinct clusters using three classificatory variables based on the proportion of total household waste arising from (1) refuse collection rounds, (2) CA sites, and (3) mini-recycling centers.

In the second phase of the research, the classification results were used to select a sample of six round days, and a door-to-door questionnaire survey was conducted with over 100 households in each area. Each of the five round-day clusters was represented in the sample. and a total of 615 questionnaire responses were obtained in August and September 1999. The questionnaire survey elicited information on householders' attitudes toward recycling activities and their use of local waste management facilities. In addition, an assessment of home composting activity was conducted in which bins or heaps found at 200 of the sampled households were examined, and the basic relationships between garden characteristics and composting activity were investigated. Findings from this research are described elsewhere (Wheeler and Parfitt, 2002).

TABLE 13.1

South Norfolk Recycling Performance
and Future Targets

Year	Household Waste Recycling Rate
1998–1999 (base year)	10%
1999–2000	11%
2003–2004	20%
2005–2006	30%

Source: From the Department for Environment, Food
and Rural Affairs, in *A Consultation Document on the
Distribution of the £140 Million Waste Minimisation and
Recycling Fund in England*, DEFRA, London, 2001
(Annexe B).

The final phase of the project used the database generated by the local
reconstruction of the household waste stream to examine the extent to
which recycling performance in South Norfolk could be improved by establishing new bring sites or introducing kerbside-collection schemes for
dry recyclables or green waste in part or all of the district. This scenario
work required a number of assumptions about levels of participation in
kerbside collections and the material composition of the household waste
stream. The former were based on data from a DETR survey of local
authority recycling performance (Parfitt et al., 2001), while the latter used
results from research conducted for Project Integra in Hampshire (see
http://www.integra.org.uk/about/research) and the Environment Agency
(Parfitt et al., 1999; Parfitt, 2000). Predictions from the scenario work were
subsequently compared against the statutory recycling targets for South
Norfolk as shown in Table 13.1.

13.3   Results

Considerable variability was found in the average weights of refuse collected for individual round days. Figure 13.7 maps the average kilogram per
household per week weights (hereafter kg/hh/wk) for the year March 1999–
February 2000 and shows a rather mixed pattern with, for example, no
simple tendency for values to be higher in more urbanized neighborhoods.
Equally, correlations with the percentage of households in particular Council Tax bands were relatively weak. These results may well reflect the
complex boundaries and diverse socioeconomic composition of some
round days (see Section 13.2). However, clearer trends were apparent in
the use of bring sites, especially when the round days were grouped into
the five clusters rather than analyzed individually. For instance, greater

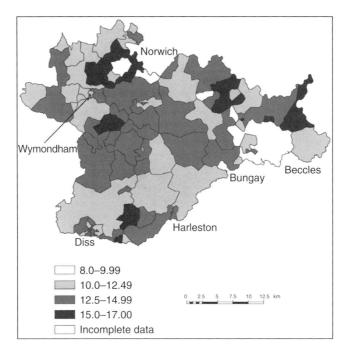

FIGURE 13.7
Average refuse collection weights (kg/hh/wk) for round days, March 1999–February 2000.
(© Crown Copyright/database right 2007. An Ordnance Survey/EDINA supplied service.)

quantities of waste were taken to CA sites by residents in the northern half
of the district where average travel times to such facilities were lowest.
By contrast, householders resident in the most southerly cluster of round
days (Diss, Harleston, and Long Stratton) had the lowest levels of accessibility to CA facilities and the highest mean quantities of refuse collection
round waste. Results from the questionnaire survey indicated that many of
these residents were driving at least 15 min to their nearest CA site, which
confirmed the results of the GIS-based calculations shown in Figure 13.5.
There was also a good level of agreement between the GIS-predicted site
catchments (e.g., Figure 13.6) and the facilities that questionnaire respondents reported using. The main difference between actual and predicted
behaviors was that some respondents used sites on the way to (or from)
Norwich or other larger towns rather than those that they resided nearest.
Nevertheless, the extent of such discrepancies was sufficiently small to
provide confidence in the procedure used to allocate the bring site data
to round days.

Figure 13.8 summarizes the results of the waste stream reconstruction exercise for the five clusters of round days. The boundaries on this map are those
of the 40 refuse collection round days grouped into the clusters, whereas the
grayscale shading depicts the average total recycling rate each week. For each
cluster there is also a proportional symbol whose size represents the total

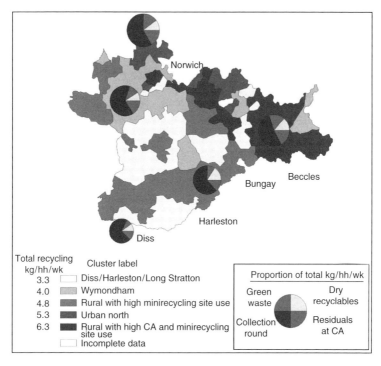

FIGURE 13.8
Variations in waste generation and recycling activity by round-day cluster. (© Crown Copyright/database right 2007. An Ordnance Survey/EDINA supplied service.)

weight of waste per household each week and the subdivisions show the proportions of the waste stream going to different outlets.

A feature of Figure 13.8 is that some of the clusters are quite geographically concentrated, even though there was no explicitly spatial factor in the definition of groups. Such an outcome, however, can be attributed to the contrasts in the accessibility and use of bring sites mentioned above. The highest recycling rates were associated with round days where residents had both CA sites and minirecycling centers easily accessible, but there was considerable variability in performance with the best areas achieving over 30% and the lowest less than 15% (these figures include estimates of materials taken to outside the district).

Analysis of the questionnaire responses showed that the amount of material put out for the refuse collection increased with household size and declined with involvement in recycling activities or composting. When asked what would encourage them to recycle more, around 50% of respondents mentioned the provision of kerbside collections, whereas another 20% suggested additional local bring sites. These options were investigated further in a series of scenarios regarding additional recycling facilities. The results of this exercise soon indicated that further bring-site provision was unlikely to be of major

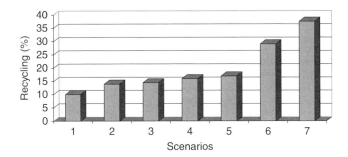

FIGURE 13.9
Predicted outcomes of introducing different kerbside schemes on district recycling rates. *Key to scenarios*: (1) baseline, 1998/99; (2) dry recyclables kerbside, Round 1; (3) dry recyclables kerbside, Round 3; (4) dry + green waste kerbside, Round 1; (5) dry + green waste kerbside, Round 3; (6) dry recyclables kerbside, all rounds; and (7) dry + green waste kerbside, all rounds.

benefit in helping South Norfolk to meet its statutory targets and therefore attention was focused on the impact of new kerbside schemes.

Figure 13.9 illustrates the predicted outcomes of a series of scenarios concerning the introduction of new kerbside schemes. The GIS database was used as part of this exercise to help determine where the introduction of kerbside-collection schemes would have the greatest impact on recycling performance. In addition, options for meeting the statutory targets with multimaterial dry recyclables and green waste collections were considered. The results suggested that introducing kerbside collections of dry recyclables in Round 1 (around the edge of Norwich) and Round 3 (Wymondham and surrounding villages) would certainly boost recycling rates, but that coverage of more than these 10 round days would be required to meet the 2003–2004 target of 20%. Looking further ahead, the predictions indicated that the 2005–2006 target of 30% might be achievable if the kerbside collection of dry recyclables occurred across South Norfolk (Scenario 6), but that this would be dependent on achieving high participation rates. If this condition was not achieved, then a district-wide kerbside collection of green waste (Scenario 7) would be required to meet the 2005–2006 target.

13.4 Conclusions

The GIS-based methodology developed in this research provides a means of integrating site and round weight data to develop a better understanding of household recycling performance at a subdistrict level. Reconstructing the household waste stream in this way also provides a valuable platform for evaluating the likely impact of new recycling services. In the case of South Norfolk, a trial kerbside-collection of dry recyclables was introduced in the villages of Mulbarton and Bracon Ash in September 2003. This scheme involves each household having two wheeled bins, which for dry recyclables being collected 1 wk and the other for residual refuse the next. A number of other local authorities have also adopted such an alternate week collection system as a means of providing a cost-effective refuse and recycling service to their residents (DEFRA, 2001, 2002).

The recycling rates achieved by the kerbside trial in South Norfolk have exceeded expectations and a similar service is now being gradually rolled out to other parts of the district. If the current level of participation is maintained then South Norfolk should certainly meet its target for 2003–2004 and will be well on the way to achieving that for 2005–2006. However, there is likely to be a need to attain still higher levels of recycling at both local and national levels in the future (Cabinet Office Strategy Unit, 2002; Norfolk Waste Partnership, 2002). In these circumstances, the role of GIS in the planning and monitoring of household waste management services could be invaluable. At present, such applications are relatively rare in local authorities and it is likely to require improvements in several aspects of data quality and storage before they become more common. Nevertheless, research such as that presented in this chapter demonstrates the contribution that the use of GIS can make to the provision of information from which effective household waste management policies can be formulated.

Acknowledgments

This research was funded as an ENTRUST (Landfill Tax Credits Scheme) project through UEAEA with support from NEWS (Norfolk Environmental Waste Services) and South Norfolk Council. We would like to thank the staff of both organizations for their help while absolving them of any responsibility for the opinions expressed.

References

Barr, S., 2002, *Household Waste in Social Perspective: Values, Attitudes, Situation and Behaviour* (Aldershot: Ashgate).

Cabinet Office Strategy Unit, 2002, *Waste Not, Want Not: A Strategy for Tackling the Waste Problem in England* (London: Cabinet Office).

Community Recycling Network, 2002, *Maximising Recycling Rates—Tackling Residuals* (Bristol: Community Recycling Network) (http://www.crn.org.uk, accessed on March 2003).

Department of the Environment, Transport and the Regions, and National Assembly for Wales, 2000, *Waste Strategy 2000 for England and Wales Parts 1 and 2* (London: The Stationary Office).

Department for Environment, Food and Rural Affairs, 2001, *A Consultation Document on the Distribution of the £140 Million Waste Minimisation and Recycling Fund in England* (London: Department for Environment, Food and Rural Affairs).

Department for Environment, Food and Rural Affairs, 2002, *Municipal Waste Management Survey 2000/01* (London: Department for Environment, Food and Rural Affairs).

Eastham, L.S., 1993, *Council Tax* (London: Citizen Advice Notes Service Trust).

EnCams, 2002, *Waste Segmentation Research 2002* (Wigan: EnCams) (http://www.encams.org, accessed on November 2002).

Lovett, A.A., 2000, GIS and environmental management. In *Environmental Science for Environmental Management*, Second Edition, edited by O'Riordan, T., pp. 267–285 (Harlow: Prentice-Hall).

Lovett, A.A., Sünnenberg, G., and Haynes, R.M., 2003a, Accessibility to GP surgeries in South Norfolk: a GIS-based assessment of the changing situation 1997–2000. In *Socio-Economic Applications of Geographic Information Science: Innovations in GIS 9*, edited by Kidner, D., Higgs, G., and White, S., pp. 181–198 (London: Taylor & Francis).

Lovett, A.A., Parfitt, J.P., Hummel, J., Bone, A., Sünnenberg, G., and Pearce, S., 2003b, *Research for Monitoring and Evaluation of the Peterborough Recycling Cell* (Norwich: School of Environmental Sciences, University of East Anglia).

Lovett, A.A., Sünnenberg, G., and Haynes, R.M., 2004, Using GIS to assess accessibility to primary health care services. In *GIS in Public Health Practice: Opportunities and Pitfalls*, edited by Maheswaran, R. and Craglia, M. (London: Taylor & Francis).

Mansell, D., 2001, *Bristol Recycling Participation Study* (Bristol: The Recycling Consortium).

Norfolk Waste Partnership, 2002, *Municipal Waste Strategy for Norfolk*, 1st Revision (Norwich: Norfolk Waste Partnership).

Office for National Statistics, 2003, *Census 2001: Key Statistics for Local Authorities in England and Wales* (London: The Stationary Office).

Parfitt, J.P., 1997, *An Assessment of the Quality of Household Waste Statistics in Norfolk*. Report to the Norfolk Waste Management Partnership (Norwich: School of Environmental Sciences, University of East Anglia).

Parfitt, J.P., 2000, *A Study of the Composition of Collected Household Waste in the UK: with Particular Reference to Packaging Waste*, Technical Report P347 (Bristol: Environment Agency).

Parfitt, J.P., Flowerdew, R., and Pocock, R., 1999, *A Review of the United Kingdom Household Waste Arisings and Compositional Data*, Final report prepared under contract No. EPG 7/10/21 CLO 0201 for the Wastes Technical Division, Department of the Environment (Bristol: Environment Agency).

Parfitt, J.P., Lovett, A.A., and Sünnenberg, G., 2001, A classification of local authority waste collection and recycling strategies in England and Wales. *Resources, Conservation and Recycling* 32, 239–257.

Perrin, D. and Barton, J., 2001, Issues associated with transforming household attitudes and opinions into materials recovery: a review of two kerbside recycling schemes. *Resources, Conservation and Recycling* 33, 61–74.

Tucker, P., Murney, G., and Lamont, J., 1998, Predicting recycling scheme performance: a process simulation approach. *Journal of Environmental Management* 53, 31–48.

Wheeler, P. and Parfitt, J.P., 2002, Life cycle assessment of home composting. *Proceedings of Waste 2002 Conference*, Stratford.

14

Comparison of Discrete-Choice Modeling and Accessibility-Based Approaches: A Forest Recreation Study

Mette Termansen, Colin J. McClean, and Hans Skov-Petersen

CONTENTS

14.1 Introduction

The environment's ability to provide goods and services beyond those with traditional and well-studied markets, such as agriculture and forestry, has received increased attention over the last decades from environmental economists (Haab and McConnell, 2002). Policy makers, too, are now widely discussing these goods and services as they attempt to balance economic developments with environmental concerns (Forest Commission, 2001). One such environmental good is the benefit people gain from recreational use of forest areas. The value gained by recreationists using forests needs to be taken into account, along with other benefits from forests such as watershed protection and wildlife diversity, when policy makers assess land-use decisions (Bateman et al., 1997). As a step toward understanding the recreational benefits offered by forest areas, policy makers are likely to require information about the patterns of recreational use for existing forest areas as well as information about where potential reafforestation projects might be best located. This inherently spatial information is well suited to study using GIS applications. The recreational use made of forest areas is clearly dependent on the spatial distribution of those areas and the potential recreationists. Spatial analysis of policy initiatives, such as reafforestation schemes, using GIS and aimed at improving the recreational opportunities for local populations, would potentially be informative in the development of forest planning and policy proposals.

Researchers have approached modeling of recreational use from a variety of perspectives. Economists have been interested in estimating recreational demand functions to assist in the economic valuation of forest resources (McConnell, 1985; Hanley et al., 2003). The studies that fall in this category have been interested in the relationship between recreational demand and socioeconomic characteristics but have put less emphasis on the spatial pattern of recreational demand. Land-use planners on the other hand have mainly been interested in understanding the recreational activity and obtaining estimates of the spatial patterns of the use or potential use of recreational areas (De Vries and Ghossen, 2001). As expected, the different perspectives have resulted in different modeling approaches. The use of GIS, for example, has a long history in land-use planning but has been little used in economics, except the work on travel cost analysis (Bateman et al., 1999, 2002; Brainard et al., 1999).

The statistically based approaches usually used in economics can be divided into two main categories of models: continuous regression models and discrete-choice models. Continuous regression models have been adopted when focus is on estimating recreational demand for a specific site. Discrete-choice models have been used where the number of alternative sites is large and substitution between sites is important (Haab and McConnell, 2002).

The continuous regression models have been combined successfully with GIS (e.g., Brainard et al., 1999). However, little has been done to combine discrete-choice models with GIS. Furthermore, little has been done to assess the relative merits of purely GIS-based approaches and the economics-based approaches for policy analysis.

In this paper, we address these issues by using the same data set to model forest recreation in Denmark using a discrete-choice model built on GIS-processed spatial data and an accessibility-based model using methods entirely available within most GIS software. In both approaches visitor numbers are modeled as a function of distance between visitors and forest areas and forest site amenities. We assess the underlying assumptions in the two approaches and compare the two methodologies and the data requirements for the two analyses. In particular, we assess the applicability of the two approaches for identification of suitable reafforestation areas. We illustrate this using information on an existing reafforestation project, initiated by the Danish Forest and Nature Agency, Department of Environment. The project is located in a forestry poor region close to one of the main population centers in Denmark.

14.2 Methods

We first report on the data available for the study, then we turn to formulation of the modeling approaches and estimation procedures.

14.2.1 Data

Two Danish data sets are available for this study. The first is a household questionnaire assessing total demand for recreation and characteristics of the recreational activities of the individual respondents. This data set consists of 2895 completed questionnaires. The second is an on-site survey covering the most important forest recreational areas in Denmark (covering 75% of the country's total annual recreational visits). In total, 2095 survey points were sampled. Data were collected simultaneously at all sites, 22 times during the year-long period of the study (December 1996–November 1997) by the Danish Forest and Landscape Research Institute. Twenty sampling times were chosen randomly, while two were selected specifically on public holidays where high visitor numbers were expected (Jensen, 2003). At prespecified randomly selected times during those 22 days, all parked cars were counted and questionnaires where distributed to parked cars at the sampling locations. This gives two data resources available for this analysis. The number of cars counted is used in the accessibility-based model. The questionnaires where the origin of the trip could be precisely located (approximately 30,000) were used for the discrete-choice analysis. The origin of each trip was digitized using the postal address using Befordringsfradrag, a PC-software package developed by Carl Bro A/S (Carl Bro, 1997). This software assigns the postal addresses to the nearest node in the Danish road network. Traveling distances from the trip origin to the destination and to alternative sites were calculated using a 1:200,000 vector road map (VejnetDK; Kort and Matrikelstyrelsen/Vejdirektorartet, 1995). It is assumed that people have used the shortest route possible.

Potentially important site attributes have been derived from several data sources. The area information system, AIS (Miljø and Energiministeriet/ Danmarks miljøundersøgelser, 2000), contains information on the spatial distribution of forests, seminatural vegetation such as heather moorland, meadows, rough pasture, lakes, bogs and rivers, and different categories of built-up areas. The locations of marked nature trails and view points were obtained from "the interactive map of Denmark" (Kort and Matrikelstyrelsen, 2001). These refer to particularly scenic walking/hiking routes that are marked out in the terrain and described in an information leaflet by the National Forest and Nature Agency. Classification of the terrain was based on a 50×50 m digital elevation model (DEM). The mean slope calculated for a 1-km grid was available from Skov-Petersen (2002). Information on parking facilities was available from the survey. The site attributes used, how they have been measured, and their sources are given in Table 14.1.

The demographic data originates from a digital parish map of 2116 parishes with information of male and female population divided into six

TABLE 14.1

The Distance Measures and Site Attributes Used to Characterize Site Amenities

Variable[a]	Measurement	Data Source[b]
Travel distance (discrete choice)	Shortest distance through road network from the origin of the trip given by the respondents to all the sites	On-site survey; road network
Travel distance (accessibility)	Shortest distance through road network from all origins to all destination node in the road network	Road network
Forest area (discrete choice)	Size of the contiguous forest area associated with the site	AIS
Fraction of broadleaved area	Fraction of forest area classified as broadleaved	AIS
Distance to coast	Straight line distance from aggregate site to nearest coastline	AIS
Slope	The average slope index of the 1-km by 1-km area around the aggregated sites	DEM
Natural area edge fraction (discrete choice)	Length of boundary between the forest area and neighboring areas classified as wet or seminatural divided by the total length of the boundary of the forest area	AIS
Parking (discrete choice)	Dummy variable. If parking lots are available at one of the sites, the aggregated site is classified as Parking $= 1$, otherwise Parking $= 0$	On-site survey
Distance to info leaflets	Straight line distance from aggregate site to nearest marked nature trail described in an information leaflet	Interactive map of Denmark
View point	RUM: dummy variable. If the straight line distance from aggregate site to nearest view point is less than 5 km, View point $= 1$, otherwise View point $= 0$. Accessibility: the straight line distance site to nearest view point	Interactive map of Denmark

[a] The type of analysis used is indicated in brackets for a variable.
[b] Further details on data sources are given in the main text.

TABLE 14.2

Data Describing the Origin for Estimating Annual Demand of Recreational Visits

Variable[a]	Measurement	Data Source[b]
Trips (discrete choice)	Total number of car-borne trips per year	Household survey
Distance (discrete choice)	Shortest Euclidian distance from each individual's home address to a forest patch larger than 10 hectares	Road network, AIS
Time traveled (accessibility)	Proportion of people traveling to sites in different time bands away from the origin	Household survey
Population	Total adult population allocated to nodes in the road network	Danish statistics, AIS
Origin classification (accessibility)	Residential houses, summer houses, hotels, camping lots, and youth hostels	National Building and Dwelling Register, AIS
Income (discrete choice)	Average parish council income	Danish statistics

[a] The type of analysis used is indicated in brackets for a variable.
[b] For more details on data sources, refer to the main text.

age classes. Finer resolution data, allocating population segments to nodes in the road network, were available from the Danish Forest and Landscape Research Institute. The spatial disaggregation of the population was generated using an urban land-use map (100×100 m resolution) and is detailed by Skov-Petersen (2002). Information on average income was available from Danish Statistic at the parish level (Statistics Denmark, 2000). The characteristic of the origin is further identified using the national Building and Dwelling Register (EBST, 2001). Buildings were geocoded by means of interpolation of the Danish road network and address register (DAV, 2001). Information about the location of hotels, youth hostels, and camping sites were available from Miljø and Energiministeriet/Danmarks miljøundersøgelser (2000). Information available about the origin locations is given in Table 14.2. Having described the data available for the study, we now turn to a description of the two models and the estimation procedures.

14.2.2 Random Utility Model

There are a variety of well-established techniques used in transport economics and environmental economics to model individual's choices between alternative actions (Ben-Akiva and Lerman, 1985; Haab and McConnell, 2002; Train, 2003). The approach taken in this chapter is based on the random utility model (RUM) approach. In the context of recreational studies, the RUM is a discrete-choice model that focuses on the choice between substitute sites, which are mutually exclusive, for a given recreational trip on a single-choice occasion. It is assumed that the individual chooses the site that gives the maximum utility, u. The utility that an individual n obtains from visiting a site j, u_{nj}, is known to the individual not to the researcher. The researcher observes attributes of the alternative sites and attributes of the individual and specifies a function which relates these factors to the individual's utility. This function denotes the representative utility, v_{nj}. As there are aspects of utility that the researcher does not observe, then $v_{nj} \neq u_{nj}$. Therefore, the utility is composed of a deterministic and a random part, $u_{nj} = v_{nj} + \varepsilon_{nj}$, where the random part, ε_{nj}, captures the factors that influence utility but are not observed by the researcher. In the present study, there are only limited data on individuals' attributes and the deterministic utility v_j for site choice j for one-choice occasion is defined as:

$$v_j = v\left(q_j, f(d_j) \right) \tag{14.1}$$

where
q_j is a vector of characteristics of site j, described in Table 14.1
d_j is the distance to site j
$f(d)$ is therefore some function of distance

In economic applications, $f(d)$ is the cost of access either measured purely as transportation costs or including the cost of time spent in traveling. The function, $f(d)$, therefore gives some measure of the costs of obtaining the utility from the recreational visit. The random component enters through ε_j as an error associated with the site. The individual will be expected to visit the site that gives the highest utility, u_{ni}. The behavioral model is therefore that individual n will choose site i if and only if

$$v_{ni} + \varepsilon_{ni} > v_{nj} + \varepsilon_{nj}, \quad \forall j \neq i \tag{14.2}$$

where j represents the other sites being chosen from.

In order to derive probability distributions and the associated likelihood functions for the given choice, assumptions need to be made about the distribution of the error term. Making the standard assumptions, that the error terms are independently and identically distributed, extreme value (Gumbel), the logit choice probability of an individual choosing a site can be calculated (Bockstael et al., 1991; Train, 2003):

$$P_i = \frac{\exp(v_i)}{\sum_{j=1}^{J} \exp(v_j)} \tag{14.3}$$

where
P_i is the probability of an individual choosing site i
J is the number of sites in the choice set

14.2.2.1 Specification of the Representative Utility Function

The RUM is estimated using GAUSS as the necessarily large-choice sets are feasible within this software. We define the mutually exclusive recreational sites by grouping the 2095 survey locations into 581 continuous recreational areas. These were identified by local forest managers as separate recreational sites. The most centrally located questionnaire distribution point was chosen as the representative location of each recreational site.

The AIS was used to derive site attributes such as the size of contiguous forested area, distance to water features, and adjacency of the forested area with other areas of seminatural vegetation. The forest area was further classified into broadleaved forest, coniferous forest, and scrub vegetation based on LandSat TM satellite images (Miljø and Energiministeriet/ Danmarks miljøundersøgelse, 2000). This data source enabled the computation of the fraction of each forest type for each site.

We define the full-choice set for each individual as the nearest 300 sites within a distance threshold of 250 km. The complete data model is estimated using this choice set specification. The absolute choice set boundary of 250 km was chosen with reference to the empirical data as less than 0.1% of trips are further than 250 km.

We have tested various specifications of the representative utility function v_i, but only report the final specification here. The model reported was chosen based on its overall maximum likelihood score. The simplest model with good predictive capacity was chosen. Other criteria for model specification based on statistical significance proved unhelpful in this situation as even variables with a very small predictive effect become statistically significant using such a large data set.

It is expected that, the distance to sites is an important variable for modeling recreational choices. The form that this variable takes in the model was therefore analyzed testing alternative Box–Cox transformations (Greene, 2002, p. 173). The Box–Cox transformation takes the following form:

$$distance^* = \frac{distance^\lambda - 1}{\lambda} \tag{14.4}$$

where $distance^*$ is the transformed variable and $distance$ is the distance from the origin of the trip to the sites in the choice set. λ is the transformation parameter, $\lim(distance^*)$ for $\lambda \to 0$ is $Ln(distance)$ where as $\lambda = 1$ is a linear transformation. Varying the transformation parameter can therefore be used to test for nonlinearity. The transformation we used was the value of λ which gives the highest maximum likelihood score. We tested values of λ in intervals of 0.05. For more details about this analysis see Termansen et al. (2004).

14.2.2.2 Model for Total Demand

The RUM estimates the probability of visiting different sites in the choice set. It does not estimate the total number of car-borne recreation trips. For this purpose we use a count regression using the household data. We chose a negative binomial distribution, as there is evidence of overdispersion. The model enables us to predict the number of visits per individual given the socioeconomic characteristics of the individual and the distance to the nearest forest site. Applying the model at all nodes in the road network a total number of annual visits can be estimated.

14.2.2.3 Assessment of Visits to New Sites

The potential number of visits to a reafforested area is straightforward to measure using a discrete-choice model and the model for total demand. As the RUM estimates a probability distribution, it incorporates substitution effects between sites. The probability that a new site *l* which is introduced to the choice set of individual *i* would be visited is given by

$$P_i = \frac{\exp(v_l)}{\sum_{j=1}^{J} \exp(v_j) + v_l} \tag{14.5}$$

Combining this probability with the model for total demand for the individual *i* gives individual *i*'s total number of visits to site *l*. Aggregating over all individuals gives the expected number of visits to the new site.

14.2.3 Accessibility-Based Model

This model has two stages. The first stage uses the household questionnaire data to develop an accessibility index for nodes in the Danish road network, representing the demand for the recreational resources provided by Danish forests. The second stage uses the accessibility index and other recreational site characteristics to model the number of visits to forest areas. This model uses the on-site survey of recreational sites for calibration.

In the accessibility model, the decrease in demand for forest recreation with distance to the recreational resource is captured directly from the household survey data. In the questionnaire respondents were asked to state how much time they spent traveling to their recreational destination on their last visit to such a site. Here respondents were given a number of categories to choose from, e.g., less than 2 min, between 2 and 5 min, between 5 and 10 min, and so on. These time thresholds allowed an assessment of the fraction of recreational trips falling into the different segments of travel time. The distance dependence is therefore formulated using isochrone functions, where each isochrone defines the resources available within each segment of travel time. The responses were further grouped by departure point of their trip (departure from home, hotel, camping site, etc.). In effect, the survey provided information about the proportion of the Danish population willing to travel for a given amount of time from different kinds of departure points.

These proportions, *r*, were then applied to the entire Danish population distributed among the nodes of the road network with information about types of departure point at each node, calculated for Thiessen polygons around the nodes using the National Building and Dwelling Register along with the other government-collected statistics referred to in Table 14.2.

This provided the spatial pattern of demand for forest recreation across Denmark, broken down into the segments of travel time that people are willing to spend accessing the resource. For each node the potential number of recreationists leaving from a particular type of departure point k, U_{nk} was calculated as:

$$U_{nk} = \sum_m P_{mk} \cdot r_{ik} \cdot f_i(t_{nm}) \tag{14.6}$$

Here, P_{mk} is the population at origin node, m, leaving from type of origin k and r_{ik} is the proportion, which would be willing to travel to a site within the i-th isochrone. The time distance between n and m is given by t_{nm}, i denotes the isochrone and the isochrone function, $f(t_{nm})$ is defined by

$$f_i(t_{nm}) = \begin{cases} 0 & \text{for } t_{nm} \notin [\tau_i^{\min}; \tau_i^{\max}] \\ 1 & \text{for } t_{nm} \in [\tau_i^{\min}; \tau_i^{\max}] \end{cases} \tag{14.7}$$

where the time distance thresholds, $\tau_i^{\min}$ and $\tau_i^{\max}$, define the isochrones.

In the same way, using the same set of nodes and isochrones, the accessibility of supply (area) of forest recreational sites was calculated from the Danish Area Information Systems land-use map by aggregating the total area available from each node in each isochrone. Combining the demand for trips at each node for each isochrone and the area of resources available in the isochrone, an estimate of the number of trips per unit area of forest is obtained. In this way, intervening opportunities are taken into account by matching demand and supply for recreational trips.

The second stage of the model calculates forest-site characteristics around each node. The characteristics used are the same as those considered in the RUM. A regression model was then developed using the number of cars counted at each of the sites sampled in the on-site survey as the dependent variable. The independent variables included the site characteristics and the number of trips per unit area of forest calculated using the accessibility model.

The resulting regression equation provides a means of estimating an index of forest-site recreational use given a set of site characteristics and accessibility of forest users to that site. This allows the potential of reafforestation for different areas to be assessed by calculating the estimated visits to individual areas assuming that the area is reafforested and the remaining landscape kept unchanged. Sequentially calculating this for a 1-km^2 raster representation for Denmark produces a map of potential use.

14.3 Results

14.3.1 Estimated Dependence of Distance and Site Amenities

The parameter estimates for the choice model based on a complete data set of 28,947 individuals chosen between the 300 nearest sites within 250 km are

TABLE 14.3

Parameter Estimates for the RUM and the Accessibility Model

	RUM		Accessibility	
Variables	Estimates	P	Estimates	P
Travel distance[a]	−0.9976	<0.0001	n.a.	n.a.
Constant	n.a.	n.a.	5.0607	<0.0001
Accessibility	n.a.	n.a.	−0.8017	0.0012
(Accessibility)^2	n.a.	n.a.	0.0854	<0.0001
Parking	0.5060	<0.0001	n.a.	n.a.
Slope	0.0257	<0.0001	0.2817	<0.0001
Info	−0.3672	<0.0001	-3.4×10^{-5}	<0.0001
View point	0.2882	<0.0001	N.S.	
Distance to coast	−0.5819	<0.0001	-2.8×10^{-5}	<0.0001
Ln (Forest area)	0.3684	<0.0001	n.a.	n.a.
Fraction broadleaved	0.1865	<0.0001	N.S.	
Fraction natural area edge	2.6286	<0.0001	n.a.	n.a.

[a] Optimal Box–Cox transformation of distance, $\lambda = 0.3$.

shown in Table 14.3 in columns 2 and 3. The results of the accessibility model are shown in columns 4 and 5. The Table cannot be used for comparison of coefficient values but indicates whether the two approaches identify the same site characteristics as important for visitor attraction.

The discrete-choice analysis shows that travel distance is negatively related to the probability of site visit, as would be expected. The optimal Box–Cox transformation indicates that it is a nonlinear relationship with a marginally declining effect of distance as distance increases.

Several landscape attributes are shown to be related to recreational choices. Sites close to the coastline are preferred to inland sites. A natural-log specification performs better than a linear specification in the RUM, as proximity to the coast has most influence for relatively short distances. Larger forest areas are preferred to smaller. The effect of size is however nonlinear, indicating that the marginal effect of area is declining. Adjacency to seminatural areas and water features are also shown to attract visitors to the site. The topography of the landscape close to the site has a small effect indicating preference for undulating rather than flat sites. The analysis also identifies a preference for sites that are within 5 km from a view point.

We have tested two site attributes which are associated with the management of the forests for recreational use: parking facilities and marked nature trails. The analysis shows that both are positively associated with probability of visit. Sites are predicted to have more visits when parking facilities exist and when located close to an area covered by a nature trail leaflet. The natural-log transformation of distance to origin of a nature trail implies that the effect of nature trail leaflets is mainly influential close to the origin of the trails. Another site attribute influenced by forest management activities is the fraction of broadleaved forest. The analysis shows that respondents display a preference for broadleaved woodland as opposed to conifer woodland.

The accessibility index variable in the accessibility-based model is highly significant. The number of cars observed at the surveyed sites increase with the accessibility of the site. This effect seems to be nonlinear as the quadratic term is also significant. Undulating terrain is also observed to have a positive relationship with recreational use. Distance to coast has a negative relationship, forest sites further from the coast receive fewer visits than forest sites close to the coast. Furthermore, the distance to marked nature trails is also found to be an important site attribute. These results are all in agreement with the RUM results. No significant relationships are found for parking, view point, and fraction of broadleaved forest.

14.3.2 Specification of Total Demand for the Discrete-Choice Analysis

The results show that the number of annual car-borne trips per individual is decreasing with the distance to the nearest forest. Furthermore, the number of annual trips increases with income (Table 14.4). This generates a total annual number of visits of about 56 million car-borne trips.

14.3.3 Estimated Potential of Reafforestation Areas

The potential for reafforestation using the two models is illustrated in Figure 14.1. The area around Aarhus on the East coast of Jutland is shown here. The reafforestation project is marked with a triangle in Figure 14.1b.

The patterns of recreational visits are markedly different for the two predictions. This is partly because the RUM results predict visits per site (forest patch), whereas the accessibility model output is measured in visits per unit area. The accessibility model reflects closely the population centers and the topography of the area (Figure 14.1a). Site characteristics seem to have been more influential in producing the spatial pattern of visitor numbers from the RUM predictions (Figure 14.1b). The discrete-choice model

TABLE 14.4

Parameter Estimates for Count Regression Model of Trip Frequency

Variable	Coefficient	Asymptotic-z
Constant	1.5656	4.93
Income	6.06×10^{-3}	3.72
Distance to nearest forest	-1.1×10^{-4}	-3.98
Dispersion parameter	1.5213	19.92

Log-likelihood (Poisson), $-11,421.7$; log-likelihood (negative binomial), -2871.9; $N = 812$.

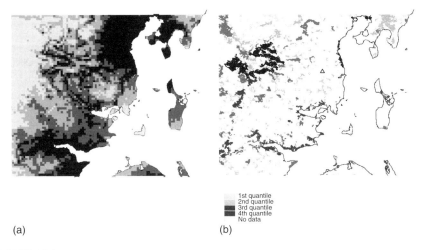

1st quantile
2nd quantile
3rd quantile
4th quantile
No data

(a)   (b)

FIGURE 14.1
Predicted indices of forest visits using (a) the accessibility model and (b) the RUM. Values are classified into quartiles for each index to allow comparison between different units used in the two maps. (Map data from Danish Forest and Landscape Research Institute.)

applied to the reafforestation project produces an estimated visitor number of 66,000 annual visits.

14.4   Discussion

The first aim of this paper was to assess the extent to which site amenities and characteristics explain the spatial distribution of recreational activity and to what extent recreational choice is purely distance dependent. As expected, the distance to the site is particularly important for modeling an individual's choice; however, other site attributes are also very influential in determining patterns of recreational use.

The estimated RUM is specified using a nonlinear specification of distance. The analysis highlighted that the linear specification was not well supported by the data and suggests that the standard practice in economics using a linear specification may deserve more careful justification. The interpretation of our results is that recreationists distinguish more between the distance to sites 10 and 20 km away than sites 50 and 60 km away. This makes intuitive sense, in that once committed to traveling 50 km an extra kilometer does not deter the recreationist to the same extent as it would if the trip was short. A nonlinear specification is embedded in the isochrones of the accessibility-based modeling approach as the number of trips generated from a node declines as the distance to the destination node increases (Skov-Petersen, 2001).

An important and policy-relevant site attribute which has been identified in this study is the size of the forested area. Both a linear and a natural-log specification have been tested. The analysis shows that a log specification is superior. This implies that there is a marginally declining improvement in visitor attraction with size. This has implications for reafforestation initiatives as it gives an indication of the trade-off between size of reafforestation project and proximity to populated areas. This effect is not tested in the accessibility model as forested areas are not treated as individual sites. This effect is however supported by research from other countries, Scarpa et al. (2000) find that forest area positively influence the willingness to pay for creation of nature reserves in Ireland.

The results also highlight that the presence of other seminatural areas adjacent to the forested area also affect recreational choices. We find that the fraction of the forest border adjacent to other seminatural areas is positively associated with visit probability. Recreationists therefore seem to prefer visiting a site with a mix of forest and other seminatural environments. These effects cannot be assessed in the accessibility model as the spatial arrangement of destination classes is not considered. Furthermore, proximity to the coast and undulating topography are also identified as being preferred by the respondents. These effects are found using both modeling approaches.

The study also identifies site attributes which have direct management implications. The species composition of the forest seems to have an impact on recreational choices, as broadleaved forest seems to be preferred to conifer forest. This is equally demonstrated in the Irish study by Scarpa et al. (2000) and in a Canadian study by Boxall et al. (1996). This effect was not identified using the accessibility model. Marked nature trails and parking facilities are other management-dependent site attributes. The discrete-choice analysis suggests that both seem to have a positive effect on visit probability. The same conclusion is derived from the accessibility model with respect to marked nature trails. The parking facilities were not tested in the accessibility model preventing any comparison.

The results from this study can potentially be valuable inputs in the policy debate on reafforestation. This is due to the direct analysis of trade-offs between accessibility captured by the distance matrix and the site attributes such as size and the presence of other landscape elements. We show how the assessment of new recreational sites is assessed in the two models. In the discrete-choice analysis, introduction of a new site into the choice set alters the probability distribution and the new site competes with the existing sites for visitors. When the accessibility-based model is used to calculate a map of recreation potential, this effect is not taken into account as each area is calculated independently.

There are therefore three main reasons why the discrete-choice framework is particularly appropriate for analysis of reafforestation initiatives. First, this framework captures the spatial arrangement of substitute sites through a distance matrix. Second, it enables a simultaneous assessment of site characteristics and to what extent they affect visitor attraction. Finally, the definition of the recreational resource as individual spatially defined recreational sites enables assessment of spatially defined site characteristics such as neighboring land uses and other landscape metrics.

The main disadvantages of the RUM approach are the detailed data requirements and the computational effort required to assess the introduction of reafforestation sites. Detailed data are required on the spatial location of the origin and the destination of each response from the on-site survey. The respondent's address from each questionnaire had to be geocoded to the Danish grid system, a much more laborious process than the use of postcode data in countries such as the United Kingdom. The accessibility-based model only required the raw data available from the household survey and the counts of cars at the on-site survey locations. Most of the data manipulation required for the accessibility-based model is achieved within GIS software. Once the accessibility-based model is built, the relative accessibility of each new potential afforestation project can be assessed. The econometric details of the RUM are specially programmed in GAUSS because of the size of the problem. For an assessment of a new forest site the expected utility provided by the site needs to be calculated. This requires calculation of the distances from all origin nodes, which would include the new site in the choice set, and the probability distribution of all the nodes need to be recalculated. This is computationally very intensive; however, intervening opportunities are fully considered in this modeling approach.

The major disadvantages of the accessibility-model are, unsurprisingly, the reverse of those found in the RUM process. There is no simultaneous assessment of the trade-offs between distance and other site characteristics. Furthermore, as all estimations are spatially aggregated there is no simple means of incorporating the wider landscape characteristics in the accessibility index.

14.5   Conclusion

This work combines GIS and RUM for modeling forest recreational behavior. Coupling such discrete-choice models and spatial databases proves effective in capturing individuals' responses to detailed landscape scale attributes in their recreational choices. Furthermore, this approach allows the inclusion of substitution effects between recreational sites. RUMs of this type allow the effect of introducing a new recreational site into an area to be assessed. We see this as potentially an extremely useful policy tool. However, the use of the RUM approach is computationally very expensive and dependent on the type of detailed data available in the Danish on-site survey. As computing power increases the RUM approach will become increasingly easy to implement, even for regions larger than Denmark.

Detailed data availability will, however, continue to rely on policy makers being willing to invest in expensive data-capture exercises.

The accessibility-based approach discussed here uses tools that are readily available in proprietary GIS software, although the computational effort involved is not trivial. It is by no means the only possible approach to studying forest recreation use by such methods, but it illustrates what can be achieved using current software and computing resources and data that might well be obtained in countries, other than Denmark, where much more limited on-site survey data may be available.

Acknowledgment

The data used in this paper were kindly made available from the Danish Forest and Landscape Research Institute. The research was funded by the Danish Socio Economic Research Council, the Danish Forest and Nature Agency, and the Danish Outdoor Council. We would like to thank Riccardo Scarpa for the assistance with the estimation of the random utility model.

References

Bateman, I.J., Dolman, P., Lovett, A., and Brainard, J., 1997, Placing the biodiversity consequences of land use change within a wider context: developing a GIS/CBA methodology for assessing conversions from agriculture to farm forestry in Wales. In *Economics of Biological Resources and Biodiversity*, edited by O'Riordan, T., pp. 34–52 (London: CSERGE, UEA and University College).

Bateman, I.J., Lovett, A.A., and Brainard, J.S., 1999, Developing a methodology for benefit transfers using geographical information systems: modelling demand for woodland recreation. *Regional Studies* 33(3), 191–205.

Bateman, I.J., Jones, A.P., Lovett, A.A., Lake, I.R., and Day, B.H., 2002. Applying geographical information systems (GIS) to environmental and resource economics. *Environmental and Resource Economics* 22, 219–269.

Ben-Akiva, M. and Lerman, S.R., 1985, *Discrete Choice Analysis: Theory and Applications to Travel Demand* (Cambridge: MIT Press).

Bockstael, N.E., McConnell, K.E., and Strand, I.E., 1991, Recreation, In *Measuring the Demand for Environmental Quality*, edited by Braden, J. and Kolstad, C., pp. 227–270 (Amsterdam: North Holland).

Boxall, P.C., Watson, D.O., and Englin, J., 1996, Backcountry recreationists' valuation of forest and park management features in wilderness parks of the western Canadian Shield. *Canadian Journal of Forest Research* 26, 982–990.

Brainard, J.S., Lovett, A.A., and Bateman, I.J., 1999, Integrating geographical information systems into travel cost analysis and benefit transfer. *International Journal of Geographical Information Systems* 13, 227–246.

Carl Bro, 1997, *Befordringsfradrag*. Carl Bro Informatik, Copenhagen, pp. 8.

DAV, 2001, Danish Road and Addresseregister (http://www.dtd.dk/dav.html).

De Vries, S. and Ghossen, M., 2001, Planning tools for the recreational function of forests and nature areas. *Paper presented at the 12th Euro Leisure Congress Cultural Events and the Leisure System*, Noordwijkerhout, the Netherlands, 18 April.

EBST, 2001, National Agency for Enterprise and Housing (http://www.ebst.dk).

Forest Commission, 2001, *Perceptions, Attitudes and Preferences in Forest and Wood-lands*. Technical Paper 018, Forest Commission, U.K.

Greene, W.H., 2002, *Econometric Analysis*, Fifth Edition (Upper Saddle River, NJ: Prentice-Hall).

Haab, T.C. and McConnell, K.E., 2002, *Valuing Environmental and Natural Resources. The Econometrics of Non-market Valuation*. New horizons in Environmental Economics (Cheltenham: Edward Elgar).

Hanley, N., Shaw, W.D., and Wright, R.E., 2003, *The New Economics of Outdoor Recreation* (Cheltenham: Edward Elgar).

Jensen, F.S., 2003, *Friluftsliv i 592 skove og andre naturområder*. The Research Serie, 32 (Denmark: Forskningscenteret for skov og landskab).

Kort and Matrikelstyrelsen/Vejdirektorartet, 1995, *Road Network for Denmark [VejnetDK]* (http://www.kms.dk).

Kort and Matrikelstyrelsen, 2001, *The Interactive Map of Denmark [Det Levende Danmarkskort]* (http://levende.kms.dk/public/)

McConnell, K.E., 1985, The economics of outdoor recreation, In *Handbook of Natural Resource and Energy Economics*, edited by Kneese, A.V. and Sweeney, J.L. (Amsterdam: North Holland).

Miljø and Energiministeriet/Danmarks miljøundersøgelser, 2000, *Areal Informations Systems* (http://www.dmu.dk/1_Viden/2_Miljoe-tilstand/3_samfund/AIS/).

Scarpa, R., Chilton, S., Hutchinson, G., and Buongiorno, J., 2000, Valuing the recreational benefits from the creation of nature reserves in Irish forests. *Ecological Economics* 33(2), 237–250.

Skov-Petersen, H., 2001, Estimation of distance-decay parameters: GIS-based indicators of recreational accessibility. In *The 8th Scandinavian Research Conference on Geographical Information Science*, edited by Bjørke, J.T. and Tveite, H. (Ås, Norway: DMS/AUN/NDRE).

Skov-Petersen, H., 2002, *GIS, Accessibility, and Physical Planning: Exemplified by Models of Recreational Activities*. Unpublished Ph.D. Thesis, Department of Geography, University of Copenhagen, Denmark.

Statistics Denmark, 2000, Gross Income by Region, Income Interval, Age and Sex (1983–2003). StatBank Denmark IF221. Copenhagen, Denmark.

Termansen, M., McClean, C.J., and Skov-Peterson, H., 2004, Recreational site choice modelling using high resolution spatial data. *Environment and Planning A*, 36, 1085–1099.

Train, K., 2003, *Discrete Choice Methods with Simulations* (Cambridge: Cambridge University Press).

Section II *Engaging with the Public*

15

Engaging Citizens: The Bradford Community Statistics Project

Derek Reeve, Erik Thomasson, Steve Scott, and Ludi Simpson

CONTENTS

15.1   Introduction

Engaging the active participation of citizens in the processes of civic governance has been a laudable, if largely unrealized goal, of local governments for decades. Recently, this goal has been much reemphasized in the United Kingdom. Almost every recent U.K. government initiative places a clear onus on local governments to collaborate with their communities. Local authorities must now establish community strategies and form local strategic partnerships to reflect community interests. In the main, however, the public have been steadfastly disinterested in participation exercises. Participation in local elections and attendance at local meetings are usually depressingly low.

The Internet has recently been seized upon as a new vehicle with which to reengage the public. Visionaries look forward to a future in which Internet-based systems will be used to involve citizens in developments in their locality, encouraging citizens to interact directly with professionals and policy makers in local decision-making processes. Because many of the issues that affect local government are land or property based, WebGIS are seen as having a major role to play within this movement towards Internet participation. There is already a significant literature describing experiments in public participation GIS (PPGIS) (Craig et al., 1998; ESF-NSF, 2001; Laurini, 2001).

Achieving the future, however, is always more difficult than the visionaries, and vendors, suggest. The rhetoric surrounding PPGIS has raced far ahead of reality as represented by present PPGIS. WebGIS packages provide the technologies by which local agencies might deliver spatial information into the homes of citizens but we still are at the very beginning of the learning curve of understanding how to design systems based on these technologies efficiently to engage the public's interest. We need to understand what information should be presented, how that information is most effectively presented, and what is required for the public to be able to use the information. It is doubtful if any present PPGIS could yet claim to have become a major channel for participation between citizens and policy makers. We are still at the stage of seeing what works, of experiments and projects.

This paper contributes to the continuing PPGIS debate by detailing the PPGIS built for the Bradford Community Statistics Project (BCSP; www. bcsp-web.org). The BCSP's Maps and Stats system is an innovative and purposeful PPGIS, the lessons from which should be of interest to both researchers and other local governments. Some PPGIS sites appear primarily to be designed to disseminate prepared mapped-based information to residents and, with such sites, the manner in which the data are presented remains largely controlled by the sites' owners. The primary purpose of the Maps and Stats system, however, is to put into the hands of residents the datasets and online tools necessary to allow them, independently, to research conditions within their communities. Our site invites users to actively engage with data, rather than passively to receive them. A further distinctive feature of the BCSP is that the Maps and Stats PPGIS has been developed as one element within a broader project to build the capacity of local communities to understand and critically appraise the statistical bases upon which decisions about their localities are being made.

15.2   Policy Context

The U.K. government has rediscovered the policy significance of urban social geography. There is presently very great concern about the spatial dimension of social exclusion, the belief being that some localities are effectively excluded from the standards of well-being which are considered the norm in the rest of society. Furthermore, there is a determination that such spatial inequalities will be reduced: "Within 10 to 20 years, no one should be disadvantaged by where they live" (Neighbourhood Renewal Unit, 2001).

A range of area-based policy initiatives designed to promote convergence of social conditions between communities have been developed, e.g., Health Action Zones, Education Action Zones, Sports Action Zones, Excellence in Cities Action Zones, Sure Start, Anticrime local partnerships, the New Deal for Communities, and the Neighbourhood Renewal Fund PACT areas. In addition to U.K. government initiatives, quasi-government funding bodies, including the National Lottery Board, target significant resources into urban areas via area-based bidding processes. The EU also uses area-based statistics to direct very large structural regeneration funds to selected areas.

Although each of these policies differs in detail, there are recurrent themes presented below:

- *Territorial*: There is an emphasis upon drawing boundaries. Such territorial delineation is seen as a means of most effectively allocating limited resources. There is, however, a clear equity issue with area-based policies: Communities within designated areas will qualify for assistance under particular initiatives, those outside will not. The onus is upon policy makers to ensure that boundaries are drawn appropriately.

- *Evidence-based*: The mechanism for justifying the delineation of boundaries depends heavily upon statistical profiling. To qualify for assistance, areas have to display specified characteristics. Once designated, statistics are used to measure the progress of areas towards target norms. A clear emphasis within recent policy making has been upon evidence-based approaches.

Actually, there is nothing particularly new about this approach to urban policy making. In the United Kingdom, there has been a tradition of small-area policy initiatives. During the 1970s for example, there were housing action areas (HAAs) and general improvement areas (GIAs), the purposes of which were to identify small pockets of need within cities and then to channel resources into the areas affected. There is also, of course, a long tradition of spatially-based multiple deprivation and territorial social indicator studies, which have used statistical techniques to identify communities in need (Knox, 1975; NCRNRD, 1998; Senior, 2002).

More novel strands within current small-area policy making might be:

- *Multiagency*: There is an expectation that local providers of services to communities—local and central government departments, health and police authorities—will coordinate their initiatives towards localities. Cross-cutting initiatives are in vogue and the joined-up approach to government is being road-tested within the local policy-making arena.

- *Community involvement*: A key feature of current local policy making is a concern that communities should be active partners in the policies that are developed for their areas. There is concern to reduce the alienation from governmental processes that are a characteristic of socially deprived areas. Within current policy initiatives, therefore, there is an emphasis upon reengaging with local communities, although in practice community and voluntary groups may well retain a degree of cynicism about being the beneficiaries of another round of top–down policy initiatives.

To realize such policy initiatives local governments and other local agencies are being required to reconsider the ways local decision-making is conducted. On a pragmatic level, there is a great need to ensure that good quality small-area statistics are available to allow the characteristics of localities to be probed. At a national level, the need for improved small-area statistics is emphasized within the National Strategy for Neighbourhood Renewal initiative: "Better information needs to be available for all involved in strategy development, service design, and delivery at the local level. This should make it more likely that problems are diagnosed and effective answers produced. It also fits well with the need to involve local people more in playing their part and holding public services to account" (Social Exclusion Unit, 2000, p. 8). And at the local authority level, there is an onus upon local agencies to share their datasets and to integrate their policy making more fully than has previously been the case.

There is also a clear expectation placed upon local government to reenergize its methods of public engagement. Rather than going through the rites of public participation, there is now an expectation that communities must be genuinely active partners in formulating the policies that affect them.

15.3 Project Context

Against this national background, the BCSP is an attempt to enhance the capacity of community groups to participate more fully, and more equally, in the local policy debates that affect their communities and, specifically, to help residents to understand the statistical manipulations involved in local area policy making and grant allocation procedures. The Maps and Stats PPGIS is also seen as providing a platform that will facilitate the efficient integration and dissemination of previously disparate datasets.

The BCSP is a joint initiative between the Research and Consultation Service of the City of Bradford Metropolitan District Council (CBMDC) and the Bradford Resource Centre (BRC), the BRC being a not-for-profit organization that provides a focal point for community groups within the district. The project was funded by a grant from the European Regional Development Fund (ERDF), with matching funds provided by the CBMDC.

In its bid for funding, the BCSP provided a succinct statement of its aims:

(a) Increase the capacity of community groups to effectively use statistical information sources

(b) Make accessible via the Internet local statistics for community groups' own areas

(c) Support the voluntary sector in making a case for statistics which are more appropriate to their needs.

To achieve objective (a) a team of workers was established within the BRC with a remit to liaise with local community groups and to foster among such groups confidence that they can make effective use of statistical data for their own purposes. The BRC team provided informal "drop-in statistical surgeries" focused on helping community activists develop skills in the critical use of statistical sources and also developed a one-day "Strength in Numbers" course to explore the issues in a deeper and more structured way. Ongoing support has also been provided to those undertaking community research. Objective (c) sprang from a concern among community researchers that official statistics often fail adequately to reflect the concerns of community and voluntary organizations and that the community should have a role not only in interpreting existing statistics but also in influencing how and which statistics are made available. The Maps and Stats WebGIS was developed to achieve objective (b), this development being undertaken primarily by officers based within the Research and Consultation Service of CBMDC, although an aspect of the capacity building activities of the BRC team has been to introduce the Maps and Stats facility as a source of relevant statistical information and to feedback users' comments to the technical development team.

In building the Maps and Stats WebGIS the intention was to provide citizens with the datasets and tools necessary to allow them to conduct their own small-area analyses and thus to develop policy and funding proposals independent of council involvement. For the first time, there would be a single, comprehensive, consistent, and maintained small-area policy dataset for the district, delivered via an easily queried online system—freely available to anyone who is interested. Whereas in the past, community groups would have needed to go to the council to obtain access to relevant statistics, increasingly it is envisaged that such statistics would be available directly via the Maps and Stats WebGIS. The BCSP recognized that the role of the local authority, and other local agencies, as gatekeepers of local information should be lessened. GIS has been criticized as a technology that further concentrates the control of knowledge within bureaucracies and excludes relatively disadvantaged citizens (Pickles, 1995). The BCSP explicitly aims to reverse this trend in Bradford.

A further noteworthy feature of the BCSP has been the partnership between Bradford Council and the BRC. Ghose (2001) and Ghose and Huxhold (2002) discuss projects involving university–community partnerships as a means of democratizing GIS. The BCSP is based upon a local government–community partnership. Rather than the council developing a system for the community, the intention has been to develop the system with the community, with the BRC providing a focus for community involvement. Experiences within the project have been valuable in exploring the relationships between a local council and its communities with regard to service design, provision, and use.

15.4   Technical Features

The Maps and Stats WebGIS (www.mapsandstats.com) is implemented using AutoDesk's MapGuide software with Microsoft's Access as a data store. Here we briefly describe some of the more innovative features developed within the system.

15.4.1   Boundary-Free Small-Area Estimates

A novel method for generating boundary-free small-area estimates lies at the core of the Maps and Stats system. This method could have very wide application as it helps to resolve a significant problem that presently hampers small-area policy analysis.

The chaotic nature of Britain's small-area geography means that producing worthwhile statistical profiles for small potential policy areas has been a perennial bugbear for analysts. Census boundaries do not coincide with postcode boundaries. Health authority and police boundaries will not coincide with local authority boundaries and so the problems of sharing data compound. Potential policy initiative areas invariably cut across data boundaries. Nothing fits! And yet the increasing reliance upon evidence-based, targeted small-area antideprivation policies means that there is an increasing need for such estimates to be made.

Presently there seems to be a considerable gulf between the small-area estimation methods devised by academics and those used by practitioners. Small-area interpolation has attracted continuing interest from academics and a number of approaches, with varying degrees of sophistication, have been proposed (e.g., Flowerdew and Openshaw, 1987; Backen and Martin, 1989). The key point here, though, is that these research-based techniques seem to have achieved little penetration into practice. As Thomasson (2000) explains, more often than not, small-area estimates within local government are likely still to be made on the basis of simple polygon overlays or crude visual estimation.

The BCSP team has devised and implemented a method of small-area estimation which is practicable within a local government context and which produces consistent, replicable results. The method uses an approach that Flowerdew and Green (1991) would classify as intelligent estimation, using knowledge of the distribution of one variable, in this case residential locations, to predict others. In essence, the stages of the method are:

(a) The locations of residential properties within the district were determined. The starting point for this was to use the Address-Point dataset and to eliminate all nonresidential properties that were indicated within the "Organizations" field of the dataset. This reduced dataset was then further refined by the use of high-resolution aerial photography to help identify residential areas and to exclude further nonresidential properties. [See Harris and Longley (2000) and Robinson et al. (2002) for further examples on the use of remote imagery in studies of urban population distribution.] Internally generated local authority residential datasets can also be cross-matched with the Maps and Stats residential properties file, so the accuracy of the file be incrementally fine-tuned.

(b) The mean center of each unit postcode across the district was calculated on the basis of the identified residential properties and the number of residential properties associated with each postcode centroid recorded. Standard distance calculations were used to highlight where the mean center would not be a good indicator of the location of properties within the manually repositioned postcodes and centroids where necessary. Sadahiro (2000) suggests that the spatial median, rather than the spatial mean, might be a more appropriate measure.

(c) When any new area-based dataset is introduced into the BCSP system, a point-in-polygon operation is conducted to identify the postcode centroids which lie within each of the new dataset areas, so constructing a postcode:data-area look-up table.

(d) The dataset's value for each area is then shared between the postcode centroids enclosed within the area, in proportion to the number of residential addresses each postcode centroid represents. In effect, the value associated with a data area is spread proportionately across the postcode centroids contained within the area.

(e) Data are held within the BCSP system as postcode centroid estimates. In effect, the postcode centroids become the common-pegs upon which data from disparate areal bases can be held.

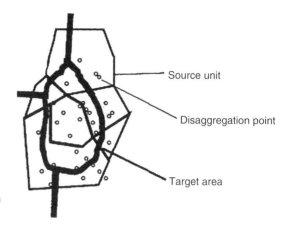

FIGURE 15.1
Source units, disaggregation
points, and target areas.

(f) When it is required to create an estimate for a newly defined target area, the estimated values for those postcode centroids contained within the target area can be simply aggregated. All datasets are held within the system as count variables, percentages and rates only being calculated from the reaggregated target data.

Figure 15.1 illustrates the process. Source data are allocated to the disaggregation points (weighted postcode centroids) and the disaggregated estimates are then available for reaggregation to target areas.

Having experimented with this method, the BCSP is aware of some limitations in the performance of this method of estimation as it is presently implemented:

- The accuracy of the system clearly depends upon adequacy of residential address locations as a predictor of distribution of the estimated variable. It is only sensible, therefore, to use the method with residentially related variables.

- There are many postcodes with only a few residential properties within them. In rural part of the district, i.e., that part outside the urban shape defined by the CBMDC's Planning Department, 60% of postcodes contain less than five residential properties. To avoid the instability of indicators calculated with low denominators, and also to preserve confidentiality, the BCSP does not return results for target areas of less than 100 residential addresses. This being the case, it is recognized that the estimating system is most appropriately used in built-up areas, where target areas are more likely to contain sufficient properties to rise above the 100 address thresholds.

- If the source area is larger than the target area, the estimates will not be sensitive to local conditions within the target area. For this reason, it is preferred to obtain source data on the smallest possible scale.

- The use of postcode centroids to represent the residential properties within a postcode area means that there is presently some
possibility of misallocation when a target boundary cuts across a
unit postcode boundary. When this happens, all the estimated
value associated with a postcode will be allocated to the target
area within which the postcode's centroid falls.

Further refinement of the method is possible. In particular, the BCSP team
is considering "spreading" source data directly to individual properties, as
this would eliminate the potentials for error associated with the use of mean
centers to represent unit postcode areas and thus misallocation problems
where target boundaries cut postcode areas. The initial decision to use
centroids was taken initially only as a pragmatic means of reducing processing loads.

15.4.2 User-Defined Target Areas

Another significant technical feature of the Map and Stats WebGIS is that it
allows users to create their own target areas. Any user can use a mouse to
define, edit, and store any area that is of interest to them. Having defined a
target area, the user then selects from a menu the dataset in which they are
interested. The Maps and Stats system then returns a report that provides a
profile of the target area, based upon the chosen dataset. Figure 15.2 shows
the Maps and Stats interface, with a user-defined target area on screen.
Figure 15.3 shows an example of the type of report that the system returns,
in this case the "Council Benefits" report. This report provides estimates of
the numbers of households claiming Housing Benefit/Council Tax Rebate

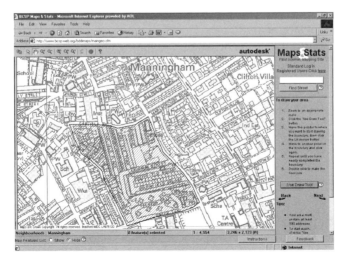

FIGURE 15.2
User-defined target area.

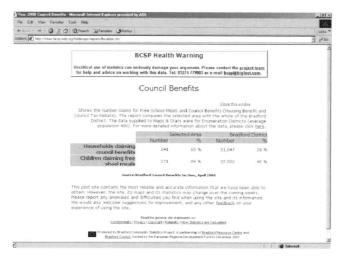

FIGURE 15.3
A target area specific report.

and Free School Meals within the user-defined target area, together with the district figures for comparison. Note that users can obtain information about the origins and reliability of the data, and how the statistics are calculated, by clicking on links embedded within the report page.

Providing the public with the facility to draw their own target areas via the WebGIS and to retrieve small-area statistics for their own defined areas gives citizens an independent ability to generate local-area policy proposals. Unlike many PPGIS, the public are not simply the passive recipients of information presented in a manner determined by the council. Rather they have the tools to form their own views and, if they wish, to enter into a dialog with the council.

In practice, the BCSP team had to balance their desire to provide open access to data against a proper concern to preserve confidentiality. It might be argued that, as the BCSP system returns estimates rather than actual observed data values, the issue of confidentiality does not arise. The view was taken however, that more stringent restraints should be put in place. The system has been designed so that it will not return profiles for user-defined areas that contain less than 100 residential addresses.

15.4.3   Information Dissemination Rather Than Analysis

As described above, the major outputs from the system at present are statistical tabulations profiling user-defined areas. The system does contain some preprepared thematic maps of selected variables, and because of the work done to identify residential locations for the estimation routine, these can be presented in dasymetric, rather than conventional choropleth formats. Figure 15.4, for example, shows the Maps and Stats site interface with

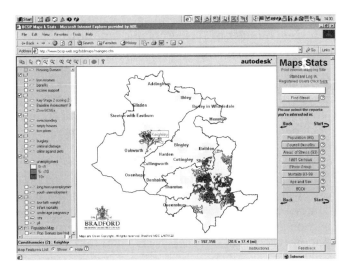

FIGURE 15.4
Dasymetric thematic mapping within the Maps and Stats site.

the left hand, thematic mapping panel opened, and a dasymetric map of the distribution of unemployment on screen.

The principal aim of the system, at least initially, however has been to develop a vehicle to allow the dissemination of statistical data, rather than to develop significant analytical functionality. The presumption is that users will download the data they require and perform any further analyses they may require independently. It is very easy to produce a wish list of additional features which might be desirable—charting, time-series analyses, user-controlled choropleth mapping, etc.—and some of these features may be added in the second phase of the project.

Each additional function, however, adds to the complexity of use of the system, and the team has had to remember that the system is intended for public use and struck a pragmatic balance between technical content and ease of use. The technical abilities of unskilled users are always likely to constrain the sophistication that can be built into the public version of the Maps and Stats system.

15.5   Data Issues

As the project progressed, the emphasis within the council's team increasingly focused on issues concerning data availability and data quality. Although developing the WebGIS site was technically interesting, its long-term usefulness will depend upon the quality and currency of the data it provides.

The project team has been successful in installing into the Maps and Stats system a large number of socioeconomic indicators. For example, the Bradford district deprivation index (BDDI), which is a superset of the DTER index, and which is a major internal planning tool for the council, has been made publicly available via the Maps and Stats site. The BDDI provides indicators under the following headings: crime, health, income, education, employment, and housing—and draws data from the census, internal council datasets, and some data from the health and police authorities. Looking a little ahead, the Maps and Stats site will clearly provide an excellent mechanism by which to ensure widespread availability of indicators derived from the 2001 census.

Experience during the BCSP, however, did raise some concerns with regard to data sources. The data loaded into the system so far has been provided on a one-off, project basis. If the system is to make a long-term contribution, however, it will be necessary to enter into formal data-sharing agreements with data suppliers, whether these suppliers are other council departments or external agencies. Ideally, a system such as Maps and Stats might hope to establish data-sharing protocols with potential suppliers to specify items such as the quality standards, frequency of delivery, formats, and conditions of use. Such data-sharing protocols are already used to regularize exchange data between central and local government, and there are already examples of protocols being negotiated at local levels. The Hertfordshire Community Profiling Partnership, for example, is using a data-sharing protocol as a basis for establishing formal agreements with its data suppliers (Z. Davies, 2002, The Hertfordshire Data Observatory, Personal correspondence). In Bradford a data-sharing group was established but, as yet, no formal data-sharing agreements have been concluded, though progress continues to be made.

The paradox that surrounds data-sharing initiatives has been widely discussed. Everyone agrees that data-sharing in principle is a good thing, and yet actually achieving such sharing on a continuing basis often appears unexpectedly difficult. (Onsrud and Rushton, 1995). There are organizational issues, cost, and legal concerns that inhibit formal data-sharing. In discovering that setting-up formal data-sharing arrangements is a slow business, the BCSP has only been experiencing a local instance of a very widely experienced difficulty. Looking into the future, however, there are grounds for optimism. First, it is reasonable to expect that now the Maps and Stats site is live, its merits will become more widely known and data providers will see the advantages of contributing their data. Second, pressures for interagency coordination from central government are encouraging local agencies to participate in initiatives such as the Maps and Stats PPGIS. Most encouragingly for the future of the Maps and Stats system, the local Health Authority and Crime Reduction Partnership has recently agreed to provide some additional resource to further the development of the system.

Data quality has been a major concern for the BCSP team. Unless datasets are very carefully prepared there is a danger that systems such as Maps and Stats could very efficiently deliver poor data. Operationally generated datasets in particular need to be cleaned prior to being used and, unless care is taken, when such datasets are combined there is a possibility that the resultant dataset will carry forward the deficiencies of both—combining two datasets each with an initial accuracy of 80% could in a worst case result in an output dataset which is only 64% accurate (INFOSHARE, 2002). A significant proportion of the council team's time has been devoted to preparing data, and the view has emerged that, in future, responsibility for data cleaning will need to be accepted by the data providers rather than the central BCSP team. This is partly for resource reasons but also reflects a principle that data cleaning should be done as close to data collection as possible, since those who are responsible for collecting data will be most likely to be able to identify and correct errors.

The hope is that CBMDC's Research and Consultation team, through the continuing development of the Maps and Stats system, will become acknowledged as a clearinghouse for small-scale data within Bradford, receiving quality assured datasets from other organizations and in return providing an efficient means of combining and disseminating data.

15.6 Community Use

There is a critical, unanswered question about PPGIS: "How will the public use them?" Indeed, is there a public demand sufficient to make the resources that go into producing them justifiable? It is an article of faith among academic geographers that interpreting spatial data and statistical mapping is difficult and requires training. Those of us who have attempted to teach rudimentary spatial analysis to undergraduates might well believe this to be true. Yet, via PPGIS, we will increasingly be making mapped data available to the general public. There is a significant need for research into the ability of the public to interpret WebGIS sites and the extent of their interest in doing so. We need to know how much demand there is for such sites; whether users can interpret the data delivered by PPGIS sites; what are the human computer interface (HCI) characteristics that will facilitate the ease of use of such systems?

The statistical emphasis within the Maps and Stats system means that these issues are particularly important. Some PPGIS sites presently adopt what is broadly look-and-see approach, allowing users to browse through prepared maps. The Maps and Stats system can function in this manner, providing access to prepared maps showing aspects of the socioeconomic conditions within Bradford. The core purpose of the system, however, is to allow users to create statistical profiles of areas of their choosing.

This requires a more active participation on the part of users. They need to be aware of the site and the facilities it offers. They need to recognize the relevance of what the site offers and to have the ability to interpret any area profiles they may create.

It is in this context that the community capacity building efforts by the BRC are most important. It was never intended that the Maps and Stats system should stand alone. The BCSP team always intended that alongside the technical development of the site, there would be a parallel effort to promote the Maps and Stats system among community groups, to explain its relevance to them, particularly in preparing funding bids, and to provide training in interpreting the statistics it can provide.*

During the BCSP, the BRC team was active in promoting the project with publicity, contacting community networks, liaising with other community initiatives, providing statistics workshops and training, and exploring the information needs of the groups they supported. The Maps and Stats Web site only became available a year after the inception of the project but the BRC records show that up to September 2001, 53 workshops focusing on aspects of the critical use of statistics within the community were delivered, 25 of which were based on the Maps and Stats site (Taylor, 2002). Valuable though these arranged meetings were, they underrepresent the interaction between the BRC and local communities during the project since many contacts were made on an ad-hoc and one-to-one basis. Over 80 community groups and 22 funding bids were assisted during the life of the initial project.

The BRC team also took responsibility for articulating community per-spectives on the Maps and Stats system, recording user feedback and emphasizing the need for clarity and ease of use. Thus, there was a continu-ing dialog within the project between the BRC team and the development team about refinement of the site. Particular issues included the fact that MapGuide requires a download that can take some minutes to install over standard Internet connections, causing some concern to inexperienced users; the area drawing and editing tools are relatively complex tasks for users not familiar with mapping tools; and the language used on the site could cause difficulty. Responding to the feedback provided from the BRC, the Maps and Stats site was continually revised to make it more usable, but it is clear that there is still much to learn about how to present statistical materials via PPGIS. The differences between community and technical perspectives on the system were not always easy to resolve.

During the BCSP concern about usability led to a discussion about whether to put the more demanding area drawing and area-editing facilities behind a password. The open site would allow everyone to have access to statistics based on standard data areas—Wards, neighborhoods, etc.—but

---

* Indeed by focusing primarily on the Maps and Stats Web site, this paper probably provides a skewed impression of the major emphasis of the overall project. The primary aim of the BCSP has been to help communities understand the statistics that are being used to take decisions about their localities. The Maps and Stats site is just a tool to facilitate this effort.

only people with a password would be allowed to use the interactive drawing tools. The Scottish Webmapping project (http://www.web-mapping. scot-homes.gov.uk/index.html) which is now developing a site similar to the BCSP's Maps and Stats system to allow local agencies in Scotland to share small-scale geographical data seems to be following this approach. Prepared maps are available directly but dynamic mapping requires users to have a password.

Within the BCSP, the decision was taken to remain true to the original intention of providing an open site: better to allow users to experiment with the statistics, and where necessary to ask for help, than to make presumptions about levels of skills within the community. Users only need to ask for a password, for administrative reasons, to store their target areas on the system's server.

15.7 Organizational Context

It was mentioned above that an intention of the project was to develop a PPGIS with the community, rather than for the community, and we believe that this is a model that other projects may wish to explore. If the council had attempted to develop the Maps and Stats site without the context and feedback provided by the BRC's community involvement, this might have resulted in a rather sterile, technological exercise.

Developing a software product in a community context is a challenging task. In an ideal world, software development takes place in a stable environment where user requirements can be well defined, objectives can be set, milestones defined, and resources allocated. In short, efficient software development requires a stable environment and clarity in decision-making. Community activity, however, thrives upon debate and inclusive decision-making. There are many varied and competing community agendas and communities of interest. These two differing styles lead to some tensions within the project. Furthermore, as neither of the parties had previous experience of such a development, it proved impossible to determine a fixed set of requirements early in the project. An attempt to adopt conventional systems development methods was discontinued as inappropriate and, in practice, an informal style of prototyping emerged as an appropriate development method. Rather than establishing an early set of requirements, the council and BRC partners negotiated progressive refinement of the site. If the development of the WebGIS had been modeled upon a conventional information systems development method, or had followed conventional local government procedures, it would certainly not have progressed as quickly as it did. Each of the major participants had to learn to accommodate the working traditions of the other.

The adoption of a fairly informal prototyping development style, however, did not negate the need for project management and quality assurance.

The project benefited from the fact that independent review was built into the original proposal. This review was useful in clarifying issues for the participants. It helped in ensuring the project was effectively steered and common objectives identified.

The ERDF grant that funded the BCSP, and within it the Maps and Stats Web site, expired in December 2001. Although further EU funding has been obtained for an expanded program of community statistical activity, including some provision for further technology development has been secured, a concern for the team will be that at some point the Maps and Stats Web site must be transformed from a project into a service, i.e., become accepted as a permanent feature of local policy making within Bradford.

In this regard, it has been encouraging that a version of the Maps and Stats site that released to Bradford Council's officers as a trial via the council's Internet proved a great success. Council officers quickly appreciated the time that could be saved by using the WebGIS to assemble area profiles compared to conventional methods. As Thomasson (2000) observes a principal reason why local authorities may previously have seemed to be unresponsive to requests for information about areas from members of the public has not been a desire to monopolize data but rather, and more prosaically, a concern about the resource implications involved in responding to such requests. In the past, collating the data sources and constructing area estimates on an ad-hoc basis have been very time consuming. With the Maps and Stats system, officers can produce area profiles for internal and external purposes with relative ease.

As a result of the internal pilot, the council agreed to purchase an authority-wide license for the MapGuide software so that a further developed version of the Map and Stats system can be made a permanent feature of the council's Intranet. This is important for the external site, as it is difficult to see how the external site could perform as a permanent feature other than by feeding off an internal service and guaranteed flows of operational data into the system.

15.8   Conclusions

We are careful not to claim too much for the BCSP. The collapse of the dot. com bubble during recent years provides a salutary warning against exaggerating the potential of Web-based technologies as means of communication with the public. If business has struggled, indeed largely failed, to develop the Web as a significant channel by which to sell goods to the public, perhaps governments and academics should be modest in the claims they make about the potential of the Web as a vehicle for promoting participation. It is doubtful if any PPGIS has yet had a major impact upon public participation rates. We are realistic about the level of demand that there may be for the Maps and Stats system among residents.

On the other hand, we do feel that the BCSP approach has elements that should be of interest to others involved in encouraging the public use of data for participation in decision-making. The Maps and Stats site now makes available to anyone who is interested a wide range of policy-relevant datasets, and this will make some contribution towards opening local government processes to scrutiny and enabling residents independently to form their opinions. A similar approach is now being adopted in at least two other local authority areas. The Open Information for Birmingham site (www.oi4b.com) has adopted a very similar approach to that developed in Bradford. The South Lanarkshire Community Plan (http://www.step.gb.com/Information_bank/) simply makes public some of the Excel spreadsheet files that are used by council officers.

We believe that developing a PPGIS within a broader program of community consultation and capacity building, as exemplified by the BCSP, provides advantages for all parties. The merits of the Maps and Stats system should not be judged solely in technical terms but as one strand within a much larger effort to engage communities in debates about their futures.

Disclaimer: The opinions expressed here are those of the authors and do not necessarily reflect the views of the agencies described.

References

Backen, I. and Martin, D., 1989, The generation of spatial population distributions from census centroid data. *Environment and Planning A* 21, 537–543.

Craig, W., Harris, T., and Weiner, D., 1998, Empowerment, marginalization and public participation GIS. *Report of the Varenius Workshop*, Santa Barbara, California, 15–17 October 1998 (http://www.ncgia.ucsb.edu/varenius/ppgis/PPGIS98_rpt.html, accessed on January 16, 2002).

ESF-NSF, 2001, Workshop on access to geographic information and participatory approaches using geographic information. *European Science Foundation–National Science Foundation Workshop*, Spoleto, Italy, 6–8 December 2001 (http://www.shef.ac.uk/~scgisa/spoleto/workshop.htm, accessed on May 23, 2002).

Flowerdew, R. and Green, M., 1991, Data integration: statistical methods for transferring data between zonal systems. In *Handling Geographical Information: Methodology and Potential Applications*, edited by Masser, I. and Blakemore, M. (Harlow: Longman).

Flowerdew, R. and Openshaw, S., 1987, *A Review of the Problems of Transferring Data from One Set of Areal Units to Another Incompatible Set*, Northern Regional Research Laboratory Research Report No. 4 (Newcastle: The University of Newcastle).

Ghose, R., 2001, Use of information technology for community empowerment: transforming geographical information systems into community information systems. *Transactions in Geographical Information Systems* 5(2), 141–163.

Ghose, R. and Huxhold, W.E., 2002, *The Role of Multi-Scalar GIS-Based Indicator Studies in Formulating Neighborhood Planning Policy* (http://www.urisa.org/Journal/

accepted/ghose/neighborhood_strategic_planning_through_gis_based_indicators. htm, accessed on May 23, 2002).

Harris, R.J. and Longley, P.A., 2000, New data and approaches for urban analysis: modelling residential densities. *Transactions in GIS* 4(3), 217–234.

INFOSHARE, 2002, *Information Is Everything* (http://www.infoshare.ltd.uk/, accessed on May 23, 2002).

Knox, P., 1975, *Social Well-Being: A Spatial Perspective* (London: Oxford University Press).

Laurini, R., 2001, *Information Systems for Urban Planning* (London: Taylor & Francis).

NCRNRD, 1998, *Social Indicators: An Annotated Bibliography on Trends, Sources and Developments, 1960–1998.* North Central Regional Center for Rural Development, Iowa State University (http://www.ag.iastate.edu/centers/rdev/indicators/contents.html, accessed on May 23, 2002).

Neighbourhood Renewal Unit, 2001, *The Vision for Neighbourhood Renewal* (http://www.neighbourhood.dtlr.gov.uk/overview/index.htm, accessed on May 23, 2002).

Onsrud, H.J. and Rushton, G. (editors), 1995, *Sharing Geographic Information* (New Brunswick: Center for Urban Policy Research).

Pickles, J. (editor), 1995, *Ground Truth: The Social Implications of Geographical Information Systems* (New York: Guildford Press).

Robinson, S., Langford, M., and Tate, N., 2002, Modelling population distribution with OS landline plus and Landsat imagery. *Proceedings of the GIS Research UK 10th Annual Conference* (Sheffield: GISRUK).

Sadahiro, Y., 2000, Accuracy of count data estimated by the point-in-polygon method. *Geographical Analysis* 32(1), 64–89.

Senior, M., 2002, Deprivation indicators. In *The Census Data System*, edited by Rees, P., Martin, D., and Williamson, P. (London: Wiley).

Social Exclusion Unit, 2000, Better Information Report of the Policy Action Team No. 18. *National Strategy for Neighbourhood Renewal* (London: HMSO).

Taylor, P., 2002, *Evaluation of the Bradford Community Statistics Project: Final Report.* Unpublished Consultant's Report. City of Bradford Resource Centre.

Thomasson, E.N., 2000, Small area statistics online. *British Urban and Regional Information Systems Association (BURISA)* 144, 2–9.

16

Public-Oriented Interactive Environmental Decision Support System

Tan Yigitcanlar

CONTENTS

16.1 Introduction

The importance of broadening community participation in environmental decision-making is widely recognized and lack of participation in this process appears to be a perennial problem. In this context, there have been calls from some academics for the more extensive use of geographical information systems (GIS) and distance learning technologies, accessible via the Internet, as a possible means to inform and empower communities. However, a number of problems exist. For instance, at present the scope for online interaction between policy makers and citizens is currently limited. Contemporary Web-based environmental information systems suffer from this lack of interactivity on the one hand and on the other hand from the apparent complexity for the lay users. This chapter explores the issue of online community participation at the local level and attempts to construct a framework for a new (and potentially more effective) model of online participatory decision-making. The key components, system architecture, and stages of such a model are introduced. This model, referred to as a Community-based Internet GIS (CIGIS), incorporates advanced information technologies, distance learning, and community involvement tools, which is applied and evaluated in the field through a pilot project in Tokyo in 2002.

With the advent of ever cheaper computers and more user-friendly software and the progressive increase in environment related data collected through various monitoring and remote sensing programs, the need to handle, store, analyze, and present data has gained increased significance. Today, it is virtually impossible to think about environmental research without relying heavily on information and communication technologies (ICTs). These ICTs affect almost every aspect of the environmental debate, such as, research, monitoring, management, and, ultimately, decision-making and public involvement.

Decision-making on environmental issues is highly complex, involving semistructured and unstructured problems, large amounts of diverse data, and the need for the application of human-value judgments. ICTs provide a means of integrating data and models into a form that is readily understood by the participants in the decision-making process. In order to achieve this objective, however, online decision support systems need to be interactive and promote knowledge sharing and exchange. Ideally, if carefully designed and implemented, online decision support systems can play an important role in promoting a more extensive discourse between experts, decision-makers, and the public, than currently exists.

It is possible to argue that the development of environmental information systems is closely connected to the increase in environmental awareness over the last three decades (Evans, 1999; Haklay, 1999). Furthermore, GIS have come to be viewed as an effective technology to aid decision-making, and they have revolutionized the way that spatial data is generated, stored, analyzed, and disseminated. This information helps people to manage what they know, by making it easy to organize, manipulate, and apply the data to solve real problems (Longley et al., 2001). This same technology, however, is of little use, if it is only available to those with technical expertise or the relevant software. Access to GIS, therefore, is a vital component of any online decision-making system.

Likewise, the Internet is a powerful medium in the move to further open up information to public scrutiny. It will function as the next generation GIS platform supporting information distribution and Web-based decision support systems. Information and analysis supported by Web-based GIS with various multimedia can potentially broaden the user base for this technology. As stated by numerous scholars (e.g., Doyle et al., 1998; Brown, 1999), recent advancements in the field of Web-based GIS have created new challenges, many of which are technical (e.g., bandwidth, Internet connection speed; development of multilevel systems to enable full access by different users; limitations of the Internet regarding multimedia; and ease of interaction) and social (e.g., technophobia; danger of creating an information underclass; lack of commercial and political will; antipathy and apathy; and lack of understanding surrounding public and personal use of the Internet). Nevertheless, the future evolution to Web-based GIS is seen as a critical component in the development of online participatory systems, which allow the visualizing and modeling of the environment in a networked virtual reality environment (Carver et al., 1995; Faust, 1995; Marmie, 1995; Shiffer, 1995; Beardsley and Quinn, 1996; Batty et al., 1998a,b; Yigitcanlar and Sakauchi, 2002).

Rapid progress with the development of the Internet has brought the possibility of computer-assisted environmental and urban planning over networks (Plewe, 1997; Okabe et al., 1998; Yigitcanlar, 2002b; Peng and Tsou, 2003). In particular, the possibility of creating three-dimensional virtual environments provides the potential to develop an intuitively comprehensive solution to both the process and the output of urban and environmental planning (Yigitcanlar, 2002a; Carver, 2003). It may also enhance increased citizen participation whereby users can share virtual space to develop, present, and discuss their ideas and concerns.

In this context, this chapter presents the framework for the development of an online participatory system that links the urban and environmental decision-making processes to environmental education and increased awareness. In the following sections, several key components of the interactive model (CIGIS) are introduced including a description of the system architecture, an outline of the key stages of the model, and its application in the pilot project.

16.2 Community-Based Internet GIS

The application of innovative online tools to support community-based decision-making offers considerable new opportunities particularly in environmental planning where institutions need to deal with the spatial, geographical, multidimensional, and complex characteristics of environmental problems. For example, adaptation of community-based decisions can be helpful in planning new facilities that are sensitive to the natural and cultural requirements of local people; in protecting the visual integrity of historic sites; in reaching decisions related to environmental impact assessments; when producing environmental statements; for integrated coastal zone management; and when estimating spatial variations in climatic change.

To adapt such community-oriented approaches, decision-makers and technicians (particularly planners) have to consider how innovative technologies can best facilitate improved synergies between development, community, and environment. As mentioned, GIS are useful tools in this context. In recent years, Web-based public participation GIS (Web-based PPGIS) has become an important research theme in the GIS and planning communities [see Carver and Openshaw (1995), Craig (1998), Kingston et al. (2000), and Boott et al. (2001) for more information on Web-based PPGIS]. However, there are currently few successful online participatory systems available (e.g., Kakiko Map, 2002; Yamato City, 2002). Therefore, existing

Web-based PPGIS applications may need to be customized, new GIS app-roaches may have to be developed, and large numbers of participatory environmental decision support systems will eventually need to be located on the Web (Yigitcanlar, 2003). Consequently, in order to explore this issue further, an innovative Web-based PPGIS approach has been developed and is presented in this chapter.

CIGIS are Web-based support systems to facilitate discussion and collab-orative decision-making. It enables various users, such as the public, tech-nicians, and politicians to interactively obtain and share information on the environment at different levels, scales, aspects, and details. It also facilitates the collaboration of these users in problem-solving throughout various decision-making stages of the community-based planning process.

CIGIS are a mechanism to support sustainable development related thinking, identify community goals, draw up planning guidelines, and collect and store data in a GIS environment. Furthermore, the main steps of decision-making, collaboration, negotiation, and consensus building are integrated into this system (Figure 16.1).

However, there are several limitations that need to be considered. For example, to be effective, a collaborative decision-making project system needs to be able to deal with various forms of participation, from individual to community empowerment and from manipulation to citizen control [see Arnstein (1969) and Heckman (1999) for more information]. This chapter will focus on the notion of community empowerment by interactive participation as we are aware of the potential misuse of this kind of ICT

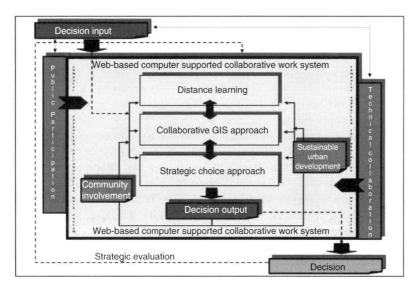

FIGURE 16.1
Community-based Internet GIS.

system in terms of promote channels for disinformation and hype. Moreover, as Karakaya (2003) states, it is also clear that while ICT tools can support interactive participation, they cannot directly alter the decision-making processes from top–down to bottom–up. Nevertheless, it is possible to argue that effective use of these tools could raise awareness and build the capacity of the public to reorient this structure.

CIGIS can also be thought of as a discussion-support tool for communities. This implies the use of GIS not just as a decision support tool, where alternatives are explored, quantified, and compared using analytical models (Densham, 1991) but more akin to the concept of "tools to think with others for the environment" (Boott et al., 2001). Although the analytical capabilities of GIS play a major role here, the main focus of the approach is not limited to this role, but rather is on the visual and contextual exploration of the problem situation and issues connected to it.

The ultimate goal of this approach is to ensure that information is made available for users to perform analyses, store, and represent their own results within the system. Users also should be able to browse the system, converse with other users, upload and download data, and move towards the design of planning and environmental policies at different scales.

16.3   System Architecture of CIGIS

The CIGIS infrastructure is designed as a computer-supported collaborative work (CSCW) system enhanced through the incorporation of an online learning environment. Furthermore, GIS represents a key tool to support information gathering and presentation. Maps are supplied from the client-side through use of an Internet map server (IMS)—an online map publishing facility. User interactivity is supported through the use of common gateway interfaces (CGIs; online forms, polling, etc.) and other means of supporting online communication (forums/discussion groups, chat rooms, etc.).

CSCW as its name suggests, is a system that supports the cooperative environmental decision-making. CSCW has numerous benefits for decision-makers (Hirscheim and Klein, 1989; Dourish and Bellotti, 1992; Ackerman and Palen, 1996; Craig, 1998; Jones, 1998; Prinz et al., 1998). Hirscheim and Klein (1989) assert that a good CSCW must not be designed in what they term as the "usual sense," but must be designed and developed within the framework of the social interactions that are embedded in the environment in which distance (collaborative) learning technology is to be incorporated. The caveat that must be stressed here is that in no sense is there an objective set of criteria that form a typology for an effective system. The system must be developed within what they term as the community's perspective. However, the CSCW system is not simply an electronic cloning or duplication of the working environment. In contrast, it is a pragmatic attempt to support the cooperative tasks of work in the context of the natural, social, and physical environment.

Web-based collaborative tools have the potential to significantly enhance the ability to train and educate the public electronically. Whether the materials are a stand-alone tutorial or a fully-fledged online workshop, the Web provides significantly new functionality in transmitting information to the user and providing forums for exchange of information. For instance, distance education and virtual classrooms are useful tools for delivering knowledge remotely, since they allow users to create their own instructional paths by using different technological alternatives (such as accessing electronic resources on the Internet, e-mail, conferences online, collaborative work, etc.). These technologies create a distance education environment which can exist beyond the traditional boundaries of a particular location and is accessible to a broad range of people, involving them in a highly interactive participatory process [see Dwyer et al. (1995), Fiorito et al. (1995), and Ibrahim and Franklin (1995) for more information]. Sharing perspectives and experiences helps people to develop ideas, and to recognize the complexities of concepts and skills. From the other side, evidence shows that if collaborative learning is effectively implemented, people can solve much more complex problems and come to far more sophisticated understandings than they could do on their own (Kolodner and Nagel, 1999).

The architecture of CIGIS, therefore, is comprised of a CSCW system with clients and servers distributed across the Internet as an open forum for all collaborators. This architecture can easily accommodate all of the relational infrastructures between planning authorities/committees, community centers, and individuals. These infrastructures should also incorporate transparent, secure, fast, and cooperative network configurations.

There are two basic approaches to deploy GIS or any other complex data driven applications on the Internet, which are server-side and client-side applications. In the case of the server-side, a Web browser is used to generate server requests and display the results by IMS. They comply with Internet standards because all the data processing is carried out by the server and the result is usually an image and a row from a database that can be wrapped in hyper text mark-up language (HTML) and sent back to the client. When a new user arrives at a server-side application site, they receive an initial response more quickly than a client-side Web-based GIS can provide.

The client-side Web-based GIS are enhanced to support GIS operations. This can be achieved by downloading a certain amount of GIS functionality to the client or involves downloading a full-blown plug-in. This enhances the user interface slightly, removing the need for processing to take place on the server, and implement solutions using vector data [see Gifford (1999) for a full review of client and server architectures].

In order to be effective, the CIGIS would need to break down the functionality of collaboration systems into smaller functions, because of the relative newness and complexity of these systems. Like in the other basic

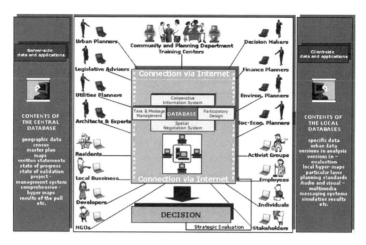

FIGURE 16.2
System architecture of CIGIS.

models with multiple phases, feedback from subsequent phases would need to be used to modify the results from the previous phases. The system constructs the trio of communication, collaboration, and coordination among all of the participants (Figure 16.2).

Turning to the discussion of how best to promote online interaction, it is important to recognize that a great deal of online interaction relies on programs called CGIs and HTML, which negotiate a dialog between the user and the Web server. The dialog goes something like this: The user types information into a form on a Web page and submits the information to the Web server. The server accepts the information and processes it in some predefined fashion, adds it to a database, e-mails it to someone, and appends it to a Web page. The server often sends a confirmation to the user, perhaps an acknowledgment of receipt or a page that displays the submission details.

Other tools to support interaction include discussion modules, the chat room, and the discussion area. The main distinction between the two is time. Participants in a chat room post and are responded to in real time. If they are not in the room while the dialog is taking place, they are not part of the discussion. The chat room model provides synchronous communication; it is like a phone conversation or a face-to-face dialog. People who post to discussion areas, by contrast, are asynchronous participants in online conversation: to contribute to a discussion they need not be online when a comment is submitted and a posting can be responded to days later. In an endeavor like the CIGIS Web site, a discussion area is likely to be more useful than a chat room. The type of user dialog sought by system administrators is better suited to the discussion format than the less structured and somewhat ephemeral chat-room model.

Having discussed the basic architecture underpinning the CIGIS model, it is now useful to explore the various stages involved in the model utilization.

16.4 Stages of CIGIS

Environmental management is a daunting task, because so many factors must be taken into account, and environmental decision-making requires that the participants understand not only the immediate impact of human activity on the environment, but also require appreciation of other factors including human health, equity, economic costs, as well as current and pending regulations. CIGIS, therefore, contains distance learning or e-learning and problem-solving processes to give the collaborators basic knowledge on relevant issues. Environmental problems, solutions, spatial and attribute data, various documents, and standards are also presented on the Internet platform by CIGIS. A schematic representation of the contents of CIGIS is shown in Figure 16.3.

Environmental and planning information systems rely heavily on GIS capabilities to work with complicated data and the ability of the expert user to understand, facilitate, and perform the analysis on the spot using the full toolbox of GIS capabilities. Moreover, CIGIS contains distance learning and problem-solving tools to give collaborators a basic background on the field and how to use the system (Figure 16.4).

The conceptual structure of CIGIS is based on the following nine functional steps of the environmental decision-making model: (1) introduce the tools and issues to the collaborators; (2) application of the tools to aid environmental decision-making; (3) identification of environmental values; (4) characterization of the environmental setting; (5) application of the tools to develop an understanding of the socioeconomic and political setting in environmental decision-making; (6) characterization of the regulatory and judicial setting; (7) integration of information; forecasting for environmental

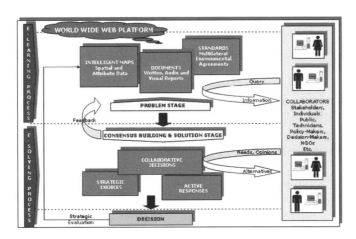

FIGURE 16.3
Contents of CIGIS.

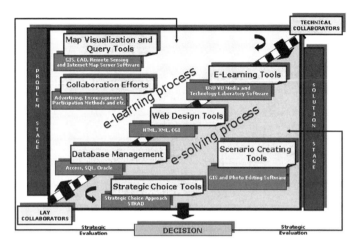

FIGURE 16.4
System overview of CIGIS.

decision-making; (8) assessment, refinement, and narrowing of options; and (9) postdecision assessment. Each step in the CIGIS model corresponds to one or more functional steps of a generic environmental decision model.

CIGIS are fundamentally comprised of two broad stages: problem and solution stages. Both stages are connected through a series of 5 sections and 14 steps. The problem stage involves the use of various e-learning tools to raise awareness of the public on the system of the environmental issues that they are facing. Furthermore, at this stage, the system endeavors to provide basic information on how to undertake environmental planning and decision-making in a collaborative manner. This stage supports community-based environmental decision-making by offering a range of new approaches for the users to select from. It assists the public in better understanding, forecasting, and visualizing the consequences of alternative scenarios for their environments. This can be accomplished through the application of ICTs, such as GIS, image, multimedia, interactive, and other computing technologies, coupled with professional planning assistance and training (Figure 16.5).

There are two sections in the first stage. The first section of CIGIS is the introduction section where collaborators get basic information about the system and log onto it, specifying a member profile (basic information such as, name, gender, age, occupation, and contact information). The CIGIS are an open platform for anyone who wants to be a part of the environmental problem-solving process. CIGIS can be seen as an open communities' network that accommodates real-time processing and establishes online relationships that strengthen environmental protection and decisions of local governments, institutions, communities, and individuals.

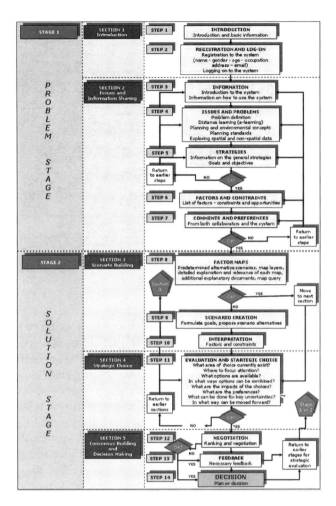

FIGURE 16.5
System flowchart of CIGIS.

The second section is the issues and information sharing section and this contains a basic introduction to the system. It incorporates text and graphic-based information on general strategies about the case studies, spatial and nonspatial data, related global multilateral environmental agreements, national and local environmental regulations, and local planning and environmental standards. Collaborators can view the list of factors, constraints and opportunities, and comment on existing preferences at the final step of the second section. If the collaborators understand the different facets of the problems and existing conditions, and have enough knowledge to be able to cope with them, then they can continue to the next section. Otherwise, they can go back to the earlier steps to gain the knowledge to enable them to continue onto further sections of the CIGIS process.

The second stage of CIGIS is the consensus building and solution stage, and it contains an interactive collaborative solving or e-solving process. This is the stage that develops and evaluates alternative scenarios to shape strategic decisions. It is important to recognize here that spatial decision support systems are often designed specifically to address semistructured problems. They often lack support for collaborative decision-making. Hence, for this reason, CIGIS accommodates collaborative GIS [see Faber (1997) and Sapient (2000) for more information] and decision tools to support the decision-making process. The solution stage of CIGIS employs new tools, which are needed to form explicit links between criteria space and the geographic space that is being managed. Environmental problems often result from the distributed and uncoordinated land-use management practices of individual decision-makers. Taken together, they cause significant environmental impacts. In developing feasible and politically acceptable solutions to such problems, it is often necessary to foster cooperation and consensus among a diverse set of special interest groups who have overlapping sets of objectives. Participants discuss their opinions on the specific cases and try to search for a solution at the CIGIS's consensus building and decision-making section.

The third section is the scenario building section. At this section, map layers can be switched on and off, zoom, pan, identify, query functions can be used, and detailed text based information can be viewed by the collaborators. Moreover, predetermined scenarios and additional explanatory documents can be studied to understand the situation and the alternative solutions available. If collaborators do not want to create their own scenarios, they can jump to step 4 and run the evaluation model for the predetermined scenarios generated by the system or by other collaborators. Otherwise, they can proceed to the next step to propose their own alternative scenarios. After they create their scenarios, they can reinterpret factors and constraints and then move on to the next section.

The fourth section is the strategic choice section, whereby scenarios are evaluated using the strategic choice approach [see Friend and Hickling (1997) and Friend (2002) for more information]. If collaborators are satisfied with the evaluated results, they can proceed to the next section. If not, they can return to the earlier sections for feedback and scenario correction.

The fifth and final section of the CIGIS is the consensus building and decision-making section. At this section, interactive negotiations between all collaborators and ranking between all chosen scenarios occur until a consensus is built between them. However, developing a consensus between collaborators can be very difficult and time consuming. If further feedback is necessary, collaborators return to the earlier stages of CIGIS for strategic evaluation and feedback. Before the final decision is approved, the entire process may need to be reviewed or in other words a postdecision assessment will be required.

At a first glance the structure of CIGIS might look quite complicated and prescriptive but in contrary it is very flexible that users may skip certain steps and include steps they are more interested in. CIGIS are continuous and dynamic decision support processes. Consequently, the decisions have to be evaluated in an iterative and cyclical way until all uncertainties are removed. This kind of approach can only facilitate better (smarter) decisions through the adoption of an open platform for all contributors, promoting consensus building, while adopting user-friendly technologies and accommodating strategic decision support systems.

As discussed by Yigitcanlar et al. (2003), collaborators, in the context of environmental problem-solving, represent several interests and bring to the negotiation table different types of training, levels of education, experience with computing technologies, and familiarity with the problems that are being addressed. This dissimilarity of knowledge can have significant effects. In spite of their differences, these people often cannot easily work together to devise solutions unless they have been trained to do so. In this context, if supported by effective user training, it is possible to argue that CIGIS could potentially evolve into an important tool to overcome the conceptual and technical barriers that inhibit the effective use of innovative technologies for environmental and urban problems.

16.5   Shibuya Community-Based Internet GIS

The CIGIS model was applied to a pilot project in Japan which commenced in 2002 called the Shibuya Community-based Internet GIS. This project is developed to raise awareness on planning and sustainable urban development (SUD) issues among the residents of Shibuya—Tokyo. The pilot project aimed to provide an easy access to urban and environmental data and to create environmental awareness among the public for achieving SUD in and around Shibuya. This project develops a comprehensive integrated Web-based information-sharing platform for the collaborators. Together with the components, it creates and promotes awareness on SUD and planning issues, relationships, virtual community, and trust online.

In this pilot project, a system for accessing the information in the database via the Internet was set up. The system uses ESRI ArcIMS technology to enable the presentation of the text information and map images. The system permits viewing the data from anywhere in the world using Web browsers. Participants can view the map of the city, perform zoom and pan operations to assist in visualization and navigation, ask such questions/queries, and then make suggestions about specific features identified from the map. All user input is stored in the Web access logs and is then used for future analysis and feedback into the planning process. In this manner a community database is created, representing a range of views and feelings about environmental and planning issues in the city.

This project was online for 6 months between 21 May and 21 November 2002. In this online exercise participants discussed sustainability issues within Shibuya which covered five major issues and problems of energy, open spaces, conservation, pollution, and urbanization/development. Only stage 1 of CIGIS was applied in this pilot study as the development of the second stage still continues. Once it is completed it will be tested on several pilot projects.

During this period, approximately 60 people logged on to this Web site. At the moderated discussion forum 36 people communicated very actively and approximately 24 participated passively. The major reason behind this low level of participation was the language barrier as the database and the Web site were developed in English instead of the local language of Japanese.

The URL of the initial Web page of the database is http://kanagawa.csis. u-tokyo.ac.jp (Figure 16.6). When users first enter the site, after an initial welcome window, they are prompted to fill in a profile. This was seen as an essential part of the system design as it could be used to build up a database of users to help validate responses and analyze the type of people who were using the system. The main page contains map, legend, tool buttons, and query frame (Figure 16.7).

Similar to what Märker et al. (2002) said about the design and conception of sociotechnical systems for online citizen participation, this pilot project has shown that public participation should not be solely technology-driven but also should be oriented towards the basic principles of cooperative planning approaches. This is known as the "new planning culture," and, among other things, allows: (1) participation at an early stage; (2) assures an equal opportunity to participate; (3) remains open with respect to both process and results; (4) assures communication and

FIGURE 16.6
Web page for accessing the Shibuya database.

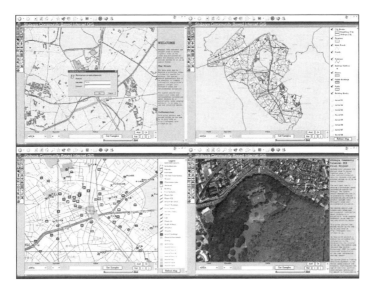

FIGURE 16.7
Shibuya Community-based Internet GIS Web site.

dialog; (5) integrates multiple perspectives; and (6) allows moderation by neutral third parties. On the other hand, the realities of city politics cannot be ignored; that is, participation procedures usually cannot achieve more than is allowed by the existing context of communication and power relationships.

The experience with the pilot study has shown that online urban information systems have the potential to catch public interest in environmental planning and also in preparation of urban development plans. The system demonstrates how it is possible to combine information and provide a greater insight into Shibuya's urban and environmental situation. The system also communicates urban and environmental information to a potentially wide ranging audience in an easy to use and intuitive manner through a variety of ways including an online discussion platform (Figure 16.8). The ability to instantaneously update the database and profile users online was seen as one of the most useful advantages of the system over the traditional techniques. The online system has a long residence time allowing people to use the system anytime and anywhere. The public do not need to attend a meeting at a particular time or place which is often the single most inhibiting factor in participating via traditional methods. The system allows faster collation of results from log files and the Web site can be used to disseminate results and feedback.

Clearly there is a need to improve public access to urban and environmental information, and encourage collaborative decision-making. Online urban and environmental information systems (UEIS) like the Shibuya Community-based Internet GIS project offer a variety of tools and

FIGURE 16.8
Shibuya Community-based Internet GIS discussion platform.

technologies to assist the management and use of urban and environmentally related data and information, and hence can be used as a tool to help to achieve SUD goals. If we are to strive to achieve the most efficient form of a sustainable information society, UEIS, their tools and results should attract the interest of as wide an audience as possible. This pilot study has shown us the role of the Internet; public participation and the development of Internet GIS are important factors in taking this potential further.

16.6 Conclusions

This chapter has proposed that spatial data related to environment should be increasingly represented virtually on the Internet in order to enhance information and decision sharing. It is argued that online participatory systems provide the opportunity to open up important national, regional, and local decision-making on environmental and other problems to a much greater audience and could actually involve the public more directly in the decisions which matter.

As in CIGIS, the integration of community-based interactive decision-making and online discussion support systems to aid planning mechanisms can help to create smart urban environments and communities. There are several major benefits of this proposed approach. CIGIS supplies equal access to data and information for users over the Internet. It endeavors to encourage individuals to share their opinions on their environments, helps communities to maintain a smart and healthy balance of urban growth, aids in conservation of natural resources, cultural identity by enhancing citizen consciousness, and also helps in bridging the gap between decision-makers, technicians, and the public. CIGIS accommodates an equitable representation of diverse views, preserves contradiction, and endorses the role of individuals and communities in creation and evaluation of development plans and strategies by promoting citizen participation in urban planning. It offers powerful tools to provide information and support decision-making in SUD, planning, and management. Finally, CIGIS supports the appropriate use of analytical tools and datasets in policy making and planning to predict and examine the possible short- and long-term outcomes on the urban areas.

As mentioned earlier the integration of community-based interactive environmental decision-making support systems to planning mechanisms has numerous benefits. However, after the application of the model was used in a pilot study it is clear that a number of issues and possible limitations still need to be explored in subsequent research projects and case studies. Those are scalability, data quality, bandwidth, Internet connection speed, related legislation, lack of public interest, accessibility, digital divide, privacy, and lack of vision of the governmental bodies.

With the current state of development of CIGIS, it is possible to assert that this approach has the potential to become a key framework for the incorporation of collaborative spatial decision-making tools at the local level. Generic ideas presented in this chapter may be of interest to people who are developing online environmental information systems and communities that are willing to integrate them.

References

Ackerman, M.S. and Palen, L., 1996, The Zephyr help instance: promoting ongoing activity in a CSCW system. *Paper Read at the ACM Conference on Human Factors in Computing Systems*, Vancouver, British Columbia, Canada, April 13–18, 1996.

Arnstein, S., 1969, A ladder of citizen participation. *Journal of American Institute of Planners* 35(7), 216–224.

Batty, M., Dodge, M., Doyle, S., and Smith, A., 1998a, *Modelling Virtual Urban Environments* (London: Centre for Advanced Spatial Analysis, UCL).

Batty, M., Dodge, M., Jiang, B., and Smith, A., 1998b, *GIS and Urban Design* (London: Centre for Advanced Spatial Analysis, UCL).

Beardsley, K. and Quinn, J.F., 1996, Information center for the environment: public access to natural resource data using an interactive query system on the World Wide Web. *Paper Read at 1996 ESRI User Conference*, Palm Springs, California, May 20–24, 1996.

Boott, R., Haklay, M., Heppell, K., and Morley, J., 2001, The use of GIS in Brownfield redevelopment. In *Innovations in GIS 8: Spatial Information and the Environment*, edited by Halls, P., pp. 241–258 (London: Taylor & Francis).

Brown, I.M., 1999, Developing a virtual reality user interface for geographic information retrieval on the Internet. *Transactions in GIS* 3(3), 207–220.

Carver, S., 2003, The future of participatory approaches using geographic information: developing a research agenda for the 21st century. *URISA Journal 15, Access and Participatory Approaches* 1, 61–71.

Carver, S. and Openshaw, S., 1995, Using GIS to explore the technical and social aspects of site selection. *Paper Read at Conference on the Geological Disposal of Radioactive Wastes*, IAEA Conference, Vienna, Austria, May 15–18, 1995.

Carver, S., Heywood, I., Cornelius, S., and Sear, D., 1995, Evaluating field-based GIS for environmental characterization, modelling and decision support. *International Journal of Geographical Information Systems* 9(4), 475–486.

Craig, W.J., 1998, The Internet aids community participation in the planning process. *Computers, Environment and Urban Systems* 22(4), 393–404.

Densham, P.J., 1991, Spatial decision support systems. In *Geographical Information Systems: Principles and Applications*, edited by Maguire, D.J., Goodchild, M.F., and Rhind, D.W., pp. 403–412 (Harlow, UK: Longman).

Dourish, P. and Bellotti, V., 1992, Awareness and coordination in shared workspace. *Paper Read at ACM Conference on Computer Supported Cooperative Work (CSCW92)*, Toronto, Canada, October 31–November 4, 1992.

Doyle, S., Dodge, M., and Smith, A., 1998, The potential of web-based mapping and virtual reality technologies for modelling urban environments. *Computers, Environment and Urban Systems* 22(2), 137–155.

Dwyer, D., Barbieri, K., and Doerr, H.M., 1995, Creating a virtual classroom for interactive education on the Web. *Paper Read at Third International World Wide Web Conference: Technology, Tools and Applications*, Darmstadt, Germany, April 10–14, 1995.

Evans, S., 1999, *A Prototype Environmental Information System for London* (*London Environment Online or LEO*). CASA Working Paper 16 (London: Centre for Advanced Spatial Analysis, UCL).

Faber, B., 1997, Active response GIS: an architecture for interactive resource modeling. *Paper Read at GIS'97 Eleventh Annual Symposium on GIS*, Vancouver, Canada, February 17–20, 1997.

Faust, N.L., 1995, The virtual reality of GIS. *Environment and Planning B: Planning and Design* 22, 257–268.

Fiorito, M., Iovane, D., and Pantano, P., 1995, An educational environment using WWW. *Paper Read at Third International World Wide Web Conference*: *Technology, Tools and Applications*, Darmstadt, Germany, April 10–14, 1995.

Friend, J., 2002, *Stradspan: New Horizons in Strategic Decision Support* (http://www.btinternet.com/~stradspan/program.htm, accessed on May, 2002).

Friend, J. and Hickling, A., 1997, Planning under pressure, *The Strategic Choice Approach*, Second Edition (Oxford: Butterworth Heinemann).

Gifford, F., 1999, Internet GIS architecture. *Client Versus Server in Mapping Awareness* 13(7), 40–42.

Haklay, M., 1999, *From Environmental Information Systems to Environmental Informatics— Evolution and Meaning* (London: Centre for Advanced Spatial Analysis, UCL).

Heckman, L.A., 1999, *Methodology Matters: Devising a Research Program for Investigating PPGIS in Collaborative Neighborhood Planning*, 2002 (http://www.ncgia.ucsb.edu/varenius/ppgis/papers/heckman/heckman.html, accessed on May, 2002).

Hirscheim, R. and Klein, K., 1989, Four paradigms of information systems development. *Social Aspects of Computing* 32(10), 1199–1216.

Ibrahim, B. and Franklin, S.D., 1995, Advanced educational uses of the World Wide Web. *Paper Read at Third International World Wide Web Conference: Technology, Tools and Applications*, Darmstadt, Germany, April 10–14, 1995.

Jones, R.M., 1998, An analysis of computer supported co-operative work systems to support decision making in regional planning. *Computers, Environment and Urban Systems* 22(4), 335–350.

Kakiko Map, 2002, *Kakiko Map for Web-based Information Sharing* (http://upstar.t. u-tokyo.ac.jp/kakiko/, accessed on May, 2002).

Karakaya, R., 2003, The use of the Internet for citizen participation: enhancing democratic local governance? *Paper Read at the 53rd Annual Conference of the Political Studies Association (PSA)*, Leicester, U.K., April 15–17, 2003.

Kingston, R., Carver, S., Evans, A., and Turton, I., 2000, Web-based public participation geographical information systems: an aid to local environmental decision-making. *Computers, Environment and Urban Systems* 24(2), 109–125.

Kolodner, J.L. and Nagel, K., 1999, The design discussion area: a collaborative learning tool in support of learning from problem-solving and design activities. In *Proceedings of the Computer Support for Collaborative Learning (CSCL) 1999 Conference*, Stanford University, Palo Alto, California, December 12–15, 1999, edited by Roschelle, C.H.J. (Mahwah, NJ: Lawrence Erlbaum).

Longley, P.A., Goodchild, M.F., Maguire, D.J., and Rhind, D.W., 2001, *Geographic Information Systems and Science* (New York: Wiley).

Märker, O., Hagedorn, H., and Trénel, M., 2002, *Internet-Based Public Consultation: Relevance–Moderation—Software* (http://www.ercim.org/publication/Ercim_News/enw48/maerker.html, accessed on February 25, 2004).

Marmie, A., 1995, Promoting and providing GIS data via the Internet. *Paper Read at Fifteenth Annual ESRI User Conference*, Palm Springs, California, May 22–26, 1995.

Okabe, A., Okunuki, K., Sagara, S., Kamachi, T., and Shiode, T., 1998, Virtual Ryoanji project: implementing a computer-assisted collaborative working environment of a virtual temple garden. *Paper Read at International Workshop on Groupware for Urban Planning*, Lyon, France, February 4–6, 1998.

Peng, Z. and Tsou, M., 2003, *Internet GIS: Distributed Geographic Information Services for the Internet and Wireless Network* (New York: John Wiley & Sons).

Plewe, B., 1997, *GIS On-Line: Information, Retrieval, Mapping and the Internet* (Santa Fe, NM: OnWord Press).

Prinz, W., Mark, G., and Pankoke-Babatz, U., 1998, Designing groupware for congruency in use. *Paper Read at Proceedings of the Conference on Computer Supported Cooperative Work*, Seattle, Washington, November 14–18, 1998.

Sapient, 2000, *Smart Places: Collaborative GIS Approach* (http://www.saptek.com/smart, accessed on May 2, 2002).

Shiffer, M.J., 1995, Interactive multimedia planning support: moving from stand-alone systems to the World-Wide-Web. *Environment and Planning B: Planning and Design* 22, 649–664.

Yamato City, 2002, *Online Urban Master Plan of the Yamato City* (http://www.city. yamato.kanagawa.jp/t-soumu/TMP/e/index.html, accessed on May 2, 2002).

Yigitcanlar, T., 2002a, Community based Internet GIS: a public oriented interactive decision support system. *Paper Read at GISRUK: GIS Research UK 10th Annual Conference*, Sheffield, U.K., April 3–5, 2002.

Yigitcanlar, T., 2002b, Placing planning online: serving maps on the Web to aid local decision making. *Paper Read at Tokyo Planning Forum*, United Nations University, Tokyo, Japan, May 11, 2002.

Yigitcanlar, T., 2003, Evaluating the potential of Australia for online public partici-patory planning. *Paper Read at 27th ANZRSAI Annual Conference 2003.*

Yigitcanlar, T. and Sakauchi, M., 2002, *Emergence of Smart Urban Ecosystems: Application of ICT for Capacity Building in Environmental Decision Making* (Tokyo: United Nations University, Institute of Advanced Studies).

Yigitcanlar, T., Baum, S., and Stimson, R., 2003, Analyzing the patterns of ICT utilization for online public participatory planning in Queensland, Australia. *Assessment Journal* 10(2), 5–21.

17

Public Participation in the Digital Age: A Theoretical Approach

Robin S. Smith

CONTENTS

17.1   Introduction

In the United Kingdom, those in practice and research often treat public participation as a simple process. In local government settings it is often felt to have distinct stages of development, partly owing to statutory planning and the temporal phases of consultation associated with it. There are many assumptions about the nature and notions of public participation, alongside a belief that those who conduct exercises and those who respond share them. In the context of U.K. local authority-initiated (or top–down) participatory activities, different elements need to be exposed and understood, in an era where greater public engagement in decision-making is proffered as a panacea to many policy or political ills.

The assumed linearity, lack of complexity, and belief in universal understandings of public participation need to be challenged; theoretical understandings of this democratic activity should be furthered; and appropriate methods be developed in line with them. This is particularly important in terms of the emerging role that information and communication technologies (ICTs) are playing through digital participation, and the hyperbole that surrounds the application of such technology to citizen engagement as a whole.

Research by Smith (2001) investigated cyberdemocracy enthusiasts' claims of ICTs for digital participation, highlighting some of the ways in which public participation has been approached and understood by various actors, both within organizations and the citizens who chose to participate. The research explored these emerging attempts through two main approaches that this chapter reports. The first involved an online baseline survey of local authority Web sites in early 1999. The second looked behind these digital façades to explore officers' approaches to, and understandings of, public participation and what levels of response or engagement these early experiments had from the public. From this material, three in-depth case studies were then conducted with pioneering organizations.

The case studies included reviewing the local authorities' literature and interviews that helped to uncover how actors understood activities and what public participation meant to them. Such ideas were then contrasted with a number of theoretical perspectives drawn from democratic and planning theories, in order to examine what components made up public participation, and what role the methods of participation were appearing to play. These not only involved digital methods such as e-mail, chat rooms, and Web sites but also more traditional methods including public meetings, writing letters, and exhibitions. As such, the research explores digital participation in an in-depth way, so that theoretical discussions can be equally applied to other participatory settings, including the emerging role of geographical information systems (GIS), and spatial information in general, for citizen engagement in decision-making.

The remainder of this chapter provides an overview of the research by reporting some of the policy contexts and the results from the survey of local authority Web sites. This is followed by an outline of the leading case studies and an exploration of the notions of participation that existed for those involved. It should be noted that a discussion of the more theoretical material stemming from this empirical research is discussed by Smith (2006) and that ideas developed in this chapter adapt some work already published in French (Smith, 2004).

17.2 E-Policy and the Web Site Survey

The policies at the time of the survey were an important backdrop to the research, both within the context of European Information Society initiatives and emerging U.K. e-government policies, at both central and local government levels. In addition, those in public services were being encouraged, through central government policies, to take increased civic leadership roles through Modernising Government (HMSO, 1999), with the introduction of new structures such as directly elected mayors and cabinet-style council committees. There was also a drive from central government to improve service delivery across the public sector through Best Value (under the 1999 Local Government Act), which aimed to continually improve and democratize services, replacing the compulsory competitive tendering approach which, arguably, did not meet customer/service-user/citizen needs. Changes in policy were very much part of the context of the local authorities' attempts to represent themselves, with Web site development offering an opportunity for experimentation, innovation, and reflection on current practice.

The survey was completed shortly after the Society for Information Technology Managers (SOCITM) had just produced their first annual survey of local authority Web sites that focused on online local government service delivery and early attempts towards transaction services, such as paying local taxes online (MAPIT, 1999). A different approach was taken with the digital participation survey in four main areas. More attention was paid to councils' activities where public participation was encouraged and supported. Of principal concern were planning and environmental matters that had a tradition of public participation, including the statutory land-use planning system and activities associated with sustainable development and Local Agenda 21. In addition to this, online interaction with elected members was also of interest, as this would offer a different avenue for a citizen to become involved in participatory activities outside consultation exercises, for example. Lastly, community-based activities were examined, particularly those that would help individuals express an opinion online for others to read and contribute, potentially helping to mobilize public opinion and allowing some exploration of bottom–up or grassroots approaches to participation initiated by communities, which may contrast or contribute to top–down ones.

Evidence of online participation was seen through two elements for these four topics. The first area was an understanding of the quantity and quality of information being provided by the authority to support public participation. The second was an understanding of the level of interaction or communication that was being offered through the Web site and/or e-mail and other related technologies. An example that was seen as more participatory involved a live exercise with a complete version of a local authority policy document that a participant could view online and fora for them to either share ideas with others, or to respond to the authority, through e-mails, online feedback forms, or surveys. The level of communication would also include information that would allow potential participants to contact the authority, from telephone numbers to addresses and directions to contact points. A less participatory Web site would only contain a fragment of the consultation document(s), guidance to where they could be found, limited contact details, and no clearly defined online consultation activities. In terms of public participation, a poor Web site would only include contact details, and a very poor one may not have any information relating to council services or any council-related information at all. It should be noted that many of the poorer examples would be classified as leading examples under other criteria, such as usability or the use of graphics.

As a basis to contrast these differing provisions in information and communication, a model was established based on the often-cited Arnstein's (1969) ladder of citizen participation (Figure 17.1).

In the lower rungs of her model, participation is restricted by limited information provision and activities that do not allow for more free and open participatory activity, in part relating to democratic principles and related notions of participation, discussed below. In the middle rungs, participation becomes more communicative and interactive between government and citizens, with increasing trust in citizens' abilities emerging in the upper sections of the model. Hunter (1999) has suggested that this model provides a good framework to examine the application of ICTs to public participation, but he only sees the top rungs as useful indicators of the type of participation taking place. It can be argued that, in top–down settings at least, this is often not the case, where exercises can occur at different rungs, at different stages and for different actors at varying times. More ideal forms of participatory democracy (at the top of the ladder) are less likely to occur in situations where elected members control budgets and make final decisions, although the idea of partnership can be often

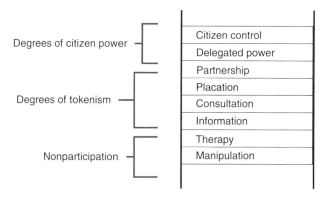

FIGURE 17.1
Arnstein's ladder of citizen participation.

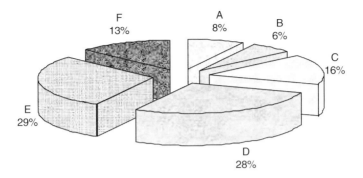

FIGURE 17.2
The level of participation present in U.K. local government Web sites in early 1999.

closely associated with current participatory activity, particularly in areas such as urban regeneration and neighborhood renewal. It should be noted that the model is not without its critics (e.g., Sharp and Connelly, 2002) but it does provide a useful introduction to some ideas relating to theories and methods of participation, such as that methods can be differentiated by their roles and relationships to power.

Web sites related to this model can only fit loosely, as the forms of participation taking place partly encompass the level of interaction, communication, and power that Arnstein represents. Six categories were defined from the digital participation survey represented in the graph (Figure 17.2).

The survey examined 287 Web sites across the whole United Kingdom, which was nearly all those available at the time. Of those where a ".gov.uk" Web site could be found, 8% could not be accessed (A), perhaps offering the greatest barrier to participation. Around 6% (B) had very limited content and were often under construction. Beyond these poorer examples, 16% appeared to be geared towards promotion through either tourism or economic development (C), which may provide some functions for service delivery but make little contribution to public participation. Very limited information could be found in the next category (D) of around 28% of the Web sites and included those with very limited information about participatory activities but they also contained online telephone directories and alphabetical lists of services, where citizens could then obtain further information, meeting some aspirations of e-government policy. If there was more content, or if there were opportunities to interact online, then the digital façade (in the sense of a frontage rather than being fake) of the Web site and the authority were seen as an information provider (E), presenting 29% of the Web sites examined. Finally, the consultative group (F) contained 13% of the examples, with the majority equivalent to the middle rungs of Arnstein's ladder, although their objectives could, arguably be more than tokenistic. This last group either contained full versions of consultation documents online, consultation exercises through Internet-based methods or more advanced digital methods, such as chat-rooms or bulletin boards. It is interesting to note that leading examples existed across all forms of U.K. local government and not just the better financed and larger English County Councils.

17.3   Case Studies

The survey found several leading examples that were worthy of further exploration through case studies. Selection was guided by features such as the forms of participation the council were involved in, the methods used, and the issues being explored, with an aim to aid comparative study. Through interviews with the officers who initiated and conducted exercises, it was found that many of the examples had very low levels of participation in terms of the proportion of the population who had responded. Looking behind the façades not only allowed exploration of the authorities' activities but also how aware an authority was of their participants. This also meant that participants could be selected (and contacted by the authority) for interview. Contrasts were also made between digital methods and other forms of engagement and activities in other policy areas within the authority. In general, where a choice existed, digital methods had far fewer responses than traditional avenues of interaction.

From these examples three local authorities, and their participants, were chosen for in-depth case study analysis. The cases sought to understand what forms and notions of participation existed for those involved, both within the organization and from the public. The selection also allowed contrasts to be made between different levels of policy making: a local example from Rushcliffe Borough Council, a strategic example from the City of Edinburgh Council, and a mixed example from the London Borough of Lewisham.

17.3.1   Case 1: The Rushcliffe Borough Council's Interim Local Plan Consultation

A number of the Web sites that ranked highly in the survey included statutory land-use planning, either as development control or planning policy consultations. The local case in Rushcliffe was particularly selected as it had the most online responses from the survey and subsequent research. Rushcliffe is an English district authority on the edge of the city of Nottingham in the East Midlands. Those officers involved with the consultation were mainly based in the planning department, with other key actors coming from information technology and public relations and media unit. A relatively wealthy borough, Rushcliffe does, however, have some pockets of poverty associated with former rural mining towns. At the time of the consultation exercise, officers felt that computer ownership was slightly above national average, which would impact on the numbers responding online. The focus for the exercise was the way that the council and the community should deal with a housing allocation from central government that was set in place by Nottinghamshire County Council through a consultation several months before the exercise of interest. The consultation exercise also sat alongside several other planning documents and activities that had tried to encourage public involvement with this politically sensitive allocation, as many did not want extra housing. For example, a more formal Issues Report (commenting on the nature and implications of the allocation and some other environmental concerns) was produced several months before the online exercise but had received limited public attention. Before a full local plan was set in place it was decided to have a less formal interim consultation that would mobilize public opinion. This was felt to allow public concerns of the allocation to be dealt with and the authority's position supported, before a deposit local plan exercise a few months later.

In terms of the methods for this interim exercise, the authority used a leaflet delivered to every resident in the borough and press releases to draw attention to the issue and, in terms of digital activities, to direct interested citizens to dedicated e-mail address and the council's Web site. The Web site itself included an online feedback form and a small amount of supplementary information. More traditional council-supplied methods included public events, comment forms (that were used at several such meetings across the borough), and a feedback slip which was part of the leaflet. The housing issue was one that many residents became engaged in and local pressure groups formed around the potential developments in their neighborhoods, a form of Not-In-My-Backyard (NIMBY)-ism. This was characterized not only by objections to developments in certain areas but also support for and suggestions of housebuilding in other parts of the borough. Many citizens chose to write letters rather than using the leaflet slip, with some including the slip as a type of coupon attached to their letters. The pressure groups organized two petitions and sent standard letters to residents in some areas, encouraging them to sign the letter that the group had drafted, add comments if they wished, and send them to the local authority. As such, this activity appears to have stimulated a relatively large number of responses compared to other activities that involve consultation on a policy document and a greater response than previous exercises that had, coincidently, not involved digital methods (Figure 17.3).

There were around 30 e-mail-based responses to the consultation, a small percentage of the total number of contributions from residents, either from those provided by the authority or pressure groups. Variation between the methods of response can be accounted for in several ways from discussions with those involved. For example, the standard letters and petitions are specific to only certain geographical areas in the borough and would reflect only a certain percentage of the population in that area and possibly a

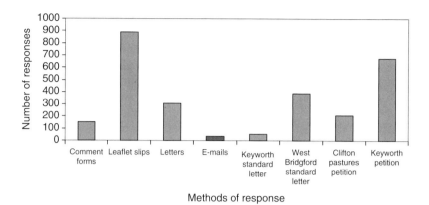

FIGURE 17.3
Rushcliffe responses. (From Rushcliffe Borough Council, *Local Plan Working Group—Borough Local Plan Review*, RBC, West Bridgford, 1999. With permission.)

certain type of resident. The leaflet slip could be seen as convenient to some, as it was specifically provided by the authority for feedback, whereas others may have thought it was the means of response that would be treated with more authority. For the same reason, others may have chosen to write letters as a possibly more acceptable and formal means of communication. In contrast, some interviewees noted choosing e-mail because it was seen as cheap and immediate option. Significantly, e-mail also opened-up increased dialog between the main consultation officers and a number of participants, helping the officers to manage their workload and to clarify many issues for the residents before any formal comments were provided. Following analysis of the responses, the authority felt that they had clearly found an issue that the public was concerned about and felt that citizens were able to respond to the topic. This can be contrasted with another example of community planning policy document consultation at the strategic level in Edinburgh.

17.3.2   Case 2: The City of Edinburgh Council's Community Plan Consultation

Two leading examples of digital participation from the survey involved community planning. This strategic form of policy making involves local actors coming together to outline visions for their area's development, drawing in policy concerns such as sustainable development, decentralization of council services, and urban regeneration. In the U.K. mode of community planning, the local authority plays a lead role with support from plan partners, including other public services, local business, and the voluntary sector, for example. Public involvement is at the heart of the process, and consultation exercises are seen as an important dimension to the development of a community plan, particularly in terms of public ownership of the policies that are developed.

Edinburgh was one of the five Scottish local authorities who piloted community planning at the time of the survey, with another leading example being Highland Council. As a unitary authority for Scotland's capital, the City of Edinburgh Council had emerged from a former regional and district authority. As such, it had inherited several buildings and bases of expertise from across the city, with community planning coming from the corporate core but drawing on these distributed actors, facilities, and other resources. Like many other urban areas, Edinburgh has a mix of geographically defined, very wealthy, and impoverished areas but often in very close proximity and with additional concerns through its city center being designated as a world heritage site.

Like Rushcliffe, officers believed that there would be a higher level of computer ownership in Edinburgh compared to other parts of the country, which officers felt would impact on the number of online responses. The Edinburgh approach to community planning involved the authority and the plan partners devising a document for public consultation and scrutiny. Public involvement was predominantly at the end of the process. As this was a new policy area, there was no tradition to influence how the exercise would take place. The policy and the exercise accompanying it were an experiment and an opportunity to try new methods of public participation, including digital ones.

Again, several methods were adopted including online versions of all the consultation documents, press releases, and advertisements in local press and several meetings with business groups, voluntary groups, and residents' associations. The authority also wanted to experiment with a call center for people to telephone their responses, which was seen useful for those who found writing English problematic. The call center was also used to host a telephone survey with their citizens' panel. These large groups contain around 1000 representatives of the community balanced by varying (geo-) demographic measures and are seen as a method to strike a balance between apathy and some of the problems of repeated engagement with the same actors, or consultation fatigue. Digital methods included a dedicated e-mail address for response to the exercise, electronic versions of the community plan's documents on the council Web site, and the development of several electronic kiosks in public areas across the city, which were additional access points to the information for a few potential participants. It is also useful to note that ICTs were part of a drive to increase information-sharing between plan partners in the context of a Web-based GIS, through Edinburgh's contribution to the Interreg IIC-funded DataShare project. This acted as another component and indicator of the evolving e-government activity in the authority.

Although aimed at the general public, the audience for the exercise included other public agencies, including national bodies (such as Scottish Natural Heritage) and nonpartner businesses in the city. The methods

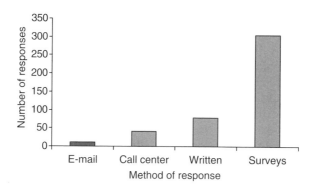

FIGURE 17.4

Responses in Edinburgh's community plan. (From City of Edinburgh Council, *City-wide Community Planning. Progress Report, June 1999*, CEC, Edinburgh, 1999. With permission.)

used in the Edinburgh case can be seen as more authority-driven than Rushcliffe's with less bottom–up/community-initiated methods (Figure 17.4).

There were only 12 e-mailed contributions in this example. A much smaller number than in Rushcliffe from a much larger population but with an overall response rate that was also much lower. The surveys gained the most responses, as they targeted citizens that were already involved and well versed in council activities. The surveys also required an immediate response, possibly contributing to greater levels of participation, as people could not delay and miss the consultation period. Written responses were also common from all groups of participants (including other government bodies and businesses), who, again, may have seen a letter as more official. The call center received just fewer than 50 responses. Although this may have offered an opportunity to some to respond, officers felt that this method had underperformed and were much more critical about it compared to the few e-mails that were sent. In this case, there was also less interaction through e-mail between officers and citizens compared to Rushcliffe. This issue of numbers of responses from the public became somewhat confused in this case, as the two participants interviewed chose to respond several times to the exercise in different guises through different methods, at residents' association meetings, by letter and also through e-mail, raising questions about their representative nature.

Before turning to the last example, it is useful to contrast several features of the methods presented in these first two cases, in terms of promotion of the activity through online and, more importantly, traditional media; dissemination of policy information online; and the promotion and support of e-mail as a means to capture public opinion. E-mails, letters, feedback slips, and comment forms obtained at meetings involved citizens framing their responses in text and in the case of the call center through similarly structured verbal responses. In contrast, surveys, standard letters, and petitions involved less commitment of time, and arguably thought, from participants, as much of the composition was provided by others. This, in part, can explain why more contributions came from this latter set of responses.

In addition, the variation in activity between the local Rushcliffe example and the strategic Edinburgh one can be attributed to two qualities. Rushcliffe involved an issue that residents could readily identify, whereas Edinburgh's strategic policy required wider interest and understanding from potential participants relating to more abstract and less immediate issues. This can be related to ideas about lower levels of response for strategic plan-making compared to local activities in U.K. town planning (Thomas, 1996). This can be seen as a geographical issue, as people appear more likely to involve themselves with activities closer to their physical areas of interest, as opposed to distant issues in terms of both location and understanding. Additionally, the connection between the publicity materials from Rushcliffe's leaflet and press releases was much more targeted on a group of participants—residents. Edinburgh had chosen to gain the views of a very broad, partially self-selecting, group and the broader publicity materials perhaps did not reach as wide a base to begin with. The relationship between traditional and digital methods is an important consideration if the numbers of people responding to an exercise is viewed as an indication of the success of the activity. This raises issues about the difference between what can be seen as statistically legitimate in a typical survey and any differences that emerge when democratic representation is being considered. Another approach occurs when participants are truly selected to be a representative community voice, as in the case of citizens' panels, found in the last study.

17.3.3 Case 3: The London Borough of Lewisham's Dialogue Project

Very few examples from the survey appeared to support more ideal forms of democracy that tended to the upper-rungs of Arnstein's ladder, in part reflected by the role of local authorities and their Web sites formulating, enacting, and displaying policy. Lewisham made attempts to introduce more grassroots engagement online. As an English unitary authority in the capital, Lewisham explored a mixed form of decision-making by having members of their citizens' panel who investigate and provide opinions on general issues that would have clear local outcomes. This took place within the corporate Equality Unit's democracy initiative—Lewisham Listens. This broad corporate activity was particularly focused on investigating and supporting elected member decision-making by a more informed and engaged public. Part of Lewisham Listens included evaluating the impact of public participation on council decisions and the methods being adopted, including digital ones. It outlined and demonstrated the open approach that the authority was taking to make more ideal forms of participation possible and how they were already reacting to central government policies such as Best Value, Modernising Government, and the emergent e-government agenda.

In terms of the survey, and their Web site as a whole, the most advanced example was Lewisham's contribution to the Dialogue Project. This pan-European Information Society Project Office (ISPO) initiative involved partners in Ronneby (Sweden) and Bologna (Italy). Unlike the other cases, this project specifically wanted to involve citizens who would not typically have the opportunity to participate in the information society, what is often referred to as the *digital have-nots*, a form of social exclusion within the digital divide. Again, as an urban area, there was a mix of poverty and wealthier areas but with relatively high unemployment and a large proportion of the population from ethnic minorities. From the authority's own surveys, it could be clearly seen that computer ownership was low. Lewisham's approach involved supplying free Internet access (including computers and modem calls) and training for around 60 members of the panel. Active members of the project participated in consultant-facilitated chat-room debates, somewhat independent of council influence except for the topics for discussion, which included strategic-to-local issues such as education, community safety, and the future needs of Lewisham's elderly. These debates occurred alongside significant group- and skill-building activities, such as support for the citizens to use e-mail in general and between panel members; bulletin boards to leave messages about topics of interest; and encouragement to explore and share other online resources, provided by the authority or through the World Wide Web. Such activities were thought to help the participants to develop their opinions to inform debates but they also had several other outcomes for those involved, including making new friends, learning new skills, and contacting family members on the other side of the world. As a result of this type of activity (that was more about communication within a small, selected, and representative group) and the context of Lewisham Listens, there are no measurements of participant response but the project's evaluation process and questionnaire recorded how useful participants found certain methods (Figure 17.5). The respondents had a slight preference for face-to-face contact (A) and e-mail (B) to the more involved, time-consuming, and technically difficult tools such as chat-rooms (C) and bulletin boards (D). Technology was a significant part of the project. Participants readily adopted the digital methods but this was not seen as a substitute to more traditional forms of interaction and communication. The digital methods provided the technical infrastructure for a forum to take place with additional outcomes alongside the citizens' advice to decision-makers. These were unexpected outcomes from the project that a strict observation of the methods alone would not uncover. The project also demonstrated that those not normally familiar with digital methods could participate online and officers felt that contributions from the citizens did impact on decision-making for a number of policies.

An example of this included a meeting (that followed online debates) where the group was invited to the council chamber to use keypads to vote on several policy options which elected members eventually adopted.

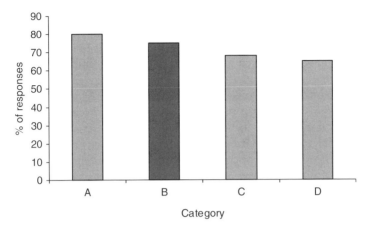

FIGURE 17.5
Participants' attitudes to methods of participation in Lewisham. (From London Borough of Lewisham, *The Dialogue Project: Consultation On-Line*, LBL, London, 1999. With permission.)

As such, the Lewisham example was the most involved, technically developed, and supported example from this research, with the authority extending the project's funding for another few months because they felt it was a success. From these contexts stem ideas about what public participation is composed of, particularly in top–down online settings.

17.4 Notions and Other Components of Participation

An important part of the research was to expose actors' notions of public participation, what they understood participation to mean, how they valued it, and what it involved. The following discussion examines interviewees' comments from each of the cases, exposing how participation was understood by those involved and some of the forms of participatory democracy that took place. As recent examinations of participation have taken a number of different theoretical approaches it is useful to outline three that were used in the research.

First, democratic theory, stemming from political science literature, provided a key to analyze digital participation. A well-established field of thought, it not only seeks to outline what public participation can be but also actors' motivations for participating in ideal and experienced situations. Democratic theories of participation also present an important relationship between different modes of political activity, representative and participatory democracy, and the roles elected members and citizens play in each.

Second, theories developed in town planning research build upon democratic theory for place-based decision-making at varying scales that compliment Rushcliffe's local example, Edinburgh's citywide context, and the mutually broad and specific activities from Lewisham. The planning literature also draws in ranging debates about participation as a practice and the methods that can be used, extending ideas and ideals to identifiable or even tangible artifacts, in terms of the methods or the outcomes of activities for all involved.

Third, these ideas and methods were focused in terms of digital participation through the social construction of technology (Bijker et al., 1987), a position that argues that technology does not exist in a technical vacuum with no social context, that technology's development is influenced by all the actors who use it (or not), and that these actors can influence its development as much as those who create and introduce it. This theoretical approach was useful in grounding the research and recognizing the importance of interviewing officers and citizens, so that both their attitudes to technology and participation could be included. Additionally, it also helped to depict the role that those viewing a participatory exercise play, both in terms of the actors inside and outside an organization.

From this theoretical background, it is possible to build a more complete picture of what actually was understood to be public participation in these cases. It should again be noted that these comments relate to leading examples of public participation from innovative local authorities at the beginnings of digital participation; that most officers and citizens interviewed saw the exercises as suitable; and that they would pursue something similar in the future. The comments demonstrate that people understand participation to have different meanings and that a lot of complimentary and divergent views emerge at different times, even from the same actor.

17.4.1 Views of Officers

In Rushcliffe, planners saw participation as "essential...to local planning" as it provided a means for the public to have a say in changes to their environment. Officers also believed that the public could feel disenfranchised if they were not being listened to or their comments not acted upon, particularly if an opportunity to have a "greater say" was being offered, or at least appeared to be. Some felt "limited numbers will get involved...even if you put stuff through their door" (Rushcliffe Officer), reflecting mixed feelings about the success of their leaflet. Another felt that less than ideal forms of participation take place:

> Much of a local plan consultation is to distract people. The process becomes a smoke screen. In particular the public have little opportunity to change things.
>
> (Rushcliffe Officer)

Although this is quite negative, the comment importantly presents participation as an impressive device that appears to tackle issues and engage citizens, when this may not be the case. Another role that participation was noted as playing was one of obligations, where the authority had "to be seen to be consulting people" (Rushcliffe Officer). This was meant in two senses, where the authority wanted to demonstrate its readiness to engage with the public and where it was important that certain audiences acknowledged that this was taking place.

Other comments emerged that more closely related to who was participating and notions of a representative community, where it was important to have a "range of views from the public" (Rushcliffe Officer) but that participation is " . . . only for those who cared, and they'd probably reply anyway," with people selecting themselves or whipping-up the interest of others. This was felt to lead to "only one side of the coin" contributing, as contented citizens were less likely to respond. The presence of familiar figures that would regularly respond to the authority's activities were also highlighted and characterized by the typical town-planning participants that Thomas (1996) identifies as middle-class middle-aged males.

Many officers desired an ideal of "true participation" and that, generally, participation was "a fundamental principle of democratic government" but that the local plan process departed from this "in a pure sense" (Rushcliffe Officer). This was made clear by one interviewee who differentiated between a wider concept of participation and consultation:

> If you could engage with the public, given resources and willingness, in a way where you could interact, for example by "planning for real" [physical models of the community that participants annotate] . . . Then this is participation in the true sense . . . so what we do is "consultation."
>
> (Rushcliffe Officer)

They also recognized its educational element and relationship with "a fundamental principle of democratic government." They suggested that the authority had "to be seen to be consulting people," thus making activities political.

In Edinburgh, officers had varying views and expectations of their approach to participation, some being quite critical, in spite of a great deal of council literature including a public participation good practice guide distributed to all officers in the authority. It was felt that although "participation is essential," there were issues of what the term meant. It is possible that the variation in corporate literature has not helped. One officer viewed participation as "critical" to service provision and a "duty," relating it to ideas of Best Value. It was suggested that participation involves the "right of individuals to be consulted" and to participate in things that affect them. As in Rushcliffe, it was felt that the public could exist in different guises, with some people wanting to participate by voting, thereby having decisions made on their behalf, whereas others may want to be consulted more often, with citizens choosing their own level of engagement. It was suggested that participation is citizens

> ... being able to express themselves to the level that they are interested in. Public transport is only public transport [i.e., not a concern] when you don't have to wait for a bus
>
> (Edinburgh Officer)

Others were more critical of what the authority were offering, highlighting "legitimization" (Edinburgh Officer) and the real impact the public could have. Participation was only "mere tokenism, it's important to have the box ticked" (Edinburgh Officer) and activities were "piecemeal, reactive and only acting to legitimate" (Edinburgh Officer). A balanced view was also expressed:

> How much they [the public] do influence things is a matter for consideration. There is a feeling that they should be involved and we should try and deliver it.
>
> (Edinburgh Officer)

Part of the variation in views and criticism will relate to the community plan focus of the research, where the authority sets the agenda and the activities for public participation. It was suggested that "organisational structures and objectives" (Edinburgh Officer) created some barriers to participation and that some staff could be frightened of participatory activities, partly because they expected responses to be negative. By comparison, others felt that there was a degree of public cynicism relating to what local government was trying to achieve and how it conducted consultations. They felt that people did not care and that there would need to be a significant change in public attitude to realize that citizens could get involved and influence decisions and that "massive education" (Edinburgh Officer) was needed both inside and outside the organization.

Officers in Edinburgh identified with the notions of participation held in corporate policy, that participation was important to how decisions were made, with a balance between rights to have a voice and citizens being duty-bound to respond to public consultations. Interviewees saw people getting involved to varying degrees, through participatory or representative structures, and that there was a spatial variation in the levels of participatory expertise across communities, possibly leading to consultation fatigue. Some saw activities as less than ideal, tokenistic, and as processes to legitimate the top–down agenda of the authority. Although public opinion was sought, its influence was unmeasured, unclear, or unreported. However, some officers felt that citizens needed more education about council activities and why it was necessary for the public to respond.

In contrast, the background of evaluating participation in Lewisham led to some distinct views of what participation meant to the authority and to its citizens. Unlike other cases, some officers focused on the varying ways that public participation can be viewed and the range of methods that can be used. In particular, Arnstein's ladder was often cited as a way of setting out the different forms of participation, as it had been used in their evaluation of Lewisham Listens:

> Public participation can be as little as reading your local news-paper, or voting, or going to the council with a positive or negative letter on services. It's not just coming along to a meeting. It's about knowing what is going on and being someone that is really involved...
> If you are on the [citizens'] panel then you could be an active citizen with an active role in the community. I view it like Arnstein's ladder, a continuum.
> The Arnstein view is the best short-hand way of defining it really... a continuum from basic information and not much of a dialogue to the potential to give control and ownership and identify resources. Some of our work is in the middle area up. We attempt to push at the limits and maximise it. We are realistic. We don't use techniques that would raise expectations. It has to be fit for purpose.
> (Lewisham Officers)

Officers also looked towards the community building aspect of participation and that public participation was, again, a "right" (Lewisham Officer) of citizens. Issues relating to Best Value policy were also drawn out as key to reinvigorating democracy.

> I have a jaundiced view. Where there is public engagement, if people felt that they had some say, then there would be participation. Pockets do occur but people are more insular. There needs to be a purpose to a consultation [we should not be consulting just for the sake of it]... it's about control and 'consultation' in the purest sense. We have to find ways of engaging, especially around regeneration with Lewisham's economy and business, and potential business, and education. Consult-ation is a key factor [but] I'm a disillusioned bureaucrat.
> (Lewisham Officer)

Officers in Lewisham saw the variation that existed, that participation was used to gather opinion through particular methods, and that this relationship between ideals and methods had varying outcomes that they frequently chose to relate to Arnstein's ladder. They realized that the public's expectations can be raised and that this can impact on the methods chosen, with an aim to be open and have transparent activities. However, officers also realized that there is a sense of power, as control of the process still lay with the authority. Participation was seen as a means to improve council activities and to present a democratic process, particularly in terms of service delivery through Best Value-led initiatives. This latter point also relates to their Lewisham Listens' democratic reforms to support elected member decision-making through increased grassroots engagement. Outcomes of exercises also appear to be given more consideration by these officers. Overall, a mixture of ideas emerged from officers in the cases. Participation was an ideal but not necessarily something achieved in practice. It was frequently viewed in wider contexts such as other policy issues, the methods used and who was involved, involving an ideal representative population and those who eventually become involved. This can be contrasted with equally varied notions of the citizens from the three cases.

17.4.2   Views of Participants

Although Thomas's "typical planning participant" was common in Rushcliffe, interviews came from many different demographic groups. Some said they had limited expectations of public participation because they had worked with local government. They chose to become involved because participation was seen as part of the democratic process, and they felt they were duty-bound to be involved. Many highlighted the need for the authority to listen and act based on the "consensus" (Rushcliffe Citizen) of public opinion but how that notion of consensus is derived and if it is desirable was not explored further. Relating to this, some felt that participation was a chance for "everyone to become involved" (Rushcliffe Citizen) at various stages in a policy process but there was also a lack of faith in this level of engagement. One interviewee noted that participation needed to be related to a topic that was an "electoral issue" for people to become involved in any numbers.

This can be contrasted with others who believed that "...most people don't bother, and the council will probably do what it wants anyway, with some sort of hidden agenda" or that "councils can do as they like or they can ask the public" (Rushcliffe Citizens). These interviewees felt that local authorities were giving out information in order to act but that a lack of response would be seen as acceptable, which can be contrasted with officers' comments about contented participants viewing consultation materials but not responding. This was underlined by another citizen who felt that the process was a "waste of time" (Rushcliffe Citizen) and that consultation exercises took place because the authority had to run them and that officers were not interested in public opinion, even though this participant did choose to respond, stating that what took place was "fair" (Rushcliffe Citizen). The level of interaction between the authority and its citizens was also muddied by another interviewee who felt that an ideal situation involved participation as both "informing" and "reassuring," which others viewed as placation and not participatory. Others believed that participatory democracy was inadequate and that "matured participation comes in terms of elected representatives" (Rushcliffe Citizen), illustrating a need to examine the relationships between representative and participatory democracy. The idea of representation also included elected representatives, where

> ...a good government will listen to what the people want...a bad one doesn't. If they are trying to consult then they are a good council...but they need to take the [party] politics out of the discussion though
> (Rushcliffe Citizen)

Citizens' representative nature was also noted by participants, feeling that the council would have to be wary of those who may overparticipate, particularly if there was "a NIMBY-type reaction" (Rushcliffe Citizen), relating to the housing issue. This was also associated with the methods used, as public meetings were felt to attract " ... guys [that] go along who are very virulent" (Rushcliffe Citizen) but that e-mail allowed "the man in the street responding far more," representing different "voices" from the community.

Consensus and a strong public voice that was "representative" were also identified by Rushcliffe's citizens. The topic under participation was seen to be influential but many felt the authority would act in a predetermined way, irrespective of public opinion, particularly relating to party politics. Being involved at a low level of interaction was reassuring to some, with appropriate approaches and methods, but others felt the low levels of engagement were worrying and that, in some sense, the participatory activity that took place was not as useful as existing representative structures.

In contrast, although several people were approached, only two citizens were available for in-depth interviews in Edinburgh. As noted above, both interviewees were very politically active in several pressure groups. One stated that, in general, when public participation was discussed then " ... you do not necessarily think of the council or planning processes" (Edinburgh Citizen). They stated that participation should " ... involve an opportunity [to participate], for it [policy] to be made visible, for opinion to be expressed and for it to be easy to use methods for feedback" (Edinburgh Citizen). Participation was seen as an opportunity to contribute to decision-making and that transparency was important, something the authority also mentioned in its literature. The second participant chose to focus on the nature of participation in local government and local democracy in general. They saw local democracy in an ideal setting, where it performed " ... a very important function of what democracy can be," before going on to note the need for adequate access to information to aid decision-making and to inform citizens.

Both interviewees acknowledged that participation is complex and one suggested that "bad government" (Edinburgh Citizen) existed where a policy or service did not involve "participation" in some sense. They felt that participation was more than the occasional voting found in representative democracy, particularly as this system " ... elected Hitler and gave him a mandate" (Edinburgh Citizen). Equally, they felt that participation was also " ... not pressure groups demanding things, [as this is] a poor complexion of democracy" (Edinburgh Citizen). However, both participants were very active in transport-related pressure groups and other community-based organizations that engaged in the community plan, to the extent that both interviewees represented different groups and used different methods on several occasions during the consultation exercise.

Openness and access were seen as important in Edinburgh in relation to information and the methods used in policy-based participation, although planning was not necessarily seen as a participatory activity. The notion of access merits a separate discussion and Smith and Craglia (2003) have outlined some ideas relating to how access can be viewed in a participatory context, building on the work of Kling (1999) who discussed social and technical ideas of access in online settings. Openness also included instances where the public needed to be involved and if they were absent then this was bad government. The interviewees saw participation's varied nature including undesirable outcomes, who was involved, how their voice was mobilized and listened to, and what the outcomes of that activity would be.

In Lewisham, unlike the other cases, the participants had been specially selected to play a part in the Dialogue Project, with some having used ICTs before, and all being heavily involved in the citizens' panel. The discussion, with around 15 participants, either involved their experience with the Dialogue Project or their views of participation in general. Again, these citizens chose to view participation in terms of ideals and more practical issues. One interviewee felt that the public involvement in local government decision-making was "fantastic and should be encouraged" but wondered if it "worked in practice?" (Lewisham Citizen). This idea became confused as they also felt that the Dialogue Project should continue and similar activities should be initiated, as it "...made you feel you were involved and could have a say...as a group" (Lewisham Citizen). Again, the involvement of several balanced opinions is raised in this case through the last comment, with the level of involvement being highlighted.

Some placed an emphasis on interaction rather than merely being questioned, and a move away from consultation, that was particularly present in the Edinburgh case. Another interviewee was critical of the public's apathy, legitimation, and overconsultation, believing that the public

> ...don't get involved enough for a start; you get the same people dragged out for everything...[and it] tends to be the same type of person...those that vote and older people.
>
> (Lewisham Citizen)

Many interviewees in Lewisham viewed participation in terms of democratic ideas, either through the involvement of a mobilized community being used to inform decision-making or the relationship that people have with their elected members. Some were concerned with the levels of public apathy that exist and that participation should allow citizens to have their say.

It's [public participation] me having shouting matches with another panel member. Getting local people involved in the democratic process of consultation and decision-making.

It's involving the people that live in the communities of the borough to have their say on what should or shouldn't happen [to their area].

The public having an input to the decision-making process of the council. People should try to be strong enough to change council ways and look at things a bit closer.

It's about participation in the democratic process. I have mixed feelings. [Sometimes participation]...leads to apathy, not wanting to vote because you feel your comments are not worthwhile. There is room for improvement and democracy is an important aspect of that, it's evolving. I want to see radical change, so we need participation; whether direct action, voting or consultation.

(Lewisham Citizens)

In contrast, another felt that there should be concerns about the involvement of older people and that when they were involved "I think it's great" (Lewisham Citizen). Typically, older people do tend to become involved in local authority exercises but this particular project tried to involve many different groups, including a wide range of ages, although their inclusion also demonstrated some elderly people's ease in using new technologies.

There was also concern that participation was a public relations' exercise but that Lewisham's officers did seem to listen. Another noted that "...the council did things we talked about. I feel we do have an input" (Lewisham Citizen), also believing participation must have a purpose, as some other activities were "...just chattering with no conclusions [but]...with Dialogue we always had conclusions" (Lewisham Citizen).

Some focused on digital participation, believing that when technology was used properly it could be "...quite good because there are those who wouldn't write or go to a council office. If it's online it's easier" (Lewisham Citizen). This was a feeling shared by many of the interviewees throughout the cases, relating notions of access to participation, as noted above.

The uncertain relationship between ideals and practice was also raised by Lewisham's citizens, although they felt that this particular project should have continued, placing emphasis on group decision-making rather than involvement at an individual level. Interaction was supported and consultation less well accepted. Criticism of the public's apathy was also raised but contrasted with the overconsultation of some groups. The elderly were identified as a particular group that were either listened to too much or did not receive enough support for participatory activities. Concerns about the authority listening to public opinion varied, with some feeling the activity was legitimation but that a collective public voice was not only listened to but also acted upon. Digital methods and participation were noted by interviewees with enthusiasm, including benefits of access for those who would not be as readily able to participate in more traditional interactive settings, although there was also criticism that the authority could do more with its Web site. Their understanding of participation was both ideal and practical, particularly reflecting their experience of a project that was geared towards supporting elected member decision-making. What was not clear was the impact that a larger number of participants would have in such activities.

17.4.3 From Notions to Theory

The notions from both sets of actors are views that focus on the philosophical democratic underpinnings of many activities, reflections on the methods employed and being aware of different types of participant, as overconsulted groups, representative voices, alongside the role of other representatives, such as elected members. Although leading examples from the survey, a lot of the criticism from both sets of actors could be related to tacit features in Arnstein's ladder that need to be exposed and understood for those involved (both within and outside organizations) to develop relationships of trust for more informed deliberative digital participation.

From these interviewees' comments and the survey it has been shown that methods of participation can vary and be used and understood to act in different ways. Public participation, itself, is neither a shared nor unique construct. Actors chose to view it in different ways, as a method, an ideal of democracy, a means to have their concerns listened to, to have ideas receive public validation, or a means to present the representative voice of citizens. These views stem from ideas brought to an activity of what participation may involve through experiences or expectation, leading to the construction of their meanings. For example, participation has been highlighted as an educational process. J.S. Mills noted that political participation, such as voting, in the workplace educated an electorate and would influence their activity at other levels of political activity (Pateman, 1970, p. 28; Holden, 1993, p. 69). What can been seen is that such educational activities exist in a participatory democracy but that they can be more than gaining a political education and experience. Involvement can also include learning more about a policy area, how decisions are made or even less expected outcomes such as how to use e-mail and Web browsers. To summarize, what emerges are five main components that construct our understanding of public participation: notions of participation, issues, audience, outcomes, and methods.

Notions of participation build on actors' understandings of what participation means to them and ideas relating to it, in a theoretical or philosophical way. The ideas that interviewees chose to express were often related to ideas or ideals of democracy. This component, in part, helps to construct the remaining four, with some characteristics of these other components occurring as notions in their own right, as well as the special case of *access*.

Issues involve those policy areas or concerns that those initiating an exercise (e.g., local government officers) are willing to present for public scrutiny. To some extent there may also be issues that the public wish to see the local authority act upon, which would often apply to more grassroots or bottom–up activities but may also occur in top–down settings if the public chose to respond with specific issues in a broad consultation context. The issue, if effectively transmitted, can influence whether responses are sent and if they are acted upon. However, often issues are expressed that do not appear relevant and it should be recognized that the ways in which these are dealt with can impact on other components.

The *audience* is seen by many as an important feature of what public participation involves. This not only includes the public, who most frequently select themselves to participate (or not) but also an internal audience within a local authority who may have expectations about an exercise, such as elected members or officers who support participatory exercises or in related departments who will be impacted upon by an activity's outcomes. The audience relates closely to the notion of representation and how some (from both inside and outside the organization) view achieving good representation as a determinant of a successful exercise; such as the "right groups" (Edinburgh Officer) having their say.

Outcomes can be actual or perceived results of exercises which will present themselves at varying stages in a process. Many consultation processes take place because they are part of an evolutionary decision-making process, such as the creation of development plans in the Rushcliffe example, but perceived linearity in the process does not mean that outcomes only emerge at the end of exercises. For the authority, perceived outcomes can influence what issues are suitable for participation, the methods that should be used to achieve these outcomes, and if an exercise should be started. Actual outcomes, such as numbers of responses or the concerns uncovered through the process, will be used to judge the impact of an exercise, in part shaping actors' perceptions of public participation and helping them to decide if any further activity is needed. As noted above, there can also be less expected outcomes, such as the educational function of an activity, in terms of learning more about participation, a particular policy area, or transferable skills, as found in the Lewisham case. Perceived outcomes will influence whether people choose to participate in the first instance or whether exercises are merely legitimation (in the Arnstein sense), possibly leading to apathy.

The final component that links the others together, by providing readily identifiable entities to study, are the methods of public participation. Different methods have different properties and are viewed and treated in various ways by actors inside and outside organizations. Methods can be traditional (often face-to-face, broadcast media, or paper-based) or digital (those which utilize ICTs) and can take three forms, those that push, advertise, or broadcast information about an exercise (ranging from newspaper adverts to Web sites), those that can be used to gain information from participants (such as letters, surveys, and feedback e-mails) and those that provide fora for multiparticipant/user communication (including public meetings and chat-rooms). Methods relate closely to notions, helping to select and measure them, both formally through policy evaluation and informally through actors' experiences.

It should be noted that the components themselves have many interrelationships and that different situations are likely to lead to components behaving differently, in terms of their influence or importance in an exercise overall or their role for the individual. It is likely, therefore, that an exploration of such contexts would be desirable if appropriate methods are being sought: what is the issue requiring citizens; who are the audiences (internally and externally); are their notions complimentary; what are the likely outcomes; and what methods should be adopted that meet these understandings, including what education and support will be required in order for the activity to operate? Blind application of any method may not only lead to disappointment but will also impact on future activities. These issues are discussed further from a theoretical perspective by Smith (2006).

17.5 Conclusion

Although public participation does have a temporal element, as exercises are initiated, it should be understood that the views of those involved will have been set before, during, and after exercises. Clearly, from the comments of both officers and citizens, there is no shared understanding of what public participation means and the way in which it is often treated as a simple process by academia and those in practice can undermine current activities and miss future opportunities for more genuine engagement between governments and their citizens. This is particularly important given the amount of consultation that governments are seeking across several policy fields.

Simultaneously, there has been increased interest and hyperbole in adopting the Internet as a means to enhance government, either through e-government service provision or for direct democracy. Digital participation, based in the top–down setting of U.K. local authorities, is still in its infancy but the general findings from this research are widely applicable in other contexts and are likely to remain true for some time. An exploration of this topic cannot solely investigate the digital façades of local authority ICTs but must also engage with the various actors who shape the activities, both within and outside organizations. Similarly, exploring digital methods alone can only offer a glimpse of how participation operates, as their context is dependent on the role and response rates of traditional methods and more inclusive activities will try to offer several, depending on the issue requiring consultation and the audiences involved. The leading examples from this research not only show the limited use of digital methods (at present) but also uncover several components that can aid theoretical understandings of public participation in general and digital participation in particular.

From a broad literature, drawing on the social construction of technology, democratic and planning theories, it is possible to begin to understand the philosophical and practice-based foundations of participation; demonstrating that participation is neither a shared, unique, nor widely understood construct or concept. The literature and actors both choose to see it as a theoretical or philosophical entity with distinct methods and from this comes five components that compose all participatory activity: notions of participation, issues, audience, outcomes, and methods.

In terms of notions, several frameworks in the literature explain the varying forms of participation that can occur. As shown in this chapter, many authors (Arnstein, 1969; Pateman, 1970; Holden, 1993) illustrate varying ways in which participation can take place, often with an element of power associated with them. In the top–down context of the research the forms of participation taking place are consultative, choosing among preset ideas that are less than the ideal. Digital methods offer particular features that link citizens with each other and government and, as noted above, it is the notions of access that relate to this argument in particular. Kling (1999) has demonstrated that access to the Internet can have social and technical components, and Smith and Craglia (2003) have suggested the need to consider the addition of political access when participatory activities are being considered, in terms of the ability to respond appropriately or for the leverage of funds.

The issues that participation can deal with occur in many ways. Scale is seen as important in a number of senses. Through the cases it is possible to see that local, strategic, and mixed policy areas all employ participation. In the Rushcliffe case, the local nature of the issue led to more responses than the strategic Edinburgh example, even though the methods used were quite similar and the population of Rushcliffe much smaller. With support, as in the Lewisham case, inexperienced participants can become actively involved in online exercises that deal with strategic policies and readily identifiable local outcomes. In the closed settings of these three cases a problem emerges relating to how the authority would deal with a response that does not meet the remit of the exercise. The public often welcome the opportunity to respond, but it is important that local authorities consider which issues to put before them, as issues can vary in meaning between individuals and at different scales (Smith, 2002).

In terms of the notions expressed, the nature of the audience appears to be important as both sets of actors within and outside the authority showed concerns that the right people should be listened to, who were also a representative voice from the community. A single public does not exist and, although Thomas's (1996) typical participant can be identified in examples of digital participation, the Dialogue Project has demonstrated an ability to contribute online from a wide range of participants.

Examining the methods of participation provides a focus for study because they are frequently the only readily identifiable components of a participatory activity that actors and researchers can identify with. As Alty and Darke (1987) indicate, all methods of participation include and exclude different audiences, and digital methods are no exception, introducing issues from the digital divide. Rushcliffe and Edinburgh had used similar methods but the outcomes in terms of online response rates differed, in part because of the issue but also because the traditional methods in the Rushcliffe case more readily engaged with a defined audience. Certain facets of the methods can be assessed through models such as Arnstein's ladder, indicating the type of participation taking place but it should be noted that a method of participation may contain more than one notion and misinterpretation of the method by actors, alongside it being adapted by the participant for their own purposes (such as learning to use computers or making new friends, as seen in the Lewisham example), can lead to misunderstandings about the nature of the exercise taking place for all involved, or not.

This research has primarily dealt with digital participation in the context of U.K. local government. Further work is needed in exploring the notions of participation that exist in other contexts and the relationships between notions of participation and access. In particular, there is also a need to explore grassroots activity in greater depth, including the role of community networks originating from citizens rather than local authorities or the emerging infrastructure and content of Web 2.0, from topical or place-based Wikis to Google Earth Community discussions, to highlight a current geographical example. Finally, participation can often relate to varying geographies and the recent research agenda on participatory approaches using geographic information (Smith, 2002) and access to geographic information (Wehn de Montalvo, 2002) should provide a useful guide to more in-depth longitudinal studies that do not rely on the flawed short-term projects that much current activity relies on, both in research and in practice. As such, a reexamination of local authority Web sites in the United Kingdom that now exist in a more mature policy context could aid not only an understanding of the role ICTs are playing for e-government but also the extent to which geographical information and participation are being adopted and supported in this context.

References

Alty, R. and Darke, R., 1987, A city centre for people: involving the community in planning for Sheffield's central area. *Planning Practice and Research* 3(1), 7–12.

Arnstein, S.R., 1969, A ladder of citizen participation. *Journal of American Institute of Planners* 35(4), 216–224.

Bijker, W.E., Hughes, T.P., and Pinch, T.J., 1987, *The Social Construction of Technological Systems* (Cambridge, MA: MIT Press).

Edinburgh Council, City of, 1999, *City-wide Community Planning. Progress Report, June 1999*, (Edinburgh: Policy and Resources Committee, City of Edinburgh Council).

HMSO, 1999, *Modernising Government* (Cm 4310) (London: The Stationery Office) (http://www.archive.official-documents.co.uk/document/cm43/4310/4310. htm, accessed on March 27, 2007).

Holden, B., 1993, *Understanding Liberal Democracy* (London: Harvester Wheatsheaf).

Hunter, D. (editor), 1999, *The Connected Council: Leading the Community in the Information Age* (Bristol: Foundation for Information Technology in Local Government).

Kling, R., 1999, Can the "next generation Internet" effectively support "ordinary citizens?" *The Information Society* 15(1), 57–63.

London Borough of Lewisham, 1999, *The Dialogue Project: Consultation On-Line* (London: London Borough of Lewisham).

Pateman, C., 1970, *Participation and Democratic Theory* (Cambridge: Cambridge University Press).

Rushcliffe Borough Council, 1999, *Local Plan Working Group—Borough Local Plan Review* (West Bridgford: Rushcliffe Borough Council).

Sharp, E. and Connelly, S., 2002, Theorising participation: pulling down the ladder. In *Planning in the UK: Agendas for the New Millennium*, edited by Rydin, Y. and Thornley, A., pp. 33–63 (Ashgate: Aldershot).

Smith, R.S., 2001, *Public Participation in the Digital Age: A Focus on British Local Government*. Ph.D. Dissertation, University of Sheffield, Sheffield.

Smith, R.S., 2002, Participatory Approaches Using Geographic Information (PAUGI): Towards a Trans-Atlantic research agenda. *5th AGILE Conference on GI Science*, Palma, Mallorca, Spain, 25th–27th April 2002, pp. 91–95 (Universitat de les Illes Balears, Palma: Mallorca).

Smith, R.S., 2004, SIG et collectivités locales: participation et accès à l'information géographique au Royaume-Uni. In *Aspects organisationnels des SIG*, edited by Roche, S. and Carron, C. (Cachan, France: Hermes).

Smith, R.S., 2006, Theories of digital participation. Chapter 3. In *GIS for Sustainable Development*, edited by Campagna, M., pp. 37–53 (Andover: CRC Press).

Smith, R.S. and Craglia, M., 2003, Digital participation and access to geographic information: a case study of UK local government. *URISA Special Public Participation GIS Volume II*, pp. 49–54.

Thomas, H., 1996, Public participation in planning. In *British Planning Policy in Transition: Planning in the 1990s*, edited by Tewdwr-Jones, M., pp. 168–188 (London: UCL Press).

Wehn de Montalvo, U., 2002, Access to geographic information: towards a trans-Atlantic research agenda. *5th AGILE Conference on GI Science*, Palma, Mallorca, Spain, 25th–27th April 2002, pp. 89–90 (Universitat de les Illes Balears, Palma: Mallorca).

Index

A

Access, 242, 249–250, 253
Accessibility, 66, 283, 307, 320
Actor network theory, 275–276
Addresses, 217–220
ADDRESS-POINT, 23–24
Adjacency, 165
Agents, 275–276
Aggregated data, 78, 83
AI, *see* Artificial Intelligence
ANZLIC, 4
Aoristic analysis, 122–123
ArcGIS, 128
ARC/INFO, 144, 161
ArcView, 76, 118, 161, 242, 249–250
Area interpolation, 304, 334–336
Arnstein's ladder, 350, 370, 383
Artificial Intelligence, 275
Automation, 242
Awareness, 114, 242–243, 270, 348, 358

B

Boundaries, 334
Boundary-free area estimates, 334

C

Cadastral map base, 14
Car ownership, 105
Carstairs index, *see* deprivation index
Cellular Automata, 275
Census
   confidentiality, 22
   use with remotely sensed data, 22
   data, 116, 302, 316
      boundaries, 334
      effect on geoemographics, 51
Centroids, 76
Cities Revealed, 36
Classification

of consumers, 43
in geodemographics, 51
Climate change, 134
Cluster(s), 121–122, 160, 223, 305, 308
   analysis, 45, 64–65, 305
   detection, 115, 160
Clustering, 70, 77–78
Community
   activists, 333
   planning, 374
COMPAS, 23–24
Confidentiality, 95, 336, 338
Consultation, 373, 386
Consumers classification, 43
Corruption, 245
Cost of NSDI, 15
Council Tax, 303
Crime, 59, 70
   event, 73
   theories of crime, 71–72
   use of geodemographics, 75
CRIMEMAP, 115
CrimeStat, 77, 115
Customers, 48

D

3D, 294–296, 349
Data
   accuracy, 152
   availability, 339
   capture, 287
   dissemination, 288–289
   quality, 11, 339
   sharing, 273–274, 286, 340
   standards, 11
Decision-making, 263, 357, 385
   and PPGIS, 347
   collaborative, 349–350, 360
Decision support, 351
DEFRA, 136